# Pragmatism and Human Genetic Engineering

# Pragmatism and Human Genetic Engineering

## Contributors

**Jorge Angel Ascacio-Martínez and Hugo Alberto Barrera-Saldaña et al.**

www.aurisreference.com

# Pragmatism and Human Genetic Engineering

Contributors: Jorge Angel Ascacio-Martínez and
Hugo Alberto Barrera-Saldaña et al.

**Published by Auris Reference Limited**

www.aurisreference.com

United Kingdom

**Copyright 2016**

**Printed in 2017 for Sale in the Indian Subcontinent**

The information in this book has been obtained from highly regarded resources. The copyrights for individual articles remain with the authors, as indicated. All chapters are distributed under the terms of the Creative Commons Attribution License, which permit unrestricted use, distribution, and reproduction in any medium, provided the original author and source are credited.

**Notice**

Contributors, whose names have been given on the book cover, are not associated with the Publisher. The editors and the Publisher have attempted to trace the copyright holders of all material reproduced in this publication and apologise to copyright holders if permission has not been obtained. If any copyright holder has not been acknowledged, please write to us so we may rectify.

Reasonable efforts have been made to publish reliable data. The views articulated in the chapters are those of the individual contributors, and not necessarily those of the editors or the Publisher. Editors and/or the Publisher are not responsible for the accuracy of the information in the published chapters or consequences from their use. The Publisher accepts no responsibility for any damage or grievance to individual(s) or property arising out of the use of any material(s), instruction(s), methods or thoughts in the book.

**Pragmatism and Human Genetic Engineering**

ISBN: 978-1-78154-954-4

British Library Cataloguing in Publication Data
A CIP record for this book is available from the British Library

Printed in the United Kingdom

Exclusively distributed by CBS Publishers & Distributors Pvt. Ltd.

Sales & Distribution Rights only for India, Pakistan, Bangladesh, Sri Lanka, Nepal and Bhutan. This book is not to be sold outside these territories.

# Contents

List of Abbreviations ............................................................................. vii

List of Contributors ................................................................................. ix

Preface ..................................................................................................... xv

Chapter 1  Genetic Engineering and
Biotechnology of Growth Hormones ..................................................... 1

Chapter 2  Plasmid-Based Genetic Modification of Human Bone
Marrow-Derived Stromal Cells: Analysis of Cell Survival
and Transgene Expression after
Transplantation in Rat Spinal Cord ....................................................... 31

Chapter 3  HLA Engineering of Human Pluripotent Stem Cells ........................... 63

Chapter 4  Use of Recombination-Mediated Genetic Engineering for
Construction of Rescue Human Cytomegalovirus Bacterial
Artificial Chromosome Clones ............................................................. 97

Chapter 5  Genetically Engineered Mouse Models for Human Lung Cancer ..... 117

Chapter 6  Speech Analysis for Diagnosis of Parkinson's Disease Using
Genetic Algorithm and Support Vector Machine ............................. 155

Chapter 7  Pragmatic Issues in Biomarker
Evaluation for Targeted Therapies in Cancer ..................................... 171

Chapter 8  Enrichment of G2/M Cell Cycle Phase in Human Pluripotent Stem
Cells Enhances HDR-Mediated Gene Repair with Customizable
Endonucleases ...................................................................................... 217

Citations ................................................................................................ 277

Index ..................................................................................................... 279

# List of Abbreviations

| | |
|---|---|
| AAV | Adeno Associated Virus |
| ARMS | Amplification Refractory Mutation System |
| BAC | Bacterial Artificial Chromosome |
| BCIRG | Breast Cancer International Research Group |
| BCS | Bovine Chorionic Somatomammotropin |
| BDNF | Brain Derived Neurotrophic Factor |
| CCM | Complete Culture Medium |
| CCSP | Clara Cell Secretory Protein |
| CFGH | Canine Growth Hormone |
| CLIA | Clinical Laboratory Improvement Amendments |
| CS | Chorionic Somatomammotropin |
| EB | Embryoid Bodies |
| ECGH | Equine Growth Hormone |
| EGF | Epidermal Growth Factor |
| EGFP | Enhanced Green Fluorescent Protein |
| ELISA | Enzyme Linked Immunosorbent Assays |
| EMA | European Medicine Agency |
| ESC | Embryonic Stem Cell |
| FACS | Fluorescence Activated Cell Sorting |
| FCGH | Feline Growth Hormone |
| FISH | Fluorescence in Situ Hybridization |
| GA | Genetic Algorithm |
| GDNF | Glial Cell Derived Neurotrophic Factor |
| HCMV | Human Cytomegalovirus |
| HCS | Human Chorionic Somatomammotropin |
| HE | Hematoxylin Eosin |
| HLA | Human Leukocyte Antigens |
| HR | Homologous Recombination |
| LDT | Laboratory Developed Tests |
| MCMV | Murine Cytomegalovirus |
| MCS | Monte Carlo search |
| MSC | Marrow Derived Stromal Cells |
| NE | Neuroendocrine |
| NGF | Nerve Growth Factor |
| NHEJ | Non Homologous End Joining |
| NSCLC | Non Small Cell Lung Cancer |
| PCR | Polymerase Chain Reaction |
| PD | Parkinson's Disease |
| PE | Phycoerythrin |
| PNN | Probabilistic Neural Network |
| PR | Progesterone Receptor |
| QP | Quadratic Programming |
| ROI | Region of Interest |
| SISH | Silver in Situ Hybridization |
| SL | Somatolactin |
| SVM | Support Vector Machine |

# List of Contributors

**Jorge Angel Ascacio-Martínez**
Department of Biochemistry and Molecular Medicine, School of Medicine, Autonomous University of Nuevo León, Monterrey Nuevo León, Av. Madero Pte. s/n Col. Mitras Centro, Monterrey, N.L., México

**Hugo Alberto Barrera-Saldaña**
Department of Biochemistry and Molecular Medicine, School of Medicine, Autonomous University of Nuevo León, Monterrey Nuevo León, Av. Madero Pte. s/n Col. Mitras Centro, Monterrey, N.L., México

**Mark W Ronsyn**
Division of Clinical Pharmacology, University of Antwerp, Antwerp, Belgium
Laboratory of Experimental Hematology, Vaccine and Infectious Disease Institute (VIDI), University of Antwerp, Antwerp, Belgium

**Jasmijn Daans**
Laboratory of Experimental Hematology, Vaccine and Infectious Disease Institute (VIDI), University of Antwerp, Antwerp, Belgium

**Gie Spaepen**
Laboratory of Physiopathology, University of Antwerp, Antwerp, Belgium

**Shyama Chatterjee**
Laboratory of Pathology, University of Antwerp, Antwerp, Belgium

**Katrien Vermeulen**
Laboratory of Experimental Hematology, Vaccine and Infectious Disease Institute (VIDI), University of Antwerp, Antwerp, Belgium

**Patrick D'Haese**
Laboratory of Physiopathology, University of Antwerp, Antwerp, Belgium

**Viggo FI Van Tendeloo**
Laboratory of Experimental Hematology, Vaccine and Infectious Disease Institute (VIDI), University of Antwerp, Antwerp, Belgium
Centre for Cell Therapy and Regenerative Medicine, Antwerp University Hospital, Antwerp, Belgium

**Eric Van Marck**
Laboratory of Pathology, University of Antwerp, Antwerp, Belgium

**Dirk Ysebaert**
Laboratory of Experimental Surgery, University of Antwerp, Antwerp, Belgium
Centre for Cell Therapy and Regenerative Medicine, Antwerp University Hospital, Antwerp, Belgium

**Zwi N Berneman**
Laboratory of Experimental Hematology, Vaccine and Infectious Disease Institute (VIDI), University of Antwerp, Antwerp, Belgium
Centre for Cell Therapy and Regenerative Medicine, Antwerp University Hospital, Antwerp, Belgium

**Philippe G Jorens**
Division of Clinical Pharmacology, University of Antwerp, Antwerp, Belgium
Centre for Cell Therapy and Regenerative Medicine, Antwerp University Hospital, Antwerp, Belgium

**Peter Ponsaerts**
Laboratory of Experimental Hematology, Vaccine and Infectious Disease Institute (VIDI), University of Antwerp, Antwerp, Belgium
Centre for Cell Therapy and Regenerative Medicine, Antwerp University Hospital, Antwerp, Belgium

**Laura Riolobos**
Department of Medicine, University of Washington, Seattle, Washington, USA

**Roli K Hirata**
Department of Medicine, University of Washington, Seattle, Washington, USA

**Cameron J Turtle**
Department of Medicine, University of Washington, Seattle, Washington, USA
Program in Immunology, Clinical Research Division, Fred Hutchinson Cancer Research Center, Seattle, Washington, USA

**Pei-Rong Wang**
Department of Medicine, University of Washington, Seattle, Washington, USA
Department of Pediatrics, Second Hospital of Shandong University, Jinan, China

**German G Gornalusse**
Department of Medicine, University of Washington, Seattle, Washington, USA

**Maja Zavajlevski**
Department of Medicine, University of Washington, Seattle, Washington, USA

**Stanley R Riddell**
Department of Medicine, University of Washington, Seattle, Washington, USA
Program in Immunology, Clinical Research Division, Fred Hutchinson Cancer Research Center, Seattle, Washington, USA
3Institute of Advanced Study, Technical University Munich, Munich, Germany

**David W Russell**
Department of Medicine, University of Washington, Seattle, Washington, USA
Department of Biochemistry, University of Washington, Seattle, Washington, USA

**Kalpana Dulal**
Department of Microbiology and Molecular Genetics, UMDNJ-NJ Medical School, 225 Warren Street, Newark, New Jersey 07101-1709, USA

**Benjamin Silver**
Department of Microbiology and Molecular Genetics, UMDNJ-NJ Medical School, 225 Warren Street, Newark, New Jersey 07101-1709, USA

**Hua Zhu**
Department of Microbiology and Molecular Genetics, UMDNJ-NJ Medical School, 225 Warren Street, Newark, New Jersey 07101-1709, USA

**Kazushi Inoue**
The Department of Pathology, Wake Forest University Health Sciences, Medical Center Boulevard, Winston-Salem, NC, USA
The Department of Cancer Biology, Wake Forest University Health Sciences, Medical Center Boulevard, Winston-Salem, NC, USA
Graduate Program in Molecular Medicine, Wake Forest University Health Sciences, Medical Center Boulevard, Winston-Salem, NC, USA

**Elizabeth Fry**
The Department of Pathology, Wake Forest University Health Sciences, Medical Center Boulevard, Winston-Salem, NC, USA

The Department of Cancer Biology, Wake Forest University Health Sciences, Medical Center Boulevard, Winston-Salem, NC, USA

**Dejan Maglic**
The Department of Pathology, Wake Forest University Health Sciences, Medical Center Boulevard, Winston-Salem, NC, USA
The Department of Cancer Biology, Wake Forest University Health Sciences, Medical Center Boulevard, Winston-Salem, NC, USA
Graduate Program in Molecular Medicine, Wake Forest University Health Sciences, Medical Center Boulevard, Winston-Salem, NC, USA

**Sinan Zhu**
The Department of Pathology, Wake Forest University Health Sciences, Medical Center Boulevard, Winston-Salem, NC, USA
Graduate Program in Molecular Medicine, Wake Forest University Health Sciences, Medical Center Boulevard, Winston-Salem, NC, USA

**Mohammad Shahbakhi**
Department of Biomedical Engineering, Dezful Branch, Islamic Azad University, Dezful, Iran

**Danial Taheri Far**
Department of Biomedical Engineering, Dezful Branch, Islamic Azad University, Dezful, Iran

**Ehsan Tahami**
Department of Biomedical Engineering, Mashhad Branch, Islamic Azad University, Mashhad, Iran

**Armand de Gramont**
New Drug Evaluation Laboratory, Centre of Experimental Therapeutics, Department of Oncology, Centre Hospitalier Universitaire Vaudois (CHUV), 1011 Lausanne, Switzerland.

**Sarah Watson**
INSERM U830, Genetics and Biology of Paediatric Tumours Group, Institut Curie, France.

**Lee M. Ellis**
Departments of Surgical Oncology, and Molecular and Cellular Oncology, University of Texas MD Anderson Cancer Center, USA.

**Jordi Rodón**

Medical Oncology, Vall d'Hebron University Hospital, Vall d'Hebron Institute of Oncology (VHIO) and Universitat Autonoma de Barcelona (UAB), Spain.

**Josep Tabernero**
Medical Oncology, Vall d'Hebron University Hospital, Vall d'Hebron Institute of Oncology (VHIO) and Universitat Autonoma de Barcelona (UAB), Spain.

**Aimery de Gramont**
Medical Oncology Department, Institut Hospitalier Franco-Britannique, France

**Stanley R. Hamilton**
Division of Pathology and Laboratory Medicine, University of Texas MD Anderson Cancer Center, USA.

**Diane Yang**
Molecular and Cellular Biology Department, Baylor College of Medicine, One Baylor Plaza, Houston, TX 77030, USA

**Marissa A Scavuzzo**
Program in Developmental Biology, Baylor College of Medicine, One Baylor Plaza, Houston, TX 77030, USA

**Jolanta Chmielowiec**
Stem Cells and Regenerative Medicine Center, Baylor College of Medicine, One Baylor Plaza, Houston, TX 77030, USA
Center for Cell and Gene Therapy, Baylor College of Medicine, Texas Children's Hospital and Houston Methodist Hospital, Houston, TX 77030, USA

**Robert Sharp**
Stem Cells and Regenerative Medicine Center, Baylor College of Medicine, One Baylor Plaza, Houston, TX 77030, USA
Center for Cell and Gene Therapy, Baylor College of Medicine, Texas Children's Hospital and Houston Methodist Hospital, Houston, TX 77030, USA

**Aleksandar Bajic**
Jan and Dan Duncan Neurological Research Institute, Texas Children's Hospital, 1250 Moursund Street, Houston, TX 77030, USA

**Malgorzata Borowiak**
Molecular and Cellular Biology Department, Baylor College of Medicine, One Baylor Plaza, Houston, TX 77030, USA
McNair Medical Institute, Houston, TX 77030, USA

# Preface

Genetic engineering is the direct manipulation of an organism's genome using biotechnology. It could be used to change physical appearance, metabolism, and even improve physical capabilities and mental faculties such as memory and intelligence. The text *Pragmatism and Human Genetic Engineering* presents overviews and pro and con viewpoints on human genes and regulation of genetic engineering. First chapter focuses on genetic engineering and biotechnology of growth hormones. The aim of second chapter is to investigate the feasibility of a plasmid-based strategy for genetic modification of human (h)MSC with enhanced green fluorescent protein (EGFP) and/or neurotrophin (NT)3. Human leukocyte antigen (HLA) engineering of human pluripotent stem cells has been discussed in third chapter. In fourth chapter, we describe the construction of HCMV BAC mutants using a homologous recombination system. In fifth chapter, we present genetically engineered mouse models for human lung cancer. Sixth chapter proposes a new algorithm for diagnosing of Parkinson's disease based on voice analysis. In seventh chapter, we review the major challenges in biomarker validation processes, including pre-analytical (sample-related), analytical, and post-analytical (data-related) aspects of assay development. Last chapter discusses how enrichment of G2/M cell cycle phase in human pluripotent stem cells enhances HDR-mediated gene repair with customizable endonucleases.

# Chapter 1

# GENETIC ENGINEERING AND BIOTECHNOLOGY OF GROWTH HORMONES

Jorge Angel Ascacio-Martínez and Hugo Alberto Barrera-Saldaña

Department of Biochemistry and Molecular Medicine, School of Medicine, Autonomous University of Nuevo León, Monterrey Nuevo León, Av. Madero Pte. s/n Col. Mitras Centro, Monterrey, N.L., México

## INTRODUCTION

In its modern conception, biotechnology is the use of genetic engineering techniques to manipulate microorganisms, plants, and animals in order to produce commercial products and processes that benefit man. These techniques, which are the backbone of the biotechnological revolution that began in the mid 1970s, have permitted the isolation and manipulation of specific genes and the development of transgenic microorganisms that produce mainly eukaryotic proteins of therapeutic use, such as vaccines, enzymes, and hormones.

Biotechnology is present in diverse areas such as food production, degradation of industrial waste, mining, and medicine. Recent achievements include drug production in transgenic animals and plants, as well as the commercial exploitation of gene sequences generated by the human genome project and similar projects of plants and animals of commercial interest that are and will be in process.

Human growth hormone was, after insulin, the second product of this new technology. This product was developed and commercialized initially by Genentech, and was used clinically for treating growth problems and dwarfism (1). Furthermore, growth hormones from different animal species have also been produced in transgenic organisms and these have been used in different examples in the aquatic animal and livestock sectors.

## THE GROWTH HORMONE (GH) FAMILY

GHs belong to a family of proteins with structural similarity and certain common functions that include prolactin (Prl), somatolactin (SL), chorionic somatomammotropin (CS), proliferin (PLF) and proteins related to Prl (PLP) (2). This family represents one of the most physiologically diverse protein groups that have evolved by gene duplication. The two most studied members of this family have been GH and Prl, which have been described from primitive fish to mammals; however, other members of the family are not so amply distributed or studied.

## Structure of Growth Hormones

GHs (see Figure 1), in general, have a molecular weight of around 22,000 Daltons (22 kDa or simply 22k) and do not require post-translational modifications. They are synthesized in somatotrophs in the hypophysis, intervening as an important endocrine factor in postnatal somatic growth and lactation.

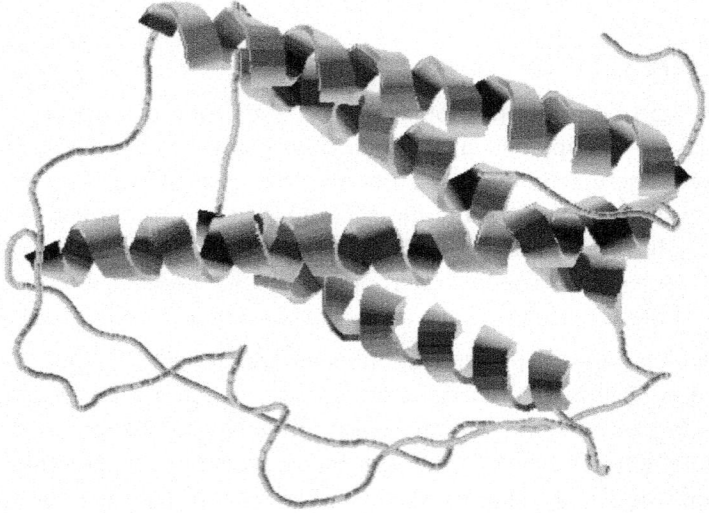

**Figure 1**. Growth hormones' consensus tridimensional structure. The GHs have in general 190 aminoacidic residues, four alpha helixes, and two sulphide bonds.

## Hormones of the Human Growth Hormone Family

### *HGH22k*

HGH22k (or HGHN) is the main product of the GH gene (hGH-N) active in the hypophysis and it is responsible for postnatal growth as well as being an important modulator of carbohydrate, lipid, nitrogen and mineral metabolism.

It is the best known hormone and the only one of the HGH family that has been commercialized.

As mentioned, besides being the cure for hypophyseal dwarfism, HGH22k postulated benefits are as an anabolic in athletics and for the treatment of trauma because of its postulated regenerative properties (3).

## HGH20k

In addition to the mRNA of HGH22k, an alternative processing pathway of the primary transcript of the hGH-N gene generates a second mRNA that is responsible for the production of the 20k isoform of HGH or HGH20k. Its smaller size is due to elimination of the first 45 nucleotides of the third exon of the mRNA and of the amino acids that correspond to positions 32-46 of the hormone, producing a protein with 176 amino acid residues (4).

This isoform comprises approximately 10% of all the GH produced in the hypophysis and although it has not been shown to be the etiological agent of any known disease, it is known that its levels are significantly higher in patients with active acromegaly and in those with anorexia nervosa (5).

The administration of exogenous HGH20k suppresses endogenous secretion of HGH22k in healthy subjects, which suggests that the regulation of secretion of both hormones is physiologically similar (6). In vitro findings suggest that both hormones can equally stimulate bone remodeling and allow anabolic effects on skeletal tissue when they are administered in vivo to laboratory animals (7).

## HGHV

Several isoforms also derive from the GH gene expressed in the placenta (hGH-V)(Table 1). The most abundant mRNA from this gene in the placenta at terminus also codifies for a 22 kDa isoform. A less abundant isoform (HGHV2) originate from a species of mRNA that retains the fourth intron and due to this, it codifies for a 26 kDa protein that anchors to the membrane and which could have a local action (8). A 25 kDa protein is also derived by glycosylation of residue 140 of asparagine from the 22 kDa isoform (9, 10). Finally, two new transcripts of this gene have recently been identified: one already known that as in the case of the HGH20k also produces a 20 kDa protein, and another novel splicing variation that results in a mRNA known as hGHV3, that traduces into a 24 kDa isoform (11).

**Table 1.** HGH-V isoforms generated by alternative splicing and processing

| Isoform | Size | Length | Characteristic |
|---|---|---|---|
| • HGH-V22k | 22kDa | 191aa | Main isoform. |
| • HGH-V25k | 25kDa | 191aa | Glycosylated version of HGH-V22. |
| • HGH-V2 | 26kDa | 230aa | Retains the fourth intron. |
| • HGH-V20K | 20kDa* | 176aa | Deletion of aa residues 32 to 46. |
| • HGH-V3 | 24kDa* | 219aa | Alternate processing at level of exon 4. |

*Only the mRNAs that codify each have been identified.

During pregnancy, while hypophyseal HGHN progressively disappears from the maternal circulation until undetectable values are reached at weeks 24 to 25, HGHV progressively increases until birth, suggesting that it has a key role during human gestation (12). It has also been found that in cases of intrauterine growth restriction, circulating levels of HGHV measured between week 31 and birth are lower than those reported in normal pregnancy (13, 14, 15).

Finally, although low levels of this hormone have been associated with intrauterine growth retardation, cases of hGH-V gene deletion have also been reported, but without an apparent pathology (16).

## Human Chorionic Somatomammotropin (HCS)

HCS is detected in maternal serum from the fourth week of gestation, increasing throughout the pregnancy in a linear fashion, and reaching high production levels of a couple of grams per day at the end of gestation. These actions result in both elevation of glucose and amino acids in the maternal circulation. These are in turn used by the fetus for his/her development. It also generates free fatty acids (by lipolytic effect), which are used as an energy source by the fetus (17, 18). Little is known about the HCS physiological role, and still is not known its action mechanism. Producing rHCS by biotechnology will help to advance these investigations.

## In Vitro Bioassays for GHs and CSHs

As stated above except for HGH22k, the functions of the rest of hormones of the human GH family have been not completely defined. Their biological activities are being studied, classifying them into at least two general categories:
- Somatogenic activities. These involve linear bone growth and alterations in carbohydrate metabolism; effects that are in part mediated by local and hepatic generation of insulinlike growth factor-I (IGF-I). The somatogenic activity of HGHV has been studied by stimulating body

weight increase in hypophysectomized rats, reporting a linear increase comparable to that produced by HGH22k (19).

- Lactogenic activities. These include stimulation of lactation and reproductive functions (20). The lactogenic activities of this hormone have been studied using a cell model (by mitogenic response to Nb2 cells) and a response that is parallel to HGH22k has been reported, although it is significantly less (19).

## The Human GH Locus

Besides the two hGH genes (normal and variant), three HCSs complement the multigenic HGH family from the human genome and these are arranged in the following order: HGHN, HCS-1, HCS-2, HGH-V y HCS-3 (21, 22) (Figure 2). While HCS-1 appears to be a pseudogene, HCS-2 and HCS-3 are very active in the placenta and interestingly; mature versions of the hormones that they codify are identical (23).

In the last few years, in our laboratory, all the hGH and HCS genes have been cloned and expressed in cell culture, and the factors that affect their levels of expression have been particularly studied (24).

In the same way, and using polymerase chain reaction (PCR) with consensus primers, several new genes and complementary DNAs (cDNAs) to the mRNA of numerous GHs have been isolated in our laboratory, mainly from mammals (unpublished results).

# GROWTH HORMONE OF ANIMAL ORIGIN

## Bovine Growth Hormone (BGH)

Bovine growth hormone (BGH) or bovine somatotropin improves the efficiency of milk production (per unit of food consumed) (25), and the production (body weight) and composition (muscle: fat ratio) of meat (26). In the case of milk cows, this permits a reduction in the number of animals needed for milk production and a subsequent savings in maintenance, feeding, water, drugs, etc. It also reduces the production of manure, and nitrogen from urine and methane (27).

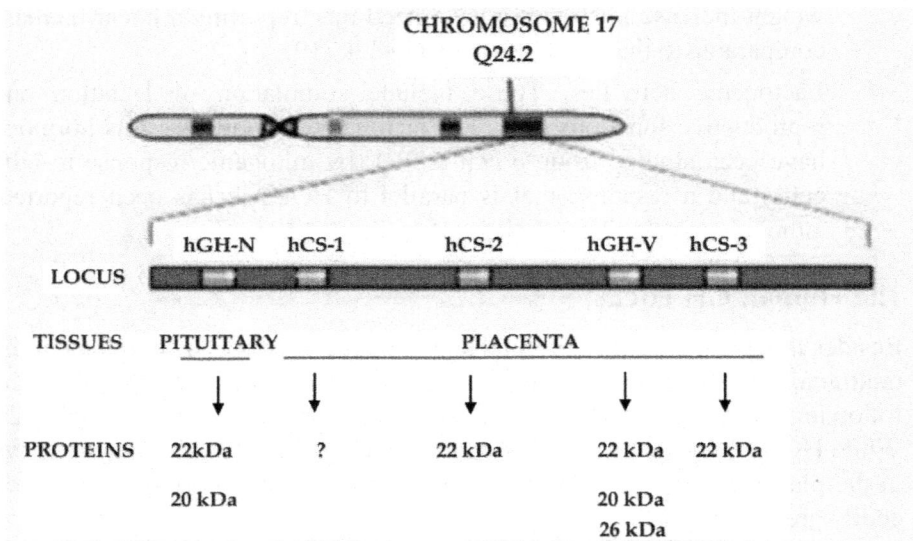

**Figure 2.** HGH-HCS multigenic complex. Located on Chromosome 17, every gene is indicated; the tissue where they are expressed and the proteic isoforms that are produced are shown.

Milk from cows treated with rBGH, does not differ from that of untreated cows (28, 29). The characteristics that have been evaluated include the freezing point, pH, thermal properties, susceptibility to oxygenation, and sensory characteristics, including taste; in fact all organoleptic properties are conserved. Also, differences have not been found in the properties necessary for producing cheese, including initial growth of the culture, coagulation, acidification, production and composition (29).

rBGH is administered subcutaneously and is dispensed as a long-acting suspension that is applied in a determined period of time. The taste of bovine meat and milk treated with rBGH is not altered, but the fat content is less.

## Caprine Growth Hormone (CHGH)

For small ruminants there are studies in lactating goats in which the administration of rBGH increased milk production 23% and stimulated mammary gland growth more than in those that were frequently milked, with it being similar to prolactin (30). However, the production of recombinant CHGH, which is identical to ovine and thus can be used in both animals, had not been reported, until we achieve its expression on the methylotrophic yeast Pichia pastoris. (See section 7.2).

## Equine Growth Hormone (ECGH)

With regard to horses, GH is used in the prevention of muscle wasting, in the repair of tendons and fractured bones, as well as for the treatment of anovulation in mares. Besides this, it is also used for repairing muscle tissue, to tonify and invigorate race horses, and for improving physical conditions in older horses by restoring nitrogen balance. It can also stimulate growth and early maturity in young horses, increase milk production in lactating mares and promote wound healing, especially of bone and cartilage (31, 32), as occurred in the case shown in Figure 3.

**Figure 3.** Uses of equine GH. The race horse "Might and Power" (right) became the winner of the Melbourne Cup in 1997. But in 1999, a tendon from one of its hooves was severely damaged. The horse was treated with ECGH, recovered and in 2000 was able to return to horse racing (32).

## Canine Growth Hormone (CFGH)

With regard to the dog (Canis familiaris), each day there is more evidence of the role that its GH (CFGH) plays in bone fracture treatment, in which the hormone helps reduce the bone restoration period (33). It is no less important in the treatment of obesity in dogs, thanks to the metabolism activation produced by the hormone in removing fatty acids, and in general, in counteracting symptoms related to the presence or absence of the same GH. Also, since this hormone is identical to pig GH (PGH) (33), its virtues are valid for the application of CFGH in the porcine industry, where it generates leaner meat (34), which is of greater value.

## Feline Growth Hormone (FCGH)

Although there is very little literature on cat GH (FCGH), the benefits identified in other GHs apply to this feline species, since these animals present the symptoms mentioned before for dogs, which are caused by the absence or low concentration of FCGH (dwarfism and alopecia, among others). Also, as referred to in the literature, biological tests of adipogenic activity in culture cells use cat serum (which contains FCGH) instead of bovine serum, because FCGH lacks adipogenicity (17). Therefore, recombinant production of this GH would be useful in the mentioned tests. It is important to point out the usefulness that recombinant FCGH would have in future research on the metabolic study and role of this hormone in this and other feline species, including of course, large wild cats in captivity.

## BIOLOGICAL POTENTIAL OF GHS

### Growth Hormones of Human Origin

Although HGH22k is widely commercialized and more functions now have been recognized to it (Table 2), the same does not occur with the other proteins and isoforms from this family; essentially the 20 kDa isoform of HGH, HGHV, also the isoform of 20 kDa of the latter (HGHV20k), and lastly, HCS. Partly because of this, many of their functions and mechanisms of action are still unknown.

**Table 2.** New functions atributed to HGH22k

| Immunization and healing | Mental function |
|---|---|
| • Resistance to common diseases<br>• Ability to heal<br>• Healing of old lesions<br>• Healing of other lesions<br>• Ulcer treatment | • Emotional stability<br>• Memory<br>• General aspect and attitude<br>• Mental energy and clarity |
| **Skin and hair** | **Muscle strength and tone** |
| • Skin elasticity<br>• Skin thickness<br>• Skin texture<br>• Growth of new hair<br>• Disappearance of wrinkles<br>• Skin hydration | • Increase in energy in general<br>• Increase muscle strength<br>• Promotion of muscle mass gain |
| **Sexual factors** | **Circulatory system** |
| • Duration of an erection<br>• Increase in libido<br>• Potential/frequency of sexual activity<br>• Regulation and control of the menstrual cycle<br>• Positive effects in the reproductive system<br>• Increase in breast-milk volume | • Improvement in circulation<br>• Stabilization of blood pressure<br>• Improvement in cardiac function |
| **Bone** | **Fats** |
| • As treatment for bone fractures<br>• Osteoporosis treatment<br>• Increases flexibility of the back and joints | • Increases "good" cholesterol (HDL) levels<br>• Reduces fat |

(Taken from Elian y cols., 1999), (3).

It is believed that some of the hormone's less abundant natural variants, such as HGH20k, could retain desirable properties of the principal hormone and lack some of its other undesirable effects, such as its diabetogenic effect, which occurs with prolonged use (35).

## Growth Hormones of Animal Origin

The biotechnological potential of GHs could be enormous, since besides its use in species of the same origin, it has been demonstrated that the GHs of mammals have activity in phylogenetically lower animals. For example, BGH and porcine GH (PGH) have been used experimentally for the treatment of hypophyseal dwarfism in dogs (36) and cats (37). Regarding farm animals, porcine, bovine, caprine and ovine livestock have been treated with exogenous GH to improve production, since it increases food conversion efficiency, growth rate, weight gain, and milk and meat production. What is surprising is the finding that BGH stimulates salmon growth, and even more interesting that bovine chorionic somatomammotropin (BCS) works even better (38).

# EXPRESSION SYSTEMS FOR GROWTH HORMONES

## The History of Human Recombinant GH

As previously mentioned, among the first cDNAs cloned and expressed in the bacteria Escherichia coli is precisely HGH (1). This expression system has been used since 1985 for the production of recombinant HGH by Genentech (protropin), which was later followed by Lily (humatrope), Biotech (biotropin), Novo Nordisk (norditropin), Serono (serostim), and others.

## Different Biotechnological Hosts

Since the recombinant protein is frequently recovered from E. coli with undesirable modifications (extra methionine, incorrect folding, aggregated forms, etc.) and contaminated with highly pyrogenic substances, toilsome purification schemes are needed to obtain it with the desired purity, structure and biological activity. For this, subsequent efforts have focused on the search for better expression systems, with production being attempted with *Saccharomyces cerevisiae* (39), *Bacillus subtilis* (40), mammal cell cultures (41), as well as transgenic animals (42). Unfortunately, these expression systems do not offer a production level greater than that of E. coli and therefore in most cases they are not profitable.

In our laboratory, we succeeded in producing HGH22k in E. coli by fusing it with maltose binding protein (rHGH-MBP) in 1994 (unpublished results). However, due to the fact that to recover the hormone, whether from the periplasm or the cytoplasm, complicated strategies were needed, together with the limitations of the bacterial systems for folding and processing foreign proteins correctly, we proposed searching for an expression system that allows synthesizing the protein, purifying it with greater ease while retaining functionality. Thus, the evaluation of different expression systems was started in our laboratories, considering the methylotrophic yeast *Pichia pastoris* as the best (43).

## Pichia pastoris as a Biotechnological Host for GHs

Yeasts offer the best of both prokaryotes and eukaryotes, since, in addition to performing some of the post-translational modifications that are common in superior organisms, they are easily grown in flasks and bioreactors, like bacteria, using simple and inexpensive culture media (44).

*P. pastoris* is a methylotrophic yeast (capable of growing in methanol as its only carbon source) that performs post-translational modifications, produces recombinant protein levels of one or two orders of magnitude above that of *Saccharomyces cerevisiae* (45), is capable of secreting heterologic proteins into the culture media (where the levels of native protein are very low), and in contrast with the latter, can be cultivated at cell densities of more than 100 g/L of dry weight (46).

# RECOMBINANT GROWTH HORMONES

In our laboratories, we identified as a scientific objective and a technological advantage, the construction and evaluation of GH protein producing *P. pastoris* strains. This as a first step in evaluating its potential in medicine as well as in animal health and productivity; searching to develop both infrastructure and experience in producing, purifying, and testing its biological activity.

Also, as previously mentioned, mammalian GHs have activity in phylogenetically inferior animals, nevertheless potentially adverse reactions to heterologic GHs can be triggered, which is why having a GH specific-species would avoid these undesirable side effects.

Regarding human hormones, we proposed constructing productive strains for the HGH22k, the HGH20k, the HGHV, and the HCS. With regard to animal GHs, we channeled our efforts into building strains to produce GHs from bovines (BGH), caprines (CHGH), ovines (OGH), equines (ECGH), canines (CFGH), porcines (PGH) and felines (FCGH); all based on the *Pichia pastoris* yeast expression system.

For this, the following experimental strategy was proposed:

- Obtain, clone, and manipulate cDNA from these hormones.
- Construct and insert into the genome of P. pastoris the hormones' expression cassettes.
- Develop the fermentation processes for each new strain.
- Implement the purification schemes of the recombinant hormones.
- Evaluate in vitro the bioactivity of the semipurified recombinant hormones.

As a result of this experimental work, we achieved the followings:

- Using different methodological approaches (RT-PCR, mutagenic PCR, subcloning, etc.) we cloned the cDNAs of the hormones of interest.
- Through genetic engineering manipulations, we converted the cloned cDNAs into expression cassettes capable of functioning in *Pichia pastoris*.
- The respective expression cassettes were integrated into the *Pichia pastoris* genome by homologous recombination.
- Through an inducible (with methanol) expression system, we were able to overproduce and recover from the culture media each of the respective recombinant hormones (rGHs/rHCSs).
- The data from the physicochemical and biological characterizations showed that the methodology described herein generates heterologous proteins that are identical to their natural counterparts and biologically active.

# TECHNOLOGICAL PLATFORM FOR THE PRODUCTION OF RECOMBINANT GHS

## Overall Strategy

As depicted in figures 4, the following are the two stages of the overall strategy in which the work was divided:

- Construction of *P. pastoris* strains carrying the hormones' expression cassettes producing rGHs/HCSs.
- Production and characterization of the recombinant hormones.

## Construction of Propagating GH cDNA Plasmids (pBS-XGHs)

Oligonucleotides for GHs cDNAs amplification by PCR were designed based on consensus nucleotide sequences of GHs (mature region) of related mammals. Extra restriction sites were added on their flanks (XhoI and AvrII) to facilitate insertion of the amplicon into the expression vector. With these, each of the hormones' cDNAs was amplified from plasmids previously constructed in our laboratory carrying the respective nucleotide sequences. Each amplicon was cloned into propagating plasmid such as the pBS II KS plasmid (+) and subsequently subcloned into the yeast expression vector pPIC9 at its multicloning site, between the restriction sites XhoI and AvrII (after previous purification of the corresponding fragment and vector), thereby giving rise to each of the pPIC9-XGH expression plasmids.

In CHGH's case, which differs from BGH in a single aa residue, a different strategy was implemented. Site-directed mutagenesis was used relying on a primer to convert codon 130 of BGH cDNA into one corresponding to CHGH. A 345 bp region containing the mutated GH cDNA was thus amplified, which was cloned in pBS and later transferred into pPIC9- BGH to converted it into pPIC9-CHGH (49).

## Construction of Expression Plasmids (pPIC9-XGHs) for Each Hormone

Preparative digestions of pBS-XGH and pPIC9 with the enzymes XhoI and AvrII were performed for all GHs except for CHGH. For CHGH, ApaI and XmaI (natural site) enzymes, which release a 133 bp fragment containing the mutagenized codon for CHGH, were used. This was purified and linked into the previously digested pPIC9-BGH vector in the same sites, replacing the fragment to originate the pPIC9-CHGH expressor vector The ligation reactions between pPIC9 and each cDNA were used to transform competent Ca++ cells of XL1-Blue Escherichia coli. PCR was used to verify that the resulting tranformants indeed carried each pPIC9-XGH, where "x" corresponds to each of the sequences of the hormone in question. The candidate clones produced by PCR with AOX1 primers for an amplicon of 1050 bp, since the expression cassette for each hormone is flanked by long regions of the AOX1 gene. While strains that were not integrated into the "cassette" gave rise to an amplicon of only 500 bp.

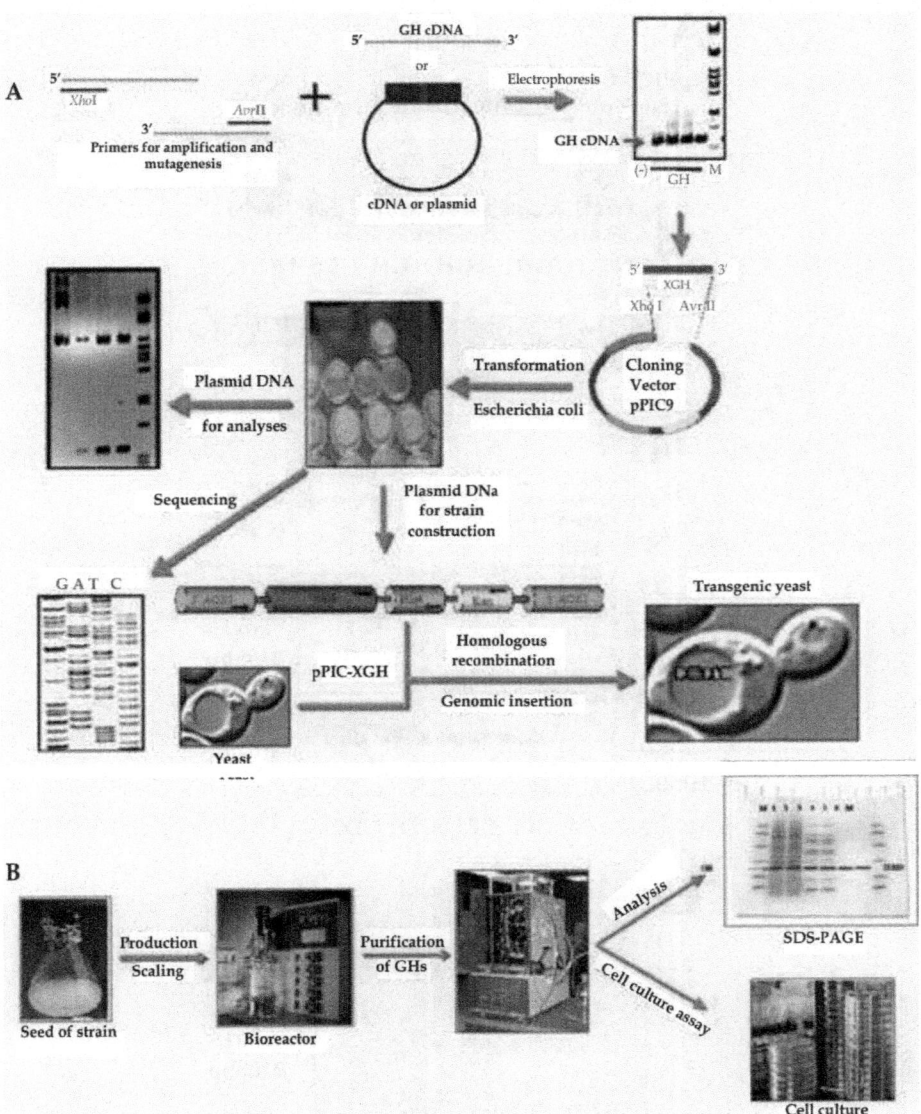

**Figure 4.** General strategy for strain construction and recombinant hormones production. (A) Genetic engineering phase. The steps followed to construct and characterize new strains of GHs and HCSs producing Pichia pastoris are shown. Protocols followed were based in different techniques (47, 48). (B) Biotechnology phase. The steps followed for the production and scaling, semipurification and bioassay of each of the recombinant hormones are shown.

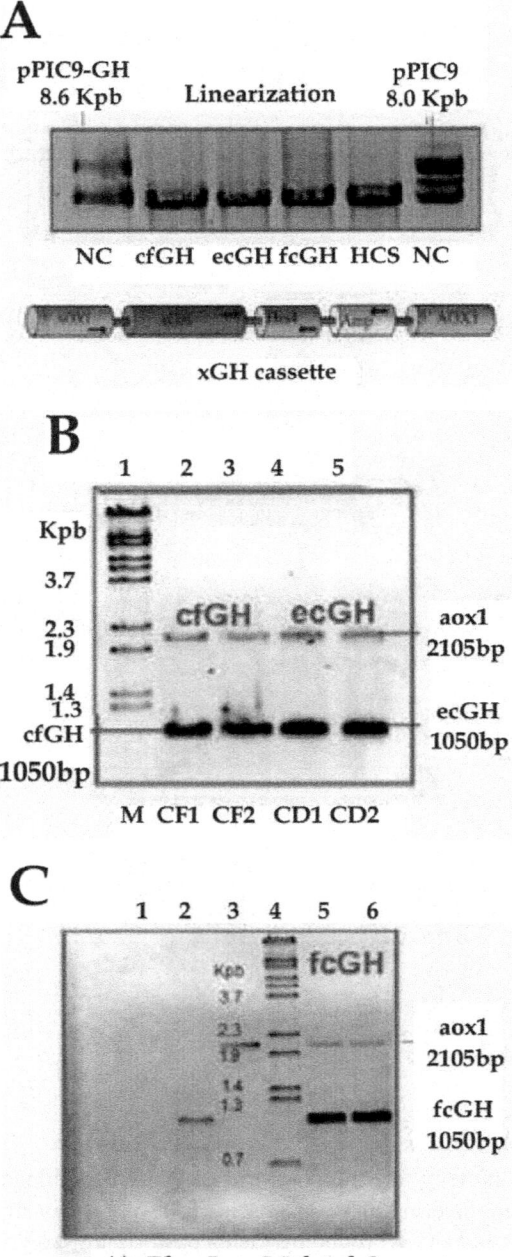

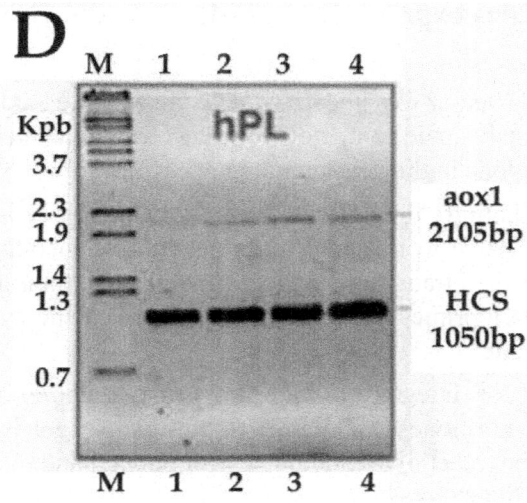

**Figure 5.** Detection of the expression "cassettes" of cfGH, ecGH, fcGH and hCS in P. pastoris genome. Analysis by PCR with AOX1 primer yeast strains transfected. In each case, the 1050 bp corresponds to the expression cassette of the recombinant hormone in question, while the 2105 bp to the AOX1 gene of the yeast itself. A) The diagram shows the linearized "cassette" of XGH and the gel products (which transfected into *Pichia pastoris* integrate the "cassette" into the genome) with SacI enzyme: cfGH = dog GH, ecGH = horse GH, fcGH = cat GH and HCS = human CS; in lane 1 NC-GH = uncut plasmid pPIC9 and in lane 6 NC = uncut pPIC9 plasmid. B) CF (1 and 2) = dog GH lanes 2 and 3 respectively, and CD (1 and 2) = horse GH lanes 4 and 5, respectively. C) (-) = negative PCR lane 1, Plas= amplification positive control lane 2, Lev = *Pichia pastoris* genomic DNA lane 3, M= pb marker lane 4 and fc (1 and 2) = cat GH lanes 5 and 6 respectively. D) Lanes 1 to 4 correspond to strains with the HCS "cassette". M = marker-bp λ BsteII. The gels correspond to 1% agarose.

## CONSTRUCTION OF GHS PRODUCING *PICHIA PASTORIS* STRAINS

The *Pichia pastoris* GS115 strain has a mutation in the histidinol dehydrogenase (his4) gene, which prevents it from synthesizing histidine. The class of plasmids used to transform it contains this gene (his4). The transformants are selected for their ability to restore growth in a medium lacking histidine. The plasmid vectors of the pPIC series and those constructed to express the GHs are of this class.

## Insertion of GHs Expression Cassettes into the Genome of *P. pastoris*

Each pPIC-XGH vector was linearized with the enzyme SacI, transformed into the yeast previously made competent for transformation and left exposed to the homologous regions in the yeast genome necessary for recombination.

After incubating the DNA with competent cells, transformation reactions were plated to recover clones needing no histidine to grow (HIS+ transformants). Then transformants were analyzed on their genomic DNAs by PCR using AOX1 primers to verify the presence of the transgenic hormone expressing cassette.

Verification of integration into the yeast genome of the expression cassette of the hormones was achieved in agarose gel by confirming that the amplification reaction rendered a prominent band of 1050 bp, which corresponds to the expression cassette of the hormone involved in each case and another of 2105 bp corresponding to the endogenous gene AOX1 of the yeast genome (Figure 5). In addition, each hormone "cassette" was subjected to nucleotide sequencing to verify that all they corresponded to the expected growth hormones.

## Analysis of New *Pichia pastoris* Strains´ Phenotypes

*Pichia pastoris* strains were grown and the biomass was adjusted to low cell density (0.5 u at 600 nm).

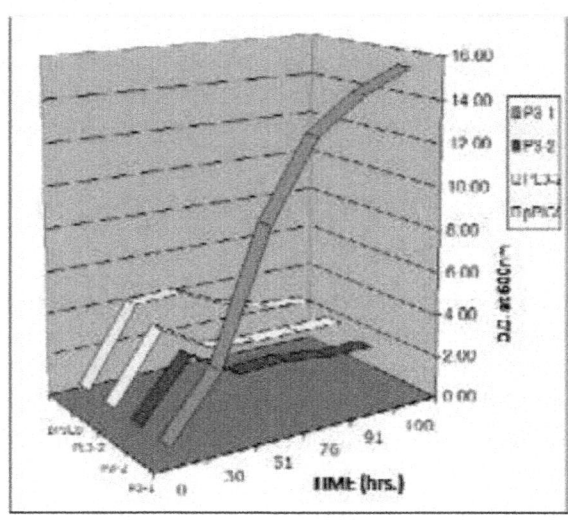

(A)

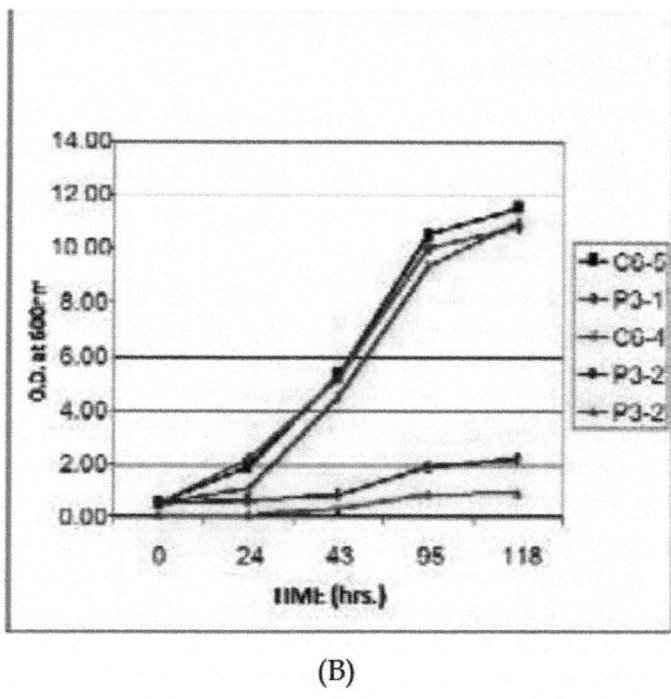

(B)

**Figure 6.** Mut phenotype characterization in Pichia pastoris strains. Growth kinetics in minimal medium using methanol as sole carbon source. Induction was started at the density of 0.5 U and ended after about 100 hrs. (A) Plot of the samples P3-1 and 2 = dog GH 1 and 2 strains, CS3 CS-2 = 2 and strain human strains pPIC9 = "mock" with the pPIC9 plasmid. (B) Graph of the C6-5T samples = horse GH, P3-1T =dog GH, C6-4T = horse GH, GH P3-Q2 = P3-dog and dog-2B = GH.

These were transferred to induction medium with 0.5% methanol and grown for 100 hours with the addition of methanol every 24 hours to compensate for its evaporation. Biomass growth was analyzed under methanol as the sole carbon source. The Mut+ phenotype strains metabolize methanol more rapidly, achieving significantly higher cell densities than their Muts counterparts that metabolize more slowly, appreciating a slight increase in biomass under the same fermentation conditions.

An analysis of the growth of the strains after 100-hour fermentation with 0.5% methanol identified the Mut phenotype of each strain.

After fermentation, the strains found to be Mut+ reached about 15 optical units at 600 nm, while those that were Muts did not exceed 2 units (Fig. 6). In the strains that had been built previously, their Mut phenotype was inferred when these were fermented in the bioreactor.

## PRODUCTION AND ANALYSIS OF RECOMBINANT HORMONES IN THE FLASK

To test the fermentation of strains, a biomass was generated in a flask. This was inoculated with a colony of each strain in 25 ml of biomass producing culture medium (BMGY) (50). This was incubated at 30°C at 250 rpm for 24 to 48 hours for the first stage of growth until a biomass with an OD of 600 nm of 10 was reached.

For the second stage, which is the induction of the recombinant hormone production, yeasts were harvested by centrifugation and the packed cells were washed with 30 mL sterile water, then these were pelleted and resuspended in fresh cassette induction medium (with methanol) (BMMY) (50). The induction was maintained by adding methanol every 24 hours to a final concentration of 1% to compensate for loss by evaporation. The experiment lasted 96 hrs. Figure 7 shows the process that was followed.

When analyzing the polyacrylamide gels of proteins from the culture media of each strain, we observed that all constructions produced and directed the secretion into the medium of the recombinant hormone in question to a greater or lesser extent. For the particular case of CFGH it was observed that a strain of Muts phenotype displayed better production of recombinant protein than its counterpart Mut+. Strain of HGH-V proved to be the least productive (Fig. 8).

Figure 9 shows the gel with the results of the production of all recombinant strains generated in *Pichia pastoris*. They all produced different amounts of their respective hormone at the level of 22 kDa, except for the HGH20k, which, migrated below the rest of the recombinant hormones.

The percentage of each recombinant hormone in the culture medium was estimated by densitometry of each gel. For this we used the Gel-Doc software by BIO-RAD (Hercules, CA. EUA) and the ImageJ program (51). The results of estimation of the percentage of each hormone in relation to background proteins from *Pichia pastoris* were: HCS = 65%, CFGH = 60%, HGH22k = 30%, ECGH = 30%, BGH = 25%, FCGH = 25%, CHGH = 25%, HGH20k = 12% and HGHV = 8%.

Production kinetics was carried out for CHGH strain in a flask with a volume of 50 ml of rich medium. Samples were taken at 24, 48, 72, 96 and 120 hours of induction with methanol with restitution every 24 hours of 1% methanol. Bradford protein determination showed that the production of total protein secreted into the culture medium was 20μg/mL by densitometry and 60% represented CHGH giving us 12μg/mL of production of the recombinant hormone.

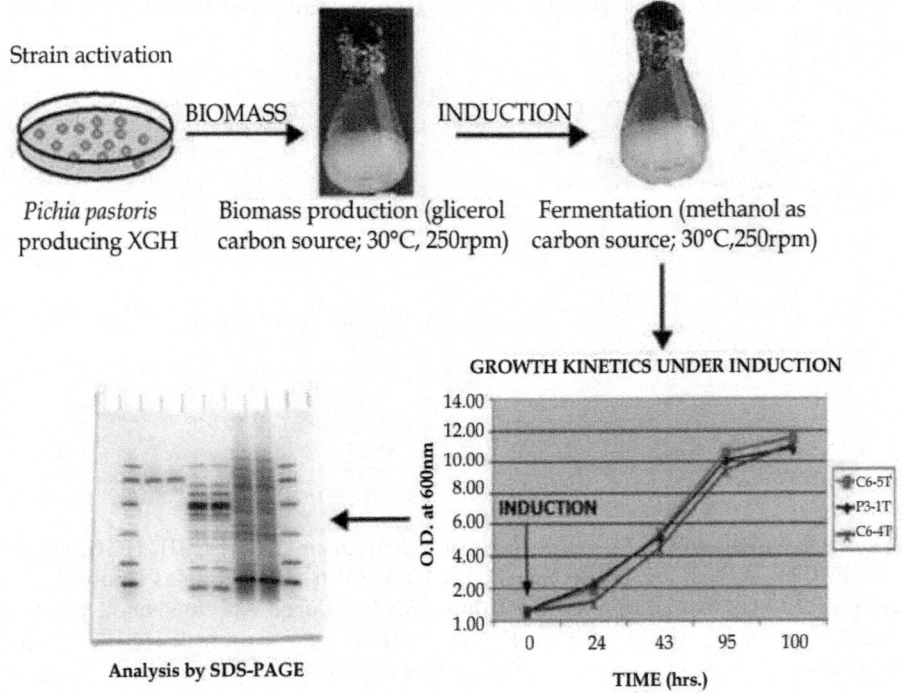

**Figure 7.** Outline of the fermentation process. General procedure for the biotechnological production of recombinant hormones by fermentation of each strain. Strains were plated to activate them, incubated in liquid medium to generate a biomass flask, and the induced transgene expression by adding methanol. The culture medium was analyzed by SDS-PAGE in search of the hormones that are migrating around 22 kDa

## PRODUCTION SCALING IN THE BIOREACTOR

When passing to a bioreactor and increasing the scale, it is possible to obtain protein concentrations 20 to 200 times greater than in flasks. In the fermentor *Pichia pastoris* reaches high cell densities greater than 100 g/L of dry weight (46).

The fermentor was Bioflo 3000 (1 liter) of New Brunswick Scientific (NBSC) (NJ. EUA). The type of fermentation conducted was in fed-batch. The parameters monitored were scheduled addition of substrates to the fermentor, pH, percentage of dissolved oxygen, agitation, temperature, and aeration. The process involved three basic steps: 1) obtaining high densities of biomass, 2) induction of the cassette expression of each hormone with methanol and 3) harvest of biomass and culture medium containing the recombinant protein.

Figure 10 shows the steps followed for the recombinant production of each hormone.

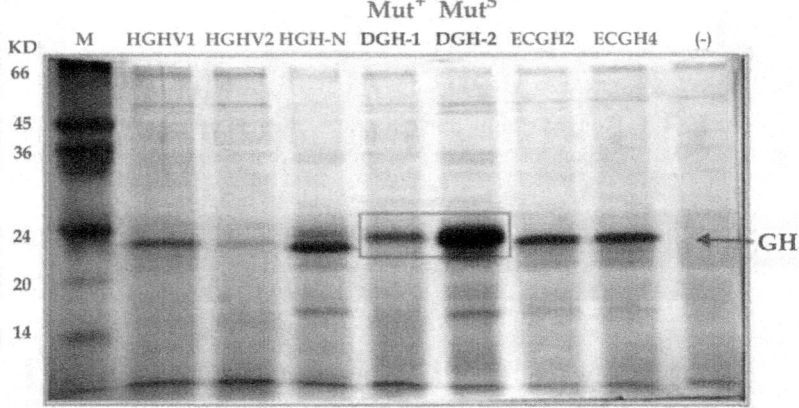

**Figure 8.** Production in *Pichia pastoris* of recombinant GHs (CFGH, HGH, HGH-V and ECGH) at flask level. The bands correspond to the GHs proteins resolved at the level of 22 kDa that come from the culture media induced with methanol. The lanes are: M = molecular weight marker, HGHV1 = HGH variant strain 1; HGHV2 = HGH variant strain 2; HGH = normal pituitary HGH of 22 kDa; DGH-1 = Dog GH strain 1; DGH-2 = dog or *Canis familiaris* GH strain 2; ECGH-2 = horse GH strain 2, ECGH-4 = horse GH strain 4 and the last lane identified as (-) = negative control of PCR. Mut+= methanol utilization plus; Muts = "methanol utilization slow". Note the prominent band of the DGH-2 corresponding to the Muts phenotype, compared to the lower intensity of Mut+. The samples correspond to 500 μL concentrates of the original media. Gel corresponds to one of 15% polyacrylamide-SDS stained with Coomassie blue.

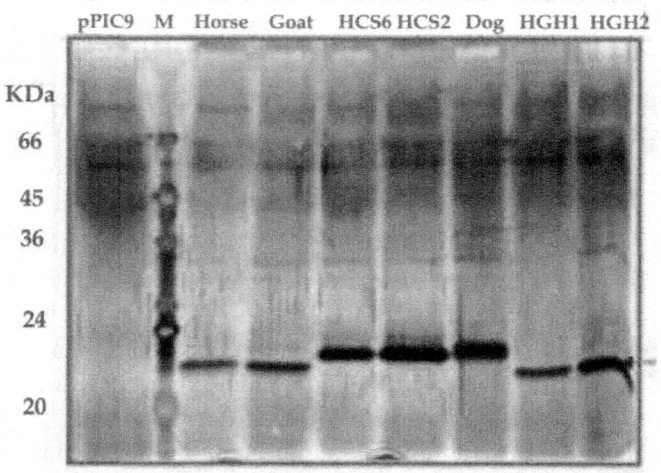

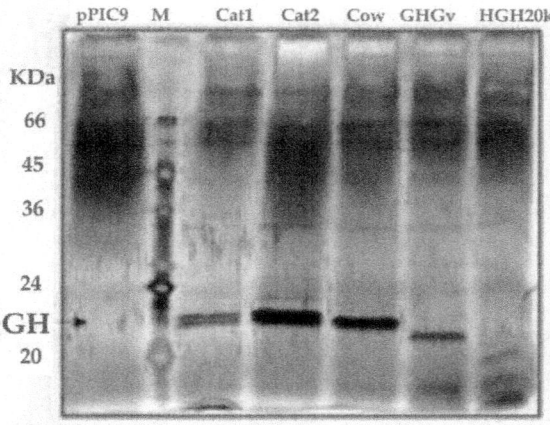

**Figure 9.** Flask production in all strains yielding recombinant hormones. In all cases a prominent band is seen (except for HGH20k) at the 22 kDa level for each hormone. They are seen in their respective lane in each case indicated by their name. The lanes of the left side gel show the proteins from the culture media with recombinant hormones: horse = ECGH, goat = CHGH, HCS (6 and 2) = Human chorionic somatomammotropin clones 6 and 2, dog = CFGH and HGH (1 and 2) = cloned human GH 1 and 2. In the right gel lanes: cat (1 and 2) = GH 1 and 2 from cat; cow = BGH; HGHv = HGH placental variant and HGH20k = isoform of 20 kDa of hGH. The pPIC9 lane refers to the "mock" strain of *Pichia pastoris*. SDS-PAGE 15% gels are silver stained.

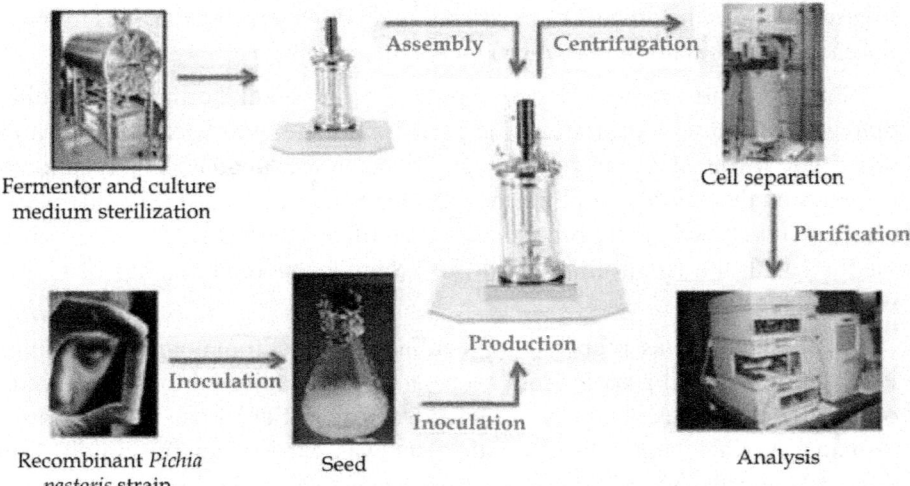

**Figure 10.** Recombinant hormone production bioreactor. This illustrates the stages of the biotechnological process of production, from the preparation of the fermentor and the medium, to the analysis of proteins in the fermented culture medium.

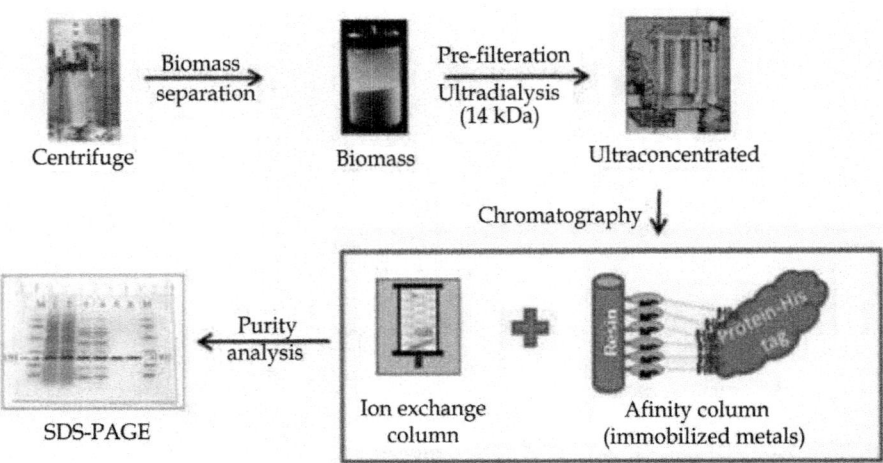

**Figure 11.** "DownStream" Process for recombinant hormones. The medium was separated from the biomass by centrifugation. This was pre-filtered with a 0.45μm membrane, and ultra-dialyzed in a cutoff membrane of 10 kDa. It was then passed through an anion exchange column (FF-QS) and/or in an affinity column by immobilized metal (IMAC). Purity was analyzed in PAGE-SDS, quantified, and finally lyophilized and stored for later use.

## SEMIPURIFICATION OF RECOMBINANT HORMONES

Figure 11 shows the process of purification or "downstream" that was followed for each recombinant hormone produced.

Each culture medium containing the recombinant hormone was ultra-dialyzed and ultraconcentrated. The pore membrane used was 14 kDa. At the same time it was ultradyalized. The ultraconcentrate obtained was lyophilized to preserve the samples and all the powder was recovered in 50 mL plastic tubes and weighed. Total protein was quantified in each lyophilized culture medium with the Bradford method (52). Samples were stored at -20°C until use.

Each sample was prepared for loading into a chromatographic column. Five mg of the total protein from each sample was adjusted to the conditions of the loading buffer. A column was loaded with Q-Sepharose fast flow anion exchange resin; column filling with resin was carried out by gravity flow. After passing the sample, washing was done with 10 mL of loading buffer. To recover the proteins from the column, 15 mL of elution buffer (loading buffer plus NaCl) was passed with an ionic strength increased sequentially with NaCl. Total protein was measured by the Bradford method (52) for each of the fractions recovered to see what percentage of the total they represented. The collected fractions were visualized on discontinuous polyacrylamide gel concentrations of 4-15% under denaturing conditions (SDS-PAGE), and stained with Coomassie and silver techniques (53). The fractions were subjected to lyophilization, the powder was recovered and weighted, total protein was measured by the Bradford method. In addition, using PAGE-SDS the percentage of purity of the monomer was determined with ImageJ software. The samples were stored at -20°C until used for further analysis or to determine their biological activity.

# TESTING THE BIOLOGICAL ACTIVITY OF RECOMBINANT HORMONES

## Lactogenic Activity Bioassay

The biological activity of hormones produced in *Pichia pastoris* recombinant was determined by their ability to promote proliferation of the Nb2 cell line, which comes from rat lymphoma (54, 55). GHs were also tested for their somatogenic activity in the adipocyte differentiation model based on the glyceraldehyde 3-phosphate dehydrogenase (GPDH) assay.

The cell-free culture medium was dialyzed and each hormone was quantified by gel densitometry. Dilutions of each recombinant hormones in the culture medium dialyzed were tested. Cell proliferation was determined by tetrazolium salt assay (MTT) (56) and was expressed as the average of three repetitions, in comparison with the positive control recombinant rat prolactin (rRPRL) and the negative control (culture medium without hormone). The concentrations tested were 0.001, 0.01, 0.1, 1, 10 and 50 nM of the following hormones, CFGH, ECGH, FCGH, HGH, HCS and RPRL. Upon completion of the testing time, we proceeded to measure the effect of hormones on cell proliferation by MTT assay (56). The activity was evaluated by color generation based on the reduction of tetrazolium salt (methyl 3- [4,5-dimethylthiazol-2-yl] -2,5-diphenyl tetrazolium) from yellow to purple forming crystals by Nb2 cell metabolism. An increase of living cells is reflected by increased metabolic activity. This increase directly correlates with the formation of absorbance monitored formazan crystals; i.e., the greater the number of cells, the greater the increase in cell metabolism with greater formazan formation and greater biological activity.

The bioassay was carried out in triplicate in a humid atmosphere with 5% $CO_2$-95% air at 37°C. After the incubation period of 3 days, we added 10μL of MTT (to a final concentration of 0.5 mg/mL) to each well. Samples were incubated for 4 hrs under the atmospheric conditions mentioned above. After this time, we added 100 uL of formazan solubilizing solution (10% SDS in 0.01 M HCl) to each of the wells. These were left incubating overnight under the same atmospheric conditions. It was verified that the precipitate of formazan purple crystals had completely dissolved and absorbance was measured at 590 nm using an ELISA plate reader.

**Somatogenic Bioassay**

To test the biological activity of goat GH, we assayed fibroblasts from cell line 3T3-F442A (pre-adipocytes) and with the glyceraldehyde 3-phosphate dehydrogenase (GPDH) assay as described in (18, 57). Cells were exposed to different concentrations of CHGH in the medium. The positive control was 10% fetal bovine serum (FBS) (v/v) the negative control was 0.25% FBS (v/v). The cells were incubated for 12 days, were harvested and proteins extracted; the supernatants were frozen in aliquots for subsequent tests. Specific activity of GPDH was measured in cell extracts by NADH oxidation measured in a spectrophotometer at 340 nm; 40 μg of protein was used. CHGH demonstrated biological activity in the essay. The same was doing with the others hormones showing biological activity.

# CONTRIBUTIONS

These are the first reports in the literature of the production by biotechnology of the recombinant GHs described here using the Pichia pastoris methylotrophic system. The new strains of Pichia pastoris constructed with GH cassettes such as canine-porcine, equine, feline, and caprine, complemented our strains previously constructed for GHs of 22 y 20 kDa, HCS and bovine GH (unpublished results), making our collections of GH clones the largest one for this protein family in the world as far as we know.

All hormones were efficiently produced, processed, and secreted into the culture media. Each hormone constituted the main proteic band among proteins secreted by Pichia pastoris analyzed by SDS-PAGE. The phenotypes Muts resulted the best for producing GHs, as was the case for CFGH (58) and for HCS (Ascacio-Martínez y Barrera-Saldaña, unpublished results).

All our constructed strains had correct processing of their heterologous secretion Saccharomyces cerevisiae alpha mating factor signal peptide in the maturation of the recombinant GHs. They were secreted into the culture media in their native and bioactive form (49, 59, 60).

Using a bioreactor increased the production of the recombinant proteins 10 to 20 times compared to an Erlenmeyer flask. Ionic chromatography was a good option in all cases for semipurification of rGHs and HCS in this system. All hormones showed biological activity in the Nb2 essay, showing that human GH had more activity than animal GHs. The same happened in the pre-adypocite system (3T3), concluding that *Pichia pastoris* produces, processes, and secretes rGHs in the bioactive form.

Biotechnological platforms were developed that made possible to move from the construction of the producer clones to the bioassay of the semipurified recombinant protein. With the technology here developed we have acquired the capacity to advance in scaling the production of veterinary and livestock rGHs for their field tests.

## PERSPECTIVES

We have developed an efficient expression system and laboratory fermentor-scale production biotechnological platform with which to partner with the productive sector to produce virtually any GH of human and animal origin with acceptable quantity, quality and activity to start field evaluations. Doing that, would allow us investigate their full potential in animal biotechnology, to then offer them as an option in veterinary treatments and to stimulate cattle production and the health of competition animals. In addition to the obvious veterinary or livestock application, their availability will also allow the discovery of unexpected biological activities for animal wellness.

## ACKNOWLEDGEMENTS

The authors want to thank J.M. Reyes, J.P. Palma, C.N. Sanchez, L.L. Escamilla, H.L. Gallardo, E.L. Cab, R.G. Padilla and M. Guerrero for their support, experiments during their thesis and valuable contributions to the information here reviewed. Authors thank Sergio Lozano for his critical reading of the manuscript.[

## REFERENCES

1. Goeddel D.V., Heyneker H.L., Hozumi T., Arentzen R., Itakura K., Yansura D.G., Ross M.J., Miozarri G., Crea R., Seeburg P. Direct expression in Escherichia coli of a DNA sequence coding for human growth hormone. Nature 281(5732): 544-548, (1979).
2. Niall H.D., Hogan M.L., Sayer R., Rosenblum I.Y., Greenwood, F.C. Sequences of pituitary and placental lactogenic and growth hormones: evolution from a primordial peptide by gene duplication. Proc. Natl. Acad. Sci. USA, 68: 866-869, (1971).
3. Elian, G., Jamieson, J., Gross, S. Staying Young: Growth Hormone and

Other Natural Strategies to Reverse the Aging Process. Age Reversal Press. First Edition pp:120, (1999).
4. De Noto F, Rutter J.W., Goodman H.M. Human growth hormone DNA sequence and mRNAstructure: possible alternative splicing. Nucleic Acid Res. 9: 3719-30, (1981).
5. Tsushima T., Katoh Y., Miyachi Y., Chihara K., Teramoto A., Irie M., Hashimoto, Y. Serum concentrations of 20K human growth hormone in normal adults and patients with various endocrine disorders. Study Group of 20K hGH. Endocr. J. 47 Suppl: S 17-21, (2000).
6. Hashimoto Y., Kamioka T., Hosaka M., Mabuchi K., Mizuchi A., Shimazaki Y., Tsuno M., Tanaka T. Exogenous 20 K growth hormone (GH) suppresses endogenous 22K GH secretion in normal men. J. Clin. Endocrinol. Metab. 85(2): 601-606, (2000).
7. Wang D.S., Sato K., Demura H., Kato Y., Maruo N., Miyachi Y. Osteo-anabolic effects of human growth hormone with 22K- and 20K Daltons on human osteoblast-like cells.Endocr. J. 46(1): 125-132, (1999).
8. Cooke N.E., Ray J., Emery J.G., Liebhaber S.A. Two distinct species of human growth hormone-variant mRNA in the human placenta predict the expression of novel growth hormone proteins. J. Biol. Chem. 263: 9001-9006, (1988).
9. Ray J., Jones B., Liebhaber S.A., y Cooke N.E. Glycosylated human growth hormone variant. Endocrinology 125: 566-568, (1989).
10. Frankenne F., Scippo M., Van Beeumen J., Igout A., Hennen G. Identification of placental human growth hormone as the growth hormone-V gene expression product. J. Clin. Endocrinol. Metab. 71: 15-18, (1990).
11. Boguszewski C.L., Svensson P.A., Jansson T., Clark R., Carlsson M.S. Carlsson B. Cloning of two novel growth hormone transcripts expressed in human placenta. J. Clin. Endocrinol. Metab. 83(8): 2878-2885, (1988).
12. Frankenne F., Closset J., Gomez F., Scippo M.L., Smal J., Hennen G. The physiology of growth hormones (GHs) in pregnant women and partial characterization of the placental GH variant. J. Clin. Endocrinol. Metab. 66(6): 1171-1180, (1988).
13. Mirlesse V., Frankenne F., Alsat E., Poncelet M., Hennen G., Evain-Brion D. Placental growth hormone levels in normal pregnancy and in pregnancies with intrauterine growht retardation. Pediatr Res. 34: 439-442, (1993).
14. Chowen J.A., Evain B.D., Pozo J., Alsat E., García Segura L.M., Argente

J. Decreased expression of placental growth hormone in intrauterine growth retardation. Pedriatr. Res. 39 (4 Pt 1): 736-739, (1996).
15. Pardi G., Marcini A.M., Cetin I. Pathophysiology of intrauterine growth retardation: rol of the placenta. Acta Pediatr. Suppl. 423: 170-172, (1997).
16. Rygaard K., Revol A., Esquivel-Escobedo D., Beck B.L., Barrera-Saldaña H.A. Absence of human placental lactogen and placental growth hormona (HGH-V) during pregnancy: PCR análisis of the deletion. Human Genet. 102(1): 87-92, (1998).
17. Morikawa M., Green H., Lewis U.J. Activity of human growth hormone and related polypeptides on the adipose convertion of 3T3 cells. Molecular and Cellular Biology 4(2): 228-231, (1984).
18. Juarez-Aguilar, E., Castro-Munozledo, F., Guerra-Rodriguez, N.E., Resendez-Perez, D., Martinez-Rodriguez, H.G., Barrera-Saldana, H.A., Kuri-Harcuch, W. Functional domains of human growth hormone necessary for the adipogenic activity of hGH/hPL chimeric molecules. J. Cell Sci. 112(18):3127-3135, (1999).
19. MacLeod J.M., Worsley I., Ray J., Friesen H.G., Liebhaber S.A., Cooke N.E. Human growth hormone-variant is a biologically active somatogen and lactogen. Endocrinol. 128(3): 1298-1302, (1991).
20. Cooke N.E. Prolactin: normal synthesis and regulation. En DeGroot L.J. (ed.) Endocrinology.Saunders, Philadelphia., 1: 384-407, (1989).
21. Chen E.Y., Liao Y.C., Smith D.H., Barrera-Saldaña H.A., Gelinas R.E., Seeburg P.H. The human growth hormone locus: nucleotide sequence, biology, and evolution. Genomics 4(4): 479-97, (1989).
22. Barrera-Saldaña H.A. Growth hormone and placental lactogen: biology, medicine and biotechnology. Gene 211: 11-18, (1998).
23. Barrera-Saldaña H.A.,Seeburg P.H., Saunders G.F. Two structurally differentgenes produce the secreted human placental lactogen hormone. J. Biol. Chem. 258: 3787- 3793, (1983).
24. Canizales-Espinosa, M., Martínez-Rodríguez, H.G., Vila, V., Revol, A., Castillo-Ureta, H., Jiménez-Mateo, O., Egly, J.M., Castrillo, J.L. and Barrera-Saldaña, H.A. Differential strength of transfected human growth hormone and placental lactogen gene promoters. J. Endocr. Genet. 4 (1): 25-36, (2005).
25. Peel C.J. Bauman D.E. Somatotropin and lactation. J. Dairy Sci. 70: 474-486, (1987).
26. Etherton T.D., Kensinger R.S. Endocrine regulation of fetal and postnatal meat animal growth.J Anim Sci. 59(2): 511-528, (1984).

27. Bauman D.E. Regulation of nutrient partitioning: homeostasis, homeorhesis and exogenous somatotropin. Keynote lecture. En: Seventh International Conference on Production Disease in Farm Animals, F.A. Kallfelz pp. 306-323, (1989).
28. Sun M. Market sours on milk hormone. Science 17: 246(4932): 876-877, (1989).
29. Juskevich J.C., Guyer C.G. Bovine growth hormone: human food safety evaluation. Science 24: 249(4971): 875-884, (1990).
30. Boutinaud M., Rulquin H., Keisler D.H., Djiane J., Jammes H. Use of somatic cells from goat milk for dynamic studies of gene expression in the mammary gland. J. Anim. Sci. 80(5): 1258-69, (2002).
31. Stewart F., Tuffnell, P.P. Cloning the cDNAfor horse growth hormone and expression in Escherichia coli. J. Mol. Endocr. 6: 189-196, (1991).
32. http://www.jockeysite.com/stories/e_melbournecup1.htm
33. Ascacio-Martínez J.A., Barrera-Saldaña, H.A. A dog growth hormone cDNA codes for mature protein identical to pig growth hormone. Gene 143: 299-300, (1994).
34. Evock, C.M., Etherton, T.D., Chung C.S., Ivy, R.E. Pituitary porcine growth hormone (pGH) and a recombinant pGH analog stimulate pig growth performance in a similar manner. J. Anim. Sci. 66(8): 1928-1941, (1988).
35. Daughaday W.H. The anterior pituitary. Williams textbook of Endocrinology. Ed. Philadelphia, WB Saunders. 568-613, (1985).
36. Eigenmann J.E. Diagnosis and treatment of dwarfism in a german shepherd dog. J. Am. Anim. Hosp. Assoc. 17: 798-804, (1981).
37. Muller G.H., Kirk R.W., Scott D.W. Small animal dermatology. Philadelphia. WB. Saunders. 4th Edition, 575-657, (1989).
38. Devlin R.H., Byatt J.C., Maclean E., Yesaki T.Y., Krivi G.G., Jaworski E.G., Clarke W.C. Bovine placental lactogen is a potent stimulator of growth and displays strong binding to hepatic receptor sites of coho salmon. General and Comparative Endocrinology, 95: 31-41, (1994).
39. Tokunaga T., Iwai S., Gomi H., Kodama K., Ohtsuka E., Ikehara M., Chisaka O., Matsubara K. Expression of synthetic human growth hormone gene in yeast. Yeast. 39: 117-120, (1985).
40. Franchi E., Maisano F., Testori S.A, Galli G., Toma S., Parente L., Ferra F.D., y Grandi G. A new human growth hormone production process using a recombinant Bacillus subtilisstrain. J. Biotechnology 18: 41-54, (1991).

41. Pavlakis G.N., Hizuka N., Gorden P., Seburg P.H., Hamer D.H. Expression of two human growth hormone genes in monkey cell infected by simian virus 40 recombinants. Proc. Natl. Acad. Sci. 78: 7398-7402, (1981).
42. Kerr D.E., Liang F., Bondioli K.R., Zhao H., Kreibich G., Wall R.J., Sun T.T. The bladder as a bioreactor: Urothelium production and secretion of growth hormone into urine. Nature Biotechnology 16: 75-78, (1997).
43. Escamilla-Treviño L.L., Viader Salvado J.M., Barrera Saldaña H., Guerrero Olazaran, M. Biosynthesis and secretion of recombinant human growth hormone. In: Pichia Pastoris. Biotechnology Letter 22: 109-114, (2000).
44. Romanos M.A., Scorer C.A., Clare. J.J. Foreign gene expression in yeast: A review. Yeast 8: 423-488, (1992).
45. Faber K.N., Harder W., Veenhuis, M. Review: Yeasts as factories for the production of foreign proteins. Yeast 11: 1131-1344, (1995).
46. Siegel R.S., Brierley R.A. Methylotrophic yeast Pichia pastoris produced in high-celldensity fermentation with high cell yields as vehicle for recombinant protein production. Biotechnol. Bioeng. 34: 403-404, (1989).
47. Ausubel, F.M., Brent, R., Kingston, R.E., Moore, D.D., Seidman, J.G., Smith, J.A., Struhl, K. Short Protocols in Molecular Biology, fourt ed., Wiley, Massachusetts, (1999).
48. Sambrook, J., Fristsch, E., Maniatis, T. Molecular Cloninig: A Laboratory Manual. Segunda Edición. Cols. Spring Harbor Laboratory Press. Cold Spring Harbor, (1989).
49. Reyes-Ruíz, J.M., Ascacio-Martínez, J.A., Barrera-Saldaña, H.A. Derivation of a growth hormone gene cassette for goat by mutagenesis of the corresponding bovine construct and its expression in Pichia pastoris. Biotechnology Letters. 28(13):1019- 25, (2006).
50. Invitrogen. Products for Gene Expression and Analysis. Instruction manual. Pichia Expression Kit. Protein Expression. A Manual of Methods for Expression of Recombinant Proteins in Pichia pastoris. Version L., (2000).
51. Rasband, W.S. ImageJ. U. S. National Institutes of Health, Bethesda, Maryland, USA, http://imagej.nih.gov/ij/, 1997-2011.
52. Bradford, M. (1976). A rapid and sensitive method for the quatitation of microgram quantities of protein utilizing the principle of protein-dye binding. Anal. Biochem., 72:248-254.
53. Merril, C.R. Gel Staining Techniques. Guide to Protein Purification:

Methods in Enzymology (Methods in Enzymology Series, Vol 182). Murray P., Deutscher John N. y Abelson. pp. 477, (1990).

54. Tanaka, T., Shiu, R.P., Gout, P.W., Beer, C.T., Noble, R.L., Friesen, H.G. A new sensitive and specific bioassay for lactogenic hormones: measurement of prolactin and growth hormone in human serum. J. Clin. Endocrinol. Metab. 51(5):1058-1063, (1980).

55. Lawson, D.M., Sensui, N., Haisenleder, D.H., Gala, R.R. Rat lymphoma cell bioassay for prolactin: observations on its use and comparison with radioimmunoassay. Life Sci. 31(26):3063-3070, (1982).

56. Gerlier, D., Thomasset, N. Use of MTT colorimetric assay to measure cell activation. J. Immunol. Methods. 94(1-2):57-63, (1986).

57. Castro-Munozledo, F., Beltran-Langarica, A., Kuri-Harcuch, W., Commitment of 3T3- F442A cells to adipocyte differentiation takes place during the first 24-36 h after adipogenic stimulation: TNF-alpha inhibits commitment. Exp. Cell Res. 284, 161- 170, 2003.

58. Ascacio-Martínez, J.A., Barrera-Saldaña H.A. Production and secretion of biologically active recombinant canine growth hormone by Pichia pastoris. Gene. 340(2):261- 266, (2004).

59. Palma-Nicolás, J.P., Ascacio-Martínez, J.A., Revol-de-Mendoza A. y Barrera-Saldaña, H.A. Production of recombinant human placental variant growth hormone in Pichia pastoris. Biotechnology Letters. 27(21):1695-1700, (2005).

60. Treerattrakool S, Eurwilaichitr L, Udomkit A, Panyim S. Secretion of Pem-CMG, a peptide in the CHH/MIH/GIH family of Penaeus monodon, in Pichia pastoris is directed by secretion signal of the alpha-mating factor from Saccharomyces cerevisiae. J Biochem. Mol. Biol. 35(5):476-81, (2002).

# Chapter 2

## PLASMID-BASED GENETIC MODIFICATION OF HUMAN BONE MARROW-DERIVED STROMAL CELLS: ANALYSIS OF CELL SURVIVAL AND TRANSGENE EXPRESSION AFTER TRANSPLANTATION IN RAT SPINAL CORD

Mark W Ronsyn[1,2], Jasmijn Daans[2], Gie Spaepen[3], Shyama Chatterjee[4], Katrien Vermeulen[2], Patrick D'Haese[3], Viggo Fl Van Tendeloo[2,6], Eric Van Marck[4], Dirk Ysebaert[5,6], Zwi N Berneman[2,6], Philippe G Jorens[1,6] and Peter Ponsaerts[2,6]

[1]Division of Clinical Pharmacology, University of Antwerp, Antwerp, Belgium

[2]Laboratory of Experimental Hematology, Vaccine and Infectious Disease Institute (VIDI), University of Antwerp, Antwerp, Belgium

[3]Laboratory of Physiopathology, University of Antwerp, Antwerp, Belgium

[4]Laboratory of Pathology, University of Antwerp, Antwerp, Belgium

[5]Laboratory of Experimental Surgery, University of Antwerp, Antwerp, Belgium

[6]Centre for Cell Therapy and Regenerative Medicine, Antwerp University Hospital, Antwerp, Belgium

## ABSTRACT

### Background

Bone marrow-derived stromal cells (MSC) are attractive targets for *ex vivo* cell and gene therapy. In this context, we investigated the feasibility of a plasmid-based strategy for genetic modification of human (h)MSC with enhanced green fluorescent protein (EGFP) and neurotrophin (NT)3. Three genetically modified hMSC lines (EGFP, NT3, NT3-EGFP) were established and used to study cell survival and transgene expression following transplantation in rat spinal cord.

## RESULTS

First, we demonstrate long-term survival of transplanted hMSC-EGFP cells in rat spinal cord under, but not without, appropriate immune suppression. Next, we examined the stability of EGFP or NT3 transgene expression following transplantation of hMSC-EGFP, hMSC-NT3 and hMSC-NT3-EGFP in rat spinal cord. While *in vivo* EGFP mRNA and protein expression by transplanted hMSC-EGFP cells was readily detectable at different time points post-transplantation, *in vivo* NT3 mRNA expression by hMSC-NT3 cells and *in vivo* EGFP protein expression by hMSC-NT3-EGFP cells was, respectively, undetectable or declined rapidly between day 1 and 7 post-transplantation. Further investigation revealed that the observed *in vivo* decline of EGFP protein expression by hMSC-NT3-EGFP cells: (i) was associated with a decrease in transgenic NT3-EGFP mRNA expression as suggested following laser capture micro-dissection analysis of hMSC-NT3-EGFP cell transplants at day 1 and day 7 post-transplantation, (ii) did not occur when hMSC-NT3-EGFP cells were transplanted subcutaneously, and (iii) was reversed upon re-establishment of hMSC-NT3-EGFP cell cultures at 2 weeks post-transplantation. Finally, because we observed a slowly progressing tumour growth following transplantation of all our hMSC cell transplants, we here demonstrate that omitting immune suppressive therapy is sufficient to prevent further tumour growth and to eradicate malignant xenogeneic cell transplants.

## Conclusion

In this study, we demonstrate that genetically modified hMSC lines can survive in healthy rat spinal cord over at least 3 weeks by using adequate immune suppression and can serve as vehicles for transgene expression. However, before genetically modified hMSC can potentially be used in a clinical setting to treat spinal cord injuries, more research on standardisation of hMSC culture and genetic modification needs to be done in order to prevent tumour formation and transgene silencing *in vivo*.

## BACKGROUND

Despite major progress in pharmacological and surgical approaches, a spinal cord injury still remains a very complex medical and psychological challenge, both for patients and their relatives as well as for involved physicians, with currently no existing curative therapy. Next to primary care using surgical osteosynthesis techniques and administration of methylprednisolone [1], further therapeutic approaches are mainly supportive and are focussed on prevention of secondary complications, like urological problems, decubitus, respiratory tract pathology, etc... However, during the past decade, significant progress has been made in animal models of spinal cord injury [2, 3], and more therapeutic strategies are likely to be discovered as the existence of an endogenous neural regenerative mechanism in the central nerve system is now generally accepted [4, 5]. In this context, a spinal cord injury should not be seen as a single event, but must be recognized as an evolving process with different stages for which different therapeutic approaches can be developed [6]. In general, functional outcome following spinal cord injury will highly depend on the severity of both primary anatomical disruption of nerve tracts (due to contusion, laceration, penetration, etc.) and secondary damage [7] caused by inevitable inflammatory reactions following the initial trauma. In brief, these secondary inflammatory responses mainly consist of an influx of peripheral inflammatory cells (macrophages, T-cells) and an activation of resident microglia. This inflammatory reaction will finally result in the formation of a central cavitation at the site of the initial trauma in the spinal cord surrounded by glial scar tissue. The latter is an important physical and chemical barrier for endogenous regeneration of ascending and descending nerve tracts and thereby compromises functional outcome. The development of future curative treatments will therefore need to combine multiple approaches that are able to modulate secondary inflammation and to enhance endogenous regeneration.

Currently, a very promising experimental strategy for promoting neuronal survival and endogenous regeneration in injured spinal cord is local delivery of neurotrophic factors. Several neurotrophic factors, like brain-derived neurotrophic factor (BDNF), glial cell-derived neurotrophic factor (GDNF), neurotrophin (NT)3 and nerve growth factor (NGF), can stimulate neurogenesis *in vitro* and *in vivo* [8], and their importance for the development of the nervous system, for axonal pathfinding and neuronal survival has made them promising targets to augment regeneration in the injured brain and spinal cord [9, 10]. Several approaches have been reported to deliver these neurotrophic factors into injured spinal cord: direct injection [11], adenoviral vectors [12], osmotic minipumps [13–15], fibrin glue [16], hydrogels [17] and genetically modified cell transplants [9, 18–20]. Safety, efficacy and applicability of these reported methodologies highly differ between the above-referenced and other published reports, implying the need for continuous study, improvement and validation of *in vivo* delivery systems for neurotrophic factors in spinal cord.

In this study, we investigated the feasibility of a plasmid-based strategy for *in vitro* genetic modification of human bone marrow-derived stromal cells (hMSC) with enhanced green fluorescent protein (EGFP) and NT3. Three genetically modified hMSC lines (EGFP, NT3, NT3-EGFP) were established and used to study cell survival and transgene expression following transplantation in rat spinal cord. First, we present a number of optimised and easy to implement techniques for reproducible histological and molecular detection of hMSC cell transplants in rat spinal cord. Using these techniques we demonstrate that genetically modified hMSC lines can survive in healthy rat spinal cord when using adequate immune suppression and can serve as vehicles for *in vivo* transgene expression. However, transgene silencing and tumour formation by hMSC cell transplants *in vivo* are of utmost importance and should therefore be addressed in priority in future research.

# RESULTS

## In Vitro Characterization of Genetically Modified hMSC Cell Transplants

A human bone marrow-derived stromal cell line (hMSC) was genetically modified with DNA plasmids encoding either (i) enhanced green fluorescent protein (EGFP) alone, (ii) neurotrophin-3 (NT3) alone, or (iii) both EGFP and NT3, as described in the Materials and Methods section. All three plasmids used for genetic modification were based on the same backbone vector and are shown in Figure 1A. The three genetically modified hMSC lines that were established for use in this study are designated as: hMSC-EGFP, hMSC-NT3 and hMSC-NT3-EGFP. Before transplantation experiments were carried out, transgene expression of these genetically modified hMSC lines was characterized at several passages during culture by PCR, real-time PCR, ELISA and flow cytometry. Figure 1B shows a representative example for the detection of transgenic EGFP and NT3 DNA and mRNA by standard PCR and RT-PCR analysis on DNA and mRNA isolated from the different hMSC lines used in this study. In addition, Figure 1C shows a representative example for the detection of transgenic EGFP and/or NT3 mRNA by real-time RT-PCR analysis on mRNA isolated from all hMSC lines used in this study. With regard to protein expression, Figure 1D shows a representative ELISA measurement of NT3 secretion by the different hMSC lines used in this study and Figure 1E shows a representative flow cytometric analysis of EGFP expression by the EGFP expressing hMSC lines used in this study.

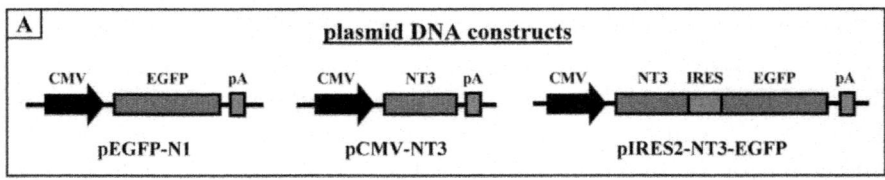

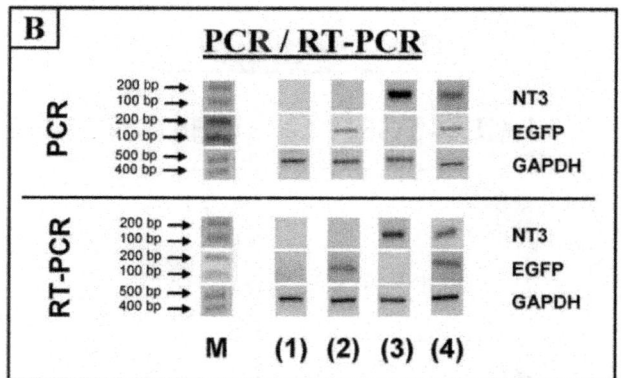
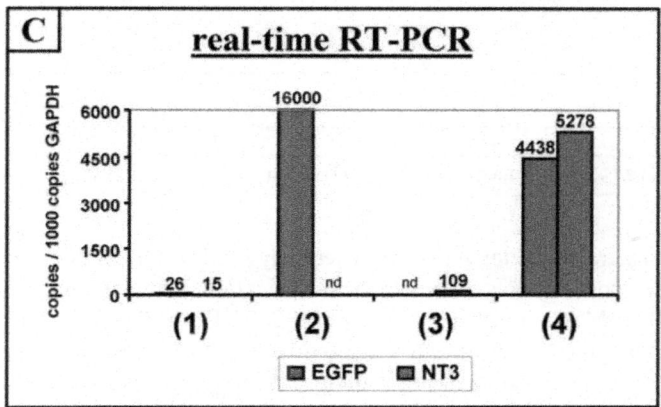
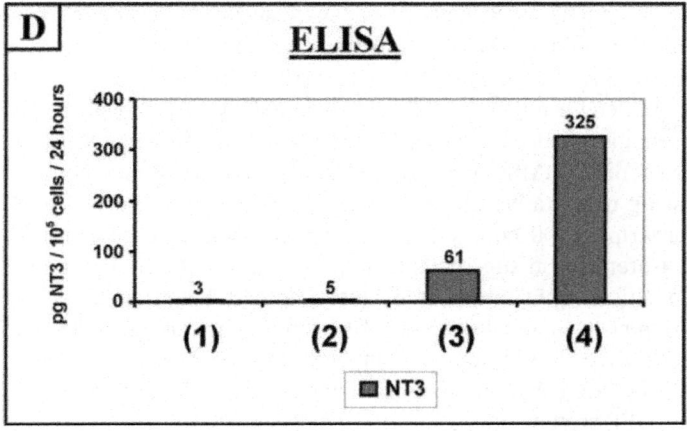

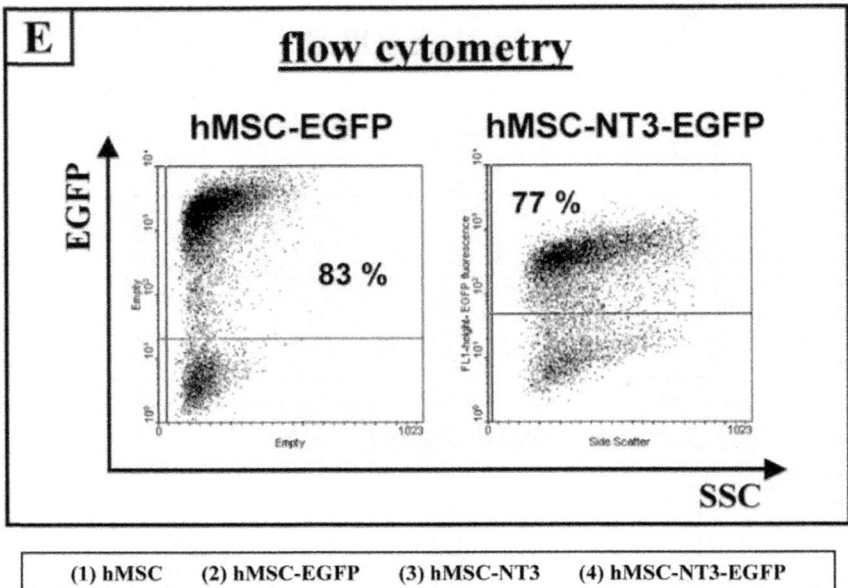

**Figure 1:** In vitro characterization of genetically modified hMSC cell transplants. (A) Plasmid DNA constructs used for genetic modification of human bone marrow-derived stromal cells (hMSC) in order to obtain hMSC-EGFP, hMSC-NT3 and hMSC-NT3-EGFP cell populations. CMV: Cytomegalovirus immediate early promotor + enhancer. EGFP: enhanced green fluorescent protein. pA: SV40 early mRNA polyadenylation signal. NT3: neurothrophin-3. IRES: internal ribosome entry site. (B) Representative standard PCR and RT-PCR analysis on DNA and mRNA isolated from the different genetically modified hMSC populations used in this study (see numbers below pictures) indicating the presence of transgenic EGFP and/or NT3 DNA and mRNA sequences. M: length marker. GAPDH: glyceraldehyde-3-phosphate dehydrogenase. (C) Representative real-time RT-PCR analysis on mRNA isolated from the different genetically modified hMSC populations used in this study (see numbers below pictures) showing quantitative differences in the level of transgenic EGFP and/or NT3 mRNA transcripts/1000 copies GAPDH; nd: no data available. (D) Representative ELISA measurement on supernatant samples from the different genetically modified hMSC populations used in this study (see numbers below pictures) showing quantitative differences in the level of NT3 secretion in picogram/$10^5$ cells/24 hours. (E) Representative flow cytometric analysis of EGFP expression by hMSC-EGFP and hMSC-NT3-EGFP populations showing quantitative differences in the level of transgenic EGFP protein expression. SSC: side scatter.

## Immunological Survival of hMSC-EGFP Cell Transplants in Rat Spinal Cord

Several reports [21–24] ascribe specific immune modulating and immune privileged features to hMSC, suggesting that these cells might serve as a universal off-the-shelf source of cells for use in regenerative medicine and can be transplanted into an allogeneic and xenogeneic host without the need for immune suppressive therapy. In order to investigate whether our established hMSC line has immune modulating properties, two series of transplantation experiments were performed using hMSC genetically modified with the EGFP reporter gene (Figure 1). First, hMSC-EGFP were transplanted in healthy rat spinal cord without systemic immune suppression. At different time points post-transplantation, animals were sacrificed and extracted spinal cords were analysed by histology for cell transplant survival, EGFP transgene expression and macrophage invasion, as presented in Figure 2A. Although on day 1 and week 1 post-transplantation hMSC-EGFP cell transplants were clearly visible by direct EGFP fluorescence and immuno-histochemical staining for EGFP, from 2 weeks post-transplantation the whole transplantation site was invaded with CD68+ macrophages and neither direct EGFP fluorescence nor immuno-histochemical staining for EGFP could indicate the immunological survival of transplanted hMSC-EGFP cells. Interestingly, despite this extensive inflammatory reaction associated with transplant rejection, no negative adverse effects were seen on general health status and locomotion of the cell transplanted animals (data not shown). Next, in order to prevent immunological rejection of hMSC-EGFP cell transplants, the same transplantation experiment was performed but with daily subcutaneous administrations of 10 mg/kg cyclosporine A starting 3 days prior to transplantation. In contrast to the results described above, efficient transplant survival was observed in all cell-transplanted animals at different time points post-transplantation. Representative examples of molecular and histological detection of hMSC-EGFP cell transplants in spinal cord are shown in Figure 2B. Following DNA and mRNA isolation from dissected spinal cord segments, PCR and RT-PCR analysis allowed to detect hMSC-EGFP cell transplants (week 3, PCR for EGFP on isolated DNA) and their transgene expression (week 3, RT-PCR for EGFP on isolated mRNA). In addition, real-time RT-PCR analysis confirmed EGFP transgene expression in spinal cord (day 1, real-time RT-PCR for EGFP on isolated mRNA). Moreover, this molecular observation of hMSC-EGFP cell transplant survival in rat spinal cord and persisting transgene expression at different time points post-transplantation was also proven by histological analysis (week 1 and 2, direct EGFP fluorescence and immuno-histochemical staining for EGFP). In summary, based on the above-described observations, all further transplantation experiments were performed under immune suppression.

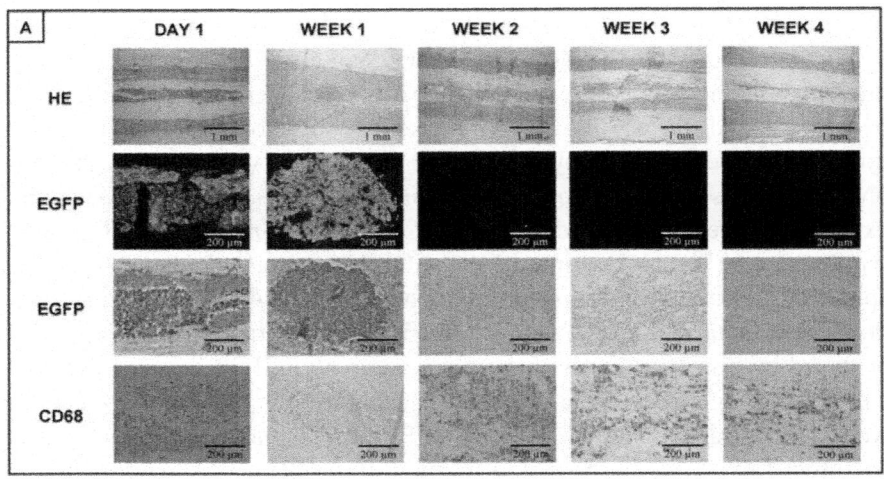

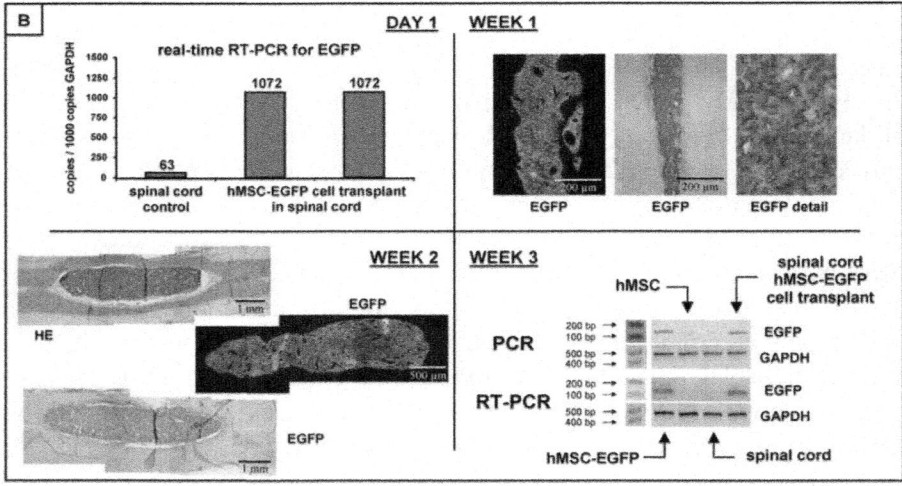

**Figure 2:** Immunological survival of hMSC-EGFP cell transplants in rat spinal cord. (A) Histological assessment of hMSC-EGFP cell transplant survival in rat spinal cord without systemic immune suppression at five time points post transplantation (day 1 and week 1–4). First row: hematoxylin-eosin (HE) staining indicating localisation and general appearance of transplantation site. Second row: direct EGFP fluorescence indicating the presence or absence of EGFP positive hMSC-EGFP cell transplants. Third row: immuno-histochemical staining for EGFP indicating the presence or absence of EGFP positive hMSC-EGFP cell transplants. Fourth row: immuno-histochemical staining for CD68 indicating macrophage infiltration into the transplantation site. All

slides were examined using a conventional light/fluorescence microscope and digital pictures were taken under magnification as indicated. Representative pictures were chosen from multiple hMSC-EGFP cell transplanted spinal cords analysed for day 1 (n = 2), week 1 (n = 2), week 2 (n = 2), week 3 (n = 2), and week 4 (n = 6) post-transplantation. (B) Molecular and histological assessment of hMSC-EGFP cell transplant survival in rat spinal cord under systemic immune suppression (subcutaneous 10 mg/kg/day cyclosporin A) at four time points post transplantation (day 1 and week 1–3). DAY 1: Real-time RT-PCR analysis on mRNA isolated from hMSC-EGFP cell transplanted spinal cords on day 1 post-transplantation (n = 2) indicating the presence of EGFP mRNA transcripts *in vivo* in spinal cord following hMSC-EGFP cell transplantation. WEEK 1: direct EGFP fluorescence and immuno-histochemical staining for EGFP indicating the presence of EGFP positive hMSC-EGFP cell transplants on week 1 post-transplantation. Representative pictures were chosen from multiple hMSC-EGFP cell transplanted spinal cords analysed 1 week post-transplantation (n = 5). WEEK 2: HE staining, direct EGFP fluorescence and immuno-histochemical staining for EGFP indicating the presence of EGFP positive hMSC-EGFP cell transplants on week 2 post-transplantation. Representative pictures were chosen from multiple hMSC-EGFP cell transplanted spinal cords analysed 2 weeks post-transplantation (n = 3). WEEK 3: Standard PCR and RT-PCR analysis on DNA and mRNA isolated from hMSC-EGFP cell-transplanted spinal cords 3 weeks post-transplantation indicating the presence of EGFP DNA sequences and EGFP mRNA transcripts *in vivo* in spinal cord following hMSC-EGFP cell transplantation.

## Transgene Expression of hMSC-NT3 Cell Transplants in Rat Spinal Cord

Following genetic modification of hMSC with a DNA plasmid encoding the EGFP reporter gene and successful *in vivo*transplantation and detection of hMSC-EGFP cells, we genetically modified hMSC with a similar DNA plasmid-encoding rat NT3 (see Figure 1A, pCMV-NT3 plasmid). In contrast to the procedures followed for obtaining the presented hMSC-EGFP line (i.e. combined antibiotics selection and cell sorting, see Materials and Methods), obtaining a pure transgene expressing hMSC-NT3 line was not possible through cell sorting, due to the lack of a fluorescent marker molecule (e.g. EGFP). Therefore, following antibiotics selection, single clones were grown and screened by ELISA for highest production of NT3 (Figure 1D, hMSC-NT3 cells). Despite significant production of NT3 by the selected hMSC-NT3 clone, only very low, but however significant, NT3 mRNA was produced by these cells (Figure 1C, hMSC-NT3 cells). However, due to this low level of NT3 mRNA produced by transgenic hMSC-NT3 cells, NT3 transgene expression following transplantation of hMSC-NT3 cells in rat spinal cord could not be detected (Figure 3). In summary, based on the above-described observations, cell sorting for an EGFP reporter gene needs to be included for further genetic modification experiments in order to obtain a transgenic cell population producing high levels of transgenic mRNA.

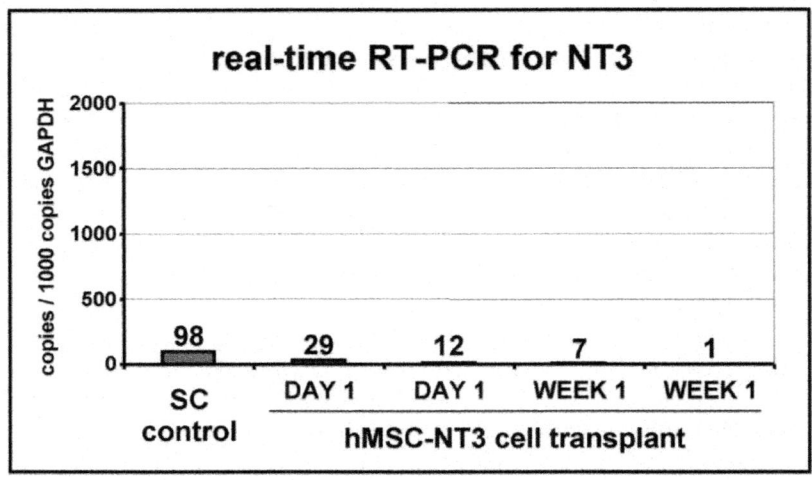

**Figure 3:** Transgene expression of hMSC-NT3 cell transplants in rat spinal cord. Molecular assessment of NT3 mRNA transcription by hMSC-NT3 cell transplants in rat spinal cord. Real-time RT-PCR analysis on mRNA isolated from hMSC-NT3 cell-transplanted spinal cords on day 1 post-transplantation (n = 2) and on week 1 post-transplantation (n = 2) demonstrating the absence of detectable exogenous NT3 mRNA transcripts. SC: spinal cord.

## Transgene Expression of hMSC-NT3-EGFP Cell Transplants in Rat Spinal Cord

In a next attempt to genetically modify hMSC to produce the NT3 neurothrophic factor, we used a DNA plasmid encoding both the NT3 and EGFP protein (Figure 1A, pIRES2-NT3-EGFP plasmid). Because transcription from this plasmid results in the production of one mRNA encoding both proteins, cell sorting for cells expressing high levels of EGFP protein was used in order to obtain an hMSC-NT3-EGFP line producing high levels of EGFP and NT3 mRNA (Figure 1C, hMSC-NT3-EGFP cells) and EGFP and NT3 protein (Figure 1D and 1E, hMSC-NT3-EGFP cells). Following transplantation in rat spinal cord under immune suppression, hMSC-NT3-EGFP cell transplants were easily detectable by morphology on HE-stained slides (Figure4A). However, while expression of EGFP was clearly demonstrated on day 1 post-transplantation by direct fluorescence microscopy (data not shown) and immuno-histochemical staining for EGFP, EGFP expression on week 1 post-transplantation, as demonstrated by immuno-histochemical staining, was strongly decreased and became almost undetectable by week 2 post-transplantation (Figure 4A and 4B). This decrease in EGFP expression was not due to lack of cell transplant survival because cell transplants could clearly be visualized on HE-stained slides (Figure 4A) and staining with an antibody against human mitochondrial antigen confirmed the human nature of the observed cell transplants (Figure 4Aand 4B). In addition, the observed EGFP transgene silencing in hMSC-NT3-EGFP cell transplants, but not in hMSC-EGFP cell transplants, was further demonstrated by laser capture microdissection (LCM) experiments.

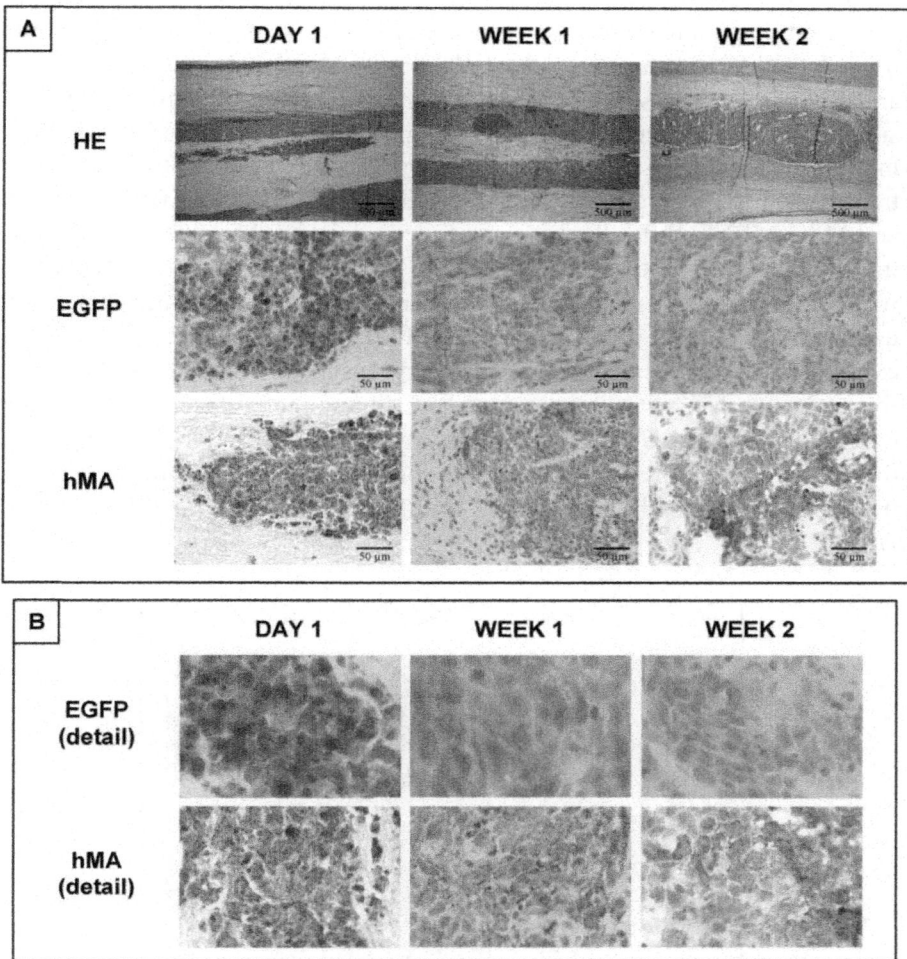

**Figure 4:** Transgene expression of hMSC-NT3-EGFP cell transplants in rat spinal cord. (A) Histological assessment of EGFP expression by hMSC-NT3-EGFP cell

transplants in rat spinal cord at three time points post transplantation (day 1, week 1 and week 2). First row: hematoxylin-eosin (HE) staining indicating localisation and general appearance of transplantation site. Second row: immuno-histochemical staining for EGFP showing a decrease in EGFP expression over time by hMSC-NT3-EGFP cell transplants. Third row: immuno-histochemical staining for human mitochondrial antigen (hMA) indicating the human nature of the hMSC-NT3-EGFP cell transplants. Representative pictures were chosen from multiple hMSC-NT3-EGFP cell-transplanted spinal cords analysed 1 day (n = 7), 1 week (n = 11) and 2 weeks (n = 4) post-transplantation. (B) Detail image of the pictures described in part (A) of this figure.

For this, hMSC-EGFP and hMSC-NT3-EGFP cells were transplanted in rat spinal cord under immune suppression. Next, at day 1 and day 7 post-transplantation, transplanted cell populations were isolated by LCM (Figure 5A) and the expression level of EGFP mRNA was analysed by real-time PCR. As suggested by the results presented in Figure 5B, while expression of EGFP mRNA by an hMSC-EGFP cell transplant decreased by 50% at day 1 post-transplantation as compared to cultured hMSC-EGFP cells, EGFP mRNA by an hMSC-NT3-EGFP cell transplant directly decreased by 80% at day 1 post-transplantation as compared to cultured hMSC-NT3-EGFP cells. However, while EGFP mRNA expression then remained stable between day 1 and day 7 post-transplantation for an hMSC-EGFP cell transplant, EGFP mRNA expression further decreased between day 1 and day 7 post-transplantation for an hMSC-NT3-EGFP cell transplant. In summary, based on the above-described results, we observed a similar degree of hMSC-NT3-EGFP cell transplant survival as described above for hMSC-EGFP cell transplants, but transgene expression by hMSC-NT3-EGFP cell transplants rapidly declined *in vivo*.

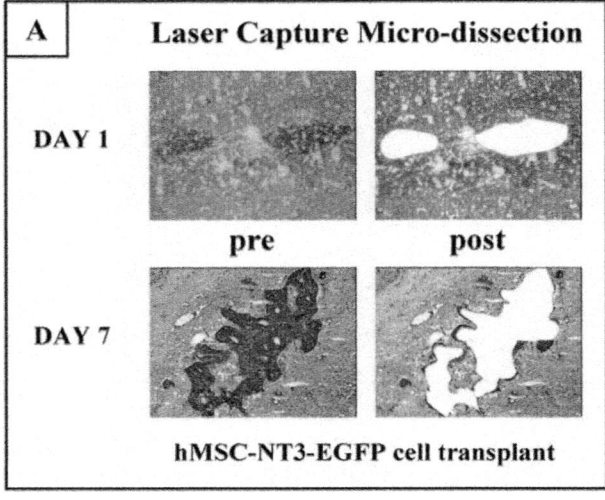

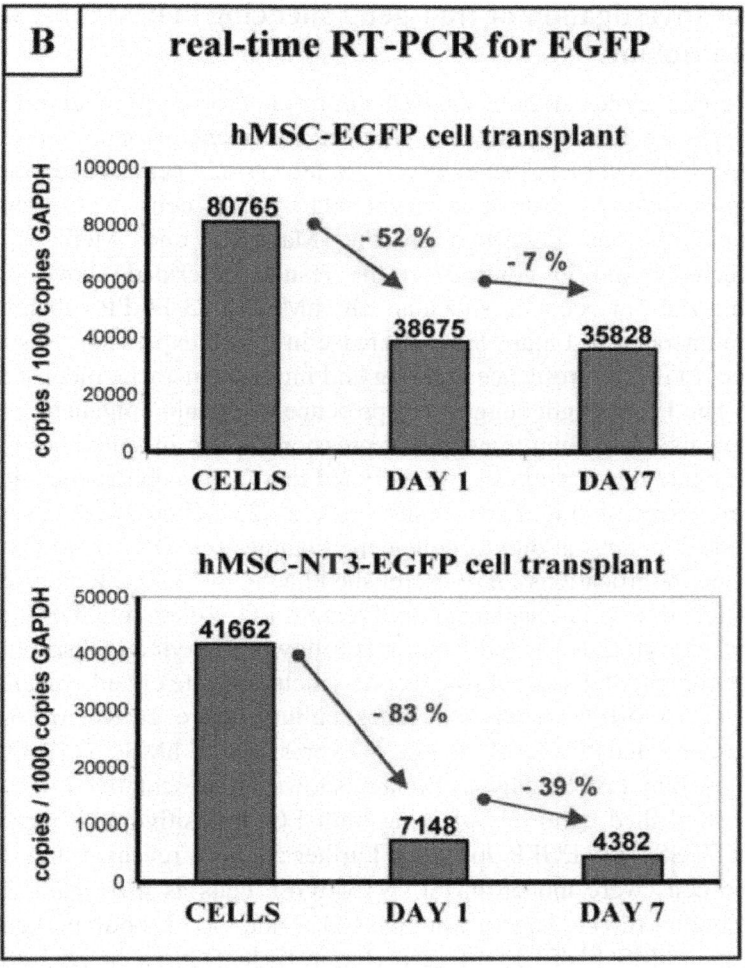

**Figure 5:** Analysis of transgene expression by hMSC-NT3-EGFP cell transplants following laser capture micro-dissection (LCM). (A) Laser Capture Micro-dissection (LCM) of hMSC-NT3-EGFP cell transplants at day 1 and day 7 post-transplantation in rat spinal cord. Pictures showing cresyl-violet stained slides pre-and post-LCM. (B) Upper graph showing the level of transgenic EGFP mRNA transcripts/1000 copies GAPDH in hMSC-EGFP cell cultures (CELLS, n = 1), on transplanted hMSC-EGFP cells at day 1 post-transplantation (DAY 1, n = 1), and on transplanted hMSC-EGFP cells at day 7 post-transplantation (DAY 7, n = 1). Lower graph showing the level of transgenic EGFP mRNA transcripts/1000 copies GAPDH in hMSC-NT3-EGFP cell cultures (CELLS, n = 1), on transplanted hMSC-NT3-EGFP cells at day 1 post-transplantation (DAY 1, n = 1), and on transplanted hMSC-NT3-EGFP cells at day 7 post-transplantation (DAY 7, n = 1).

## Further Investigation of Transgene Silencing in hMSC-NT3-EGFP Cell Transplants

Since reproducible cell transplantation into spinal cord is a relatively complex procedure, we first investigated whether subcutaneous cell transplantation might provide a valuable alternative in order to study persistence of transgene expression *in vivo*. For this, hMSC-NT3-EGFP cells were transplanted subcutaneously as described in the Materials and Methods section. Unexpectedly and in contrast to the results described above, transgene silencing did not occur in subcutaneous hMSC-NT3-EGFP cell transplants. As demonstrated in Figure 6, no decrease in EGFP expression was observed by direct EGFP fluorescence imaging and immuno-histochemical staining for EGFP. The latter might suggest the presence of certain epigenetic regulatory mechanisms controlling transgene expression *in vivo* in spinal cord. In order to investigate this hypothesis, we assumed that *in vivo* epigenetic silencing of transgene expression is an irreversible process [25]. Thus, in case the observed transgene silencing is due to epigenetic changes (eg DNA methylation and/ or histone modifications), a re-established hMSC-NT3-EGFP culture from an EGFP-negative cell transplant should remain EGFP-negative. For this, we re-established hMSC-NT3-EGFP cultures following enzymatic disruption of the cellular context of dissected spinal cords (including site of cell transplantation) at week 2 post-transplantation. When cultures were grown to confluence, the presence and the level of EGFP expression of hMSC-NT3-EGFP was investigated by FACS analysis. Figure 7shows a representative FACS analysis of an established culture containing both EGFP-positive and negative cells (Figure 7, SSC vs. EGFP dot plot). Further analysis revealed that the EGFP positive cells were indeed hMSC-NT3-EGFP cells as they stained positive for antibodies directed against human CD29 and CD73, both markers present on parental hMSC-NT3-EGFP cells (Figure 7, hCD29 vs. hCD73 dot plot on EGFP-positive cells). EGFP-negative cells present in the cultures displayed a clearly different morphology (data not shown) and presumably were rat fibroblast cells as they were not recognized by antibodies directed against human CD29 and CD73 (Figure 7, hCD29 vs. hCD73 dot plot on EGFP-negative cells). Based on the above-described results, epigenetic modification is most likely not causing the observed transgene silencing in intraspinally transplanted hMSC-NT3-EGFP cells, and a putative mechanism for this observed transgene silencing in hMSC-NT3-EGFP cell transplants in spinal cord (Figure 4), but not subcutaneously (Figure 6), remains unclear.

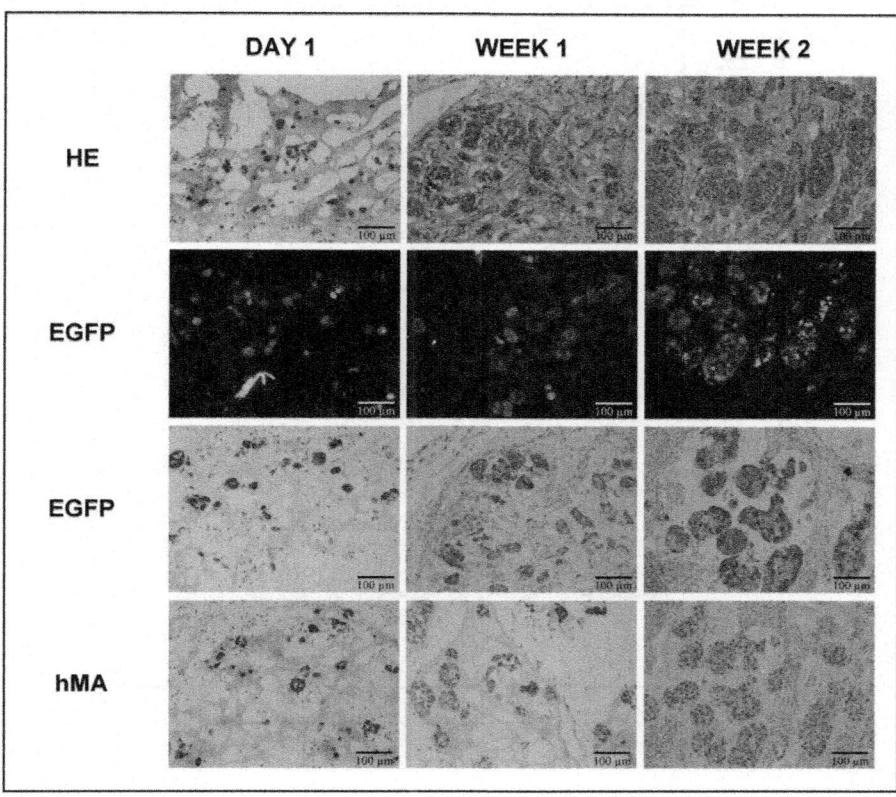

**Figure 6:** Transgene expression of subcutaneous hMSC-NT3-EGFP cell transplants. Histological assessment of EGFP expression by subcutaneous hMSC-NT3-EGFP cell transplants at three time points post transplantation (day 1, week 1 and week 2). First row: hematoxylin-eosin (HE) staining indicating the presence of nucleated cells into the subcutaneously transplanted matrigel. Second row: direct EGFP fluorescence indicating the presence of EGFP positive hMSC-NT3-EGFP cell transplants. Third row: immuno-histochemical staining for EGFP indicating the presence of EGFP positive hMSC-NT3-EGFP cell transplants. Fourth row: immuno-histochemical staining for human mitochondrial antigen (hMA) indicating the human nature of the observed EGFP positive hMSC-NT3-EGFP cell transplants. Representative pictures were chosen from multiple subcutaneous hMSC-NT3-EGFP cell transplants analysed 1 day (n = 6), 1 week (n = 5) and 2 weeks (n = 3) post-transplantation.

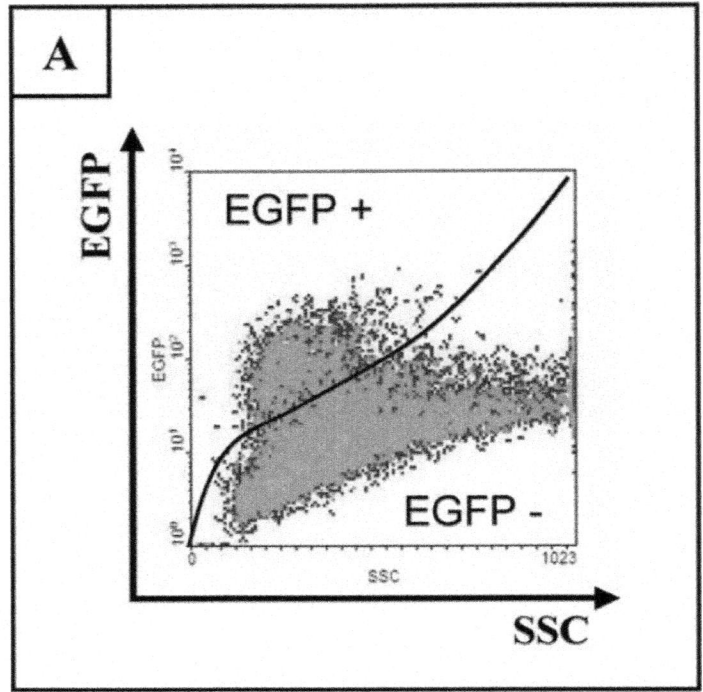

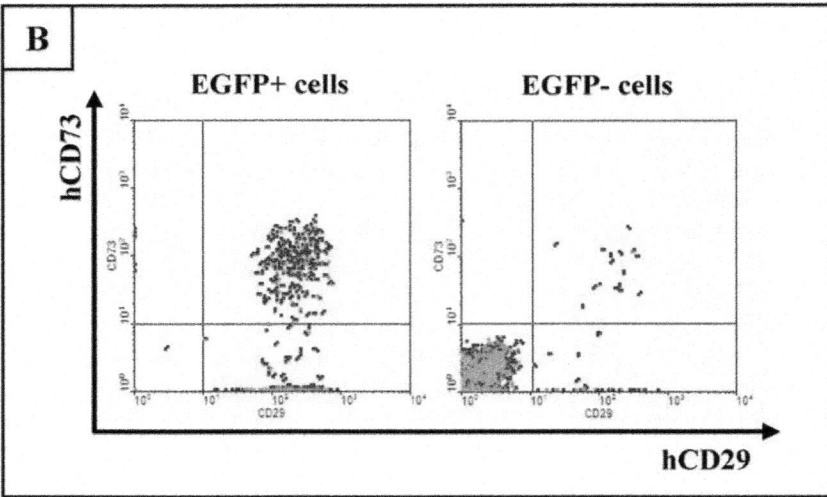

**Figure 7:** Transgene expression of re-established hMSC-NT3-EGFP cultures from hMSC-NT3-EGFP cell transplanted spinal cords. (A) Flow cytometric assessment of EGFP expression (Y-axis) by an established cell culture at passage 1 from dissected hMSC-NT3-EGFP cell transplanted spinal cords showing the presence of both EGFP positive and EGFP negative cells. (B) Flow cytometric staining for human CD29 (X-

axis) and human CD73 (= Y-axis) demonstrating that EGFP positive cells originate from hMSC-NT3-EGFP cell transplants while EGFP negative cells do not stain positive for human antibodies. Representative flow cytometric data were chosen from multiple re-established hMSC-NT3-EGFP cell cultures (n = 3).

## Tumorigenicity of hMSC Cell Transplants in Rat Spinal Cord

During progress in this study, we observed tumour growth from on week 1 post-transplantation for both hMSC-EGFP and hMSC-NT3-EGFP cell transplants in spinal cord (see histological analysis on week 2 post-transplantation in Figure 2B, Figure4 and Figure 8). Because tumour growth is a potential risk in clinical transplantation of autologous, allogeneic and xenogeneic bone marrow-derived stromal cell populations, we investigated whether omitting immune suppressive therapy would be sufficient to prevent further tumour growth and/or destroy malignant cells. As shown in Figure 8, tumorigenic growth of xenogeneic hMSC-EGFP cell transplants at week 2 post-transplantation can efficiently be controlled by the host›s immune system. Two weeks following arrest of immune suppressive therapy, the whole transplantation site is invaded by macrophages and no viable cell EGFP+ transplant could be observed.

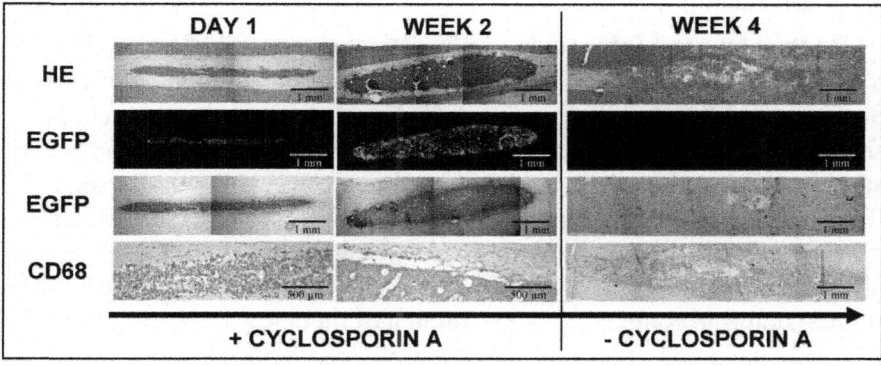

**Figure 8:** Tumorigenicity of hMSC-EGFP cell transplants. Left side: histological assessment of tumour growth by hMSC-EGFP cell transplants in rat spinal cord between day 1 and week 2 post-transplantation in immune suppressed animals (+ cyclosporine A). Right side: histological assessment of tumour regression in rat spinal cord at week 4 post-transplantation upon withdrawal of immune suppression at week 2 post-transplantation (-cyclosporine A). First row: hematoxylin-eosin (HE) staining indicating localisation and general appearance of transplantation site. Second row: direct EGFP fluorescence indicating the presence or absence of EGFP positive hMSC-EGFP cell transplants. Third row: immuno-histochemical staining for EGFP indicating the presence or absence of EGFP positive hMSC-EGFP cell transplants. Fourth row: immuno-histochemical staining for CD68 indicating macrophage infiltration into the

transplantation site. Representative pictures were chosen from multiple hMSC-EGFP cell transplanted spinal cords analysed 1 day (n = 4), 2 weeks (n = 3) and 4 weeks (n = 2) post-transplantation.

## DISCUSSION

Stem cells, both embryonic and adult, are attractive targets in the development of cell and gene therapy approaches in a variety of diseases and injuries. While embryonic stem cells have gained much interest due to their pluripotent differentiation capacity, the potential clinical use of multipotent adult stem cell populations in regenerative medicine seems to be more realistic taking into account certain immunological, practical and ethical advantages. Currently, our research is focussing on the use of adult bone marrow derived stromal cells (MSC) as a cellular minipump for neurotrophic factors in injured spinal cord in order to promote neural regeneration. In this context, the aim of this study was to investigate the feasibility of a plasmid-based strategy for genetic modification of human (h)MSC with enhanced green fluorescent protein (EGFP) and/or neurotrophin (NT)3, and to study *in vivo* cell survival and transgene expression of genetically modified hMSC following transplantation in rat spinal cord.

In the first part of this study, we investigated survival of hMSC-EGFP cell transplants in rat spinal cord in immune competent and immune suppressed rats. Survival of hMSC-EGFP in normal spinal cord of immune competent rats was significantly compromised from an early time point (day 1 post-transplantation) and resulted in total graft eradication from on week 2 post-transplantation (Figure 2A). However, suppressing the immune system with cyclosporine A resulted in successful long-term *in vivo* graft survival (Figure 2B). The observed immunological rejection of MSC in the absence of systemic immune suppression is in contrast to a large amount of literature suggesting potential *in vivo* immune modulation and immune privileged properties of MSC [21–24]. However, we believe that there is a common misunderstanding regarding the immune privileged nature of MSC. While these cells have shown immune suppressive effects on Graft-versus-Host disease when given intravenously as shown in several publications, currently, as also demonstrated by the presented results, there is no strong evidence that these cells are of such an immune tolerant nature that they can be transplanted in allogeneic and xenogeneic hosts without giving proper immune suppression. The latter was recently also suggested by Eliopoulos *et al.*[26], who clearly demonstrated that mice implanted with MHC-mismatched MSC-expressing murine erythropoietin, in contrast to mice implanted with syngeneic MSC-expressing erythropoietin, failed in causing a sustained rise in hematocrit level. In addition, Swanger *et al.*[27] demonstrated improved survival of transplanted bone marrow stromal cells in spinal cord when using high-dose cyclosporine A, indicating not only the importance of immune suppression, but also a link between cell survival and level of immune suppression. Therefore, based on these published reports and supported by our findings, we conclude that xenogeneic hMSC cell transplantation in rat is feasible and long-term transplant survival can be achieved under appropriate immune suppression.

In the second part of this study, we investigated the stability of EGFP or NT3 transgene expression by hMSC-EGFP, hMSC-NT3 and hMSC-NT3-EGFP cell transplants in rat spinal cord. While *in vivo* EGFP expression of transplanted hMSC-EGFP cells was readily detectable at different time points post-transplantation (Figure 2B), *in vivo* transgene expression by hMSC-NT3 and hMSC-NT3-EGFP cells was, respectively, undetectable (Figure 3) or declined rapidly between day 1 and 7 post-transplantation (Figure 4). *In vivo* transgene silencing is frequently reported and is currently considered as a major obstacle in the development of *ex vivo* gene therapy protocols using stem cells. One possible explanation for this transgene silencing might be epigenetic modification of introduced transgenic DNA sequences. Transgene silencing can be initiated and sustained by: (i) methylation of CpG-rich DNA sequences which inhibits RNA transcription, or (ii) deacetylation of histones which allows DNA condensation and makes DNA inaccessible for RNA transcription [28]. In our study, we used an internal ribosome entry site (IRES)-based construct for simultaneous expression of EGFP and NT3 protein starting from a single messenger RNA. Others previously reported similar transgene silencing when using IRES-based constructs, however*in vivo* transgene silencing could be reversed when animals were treated with 5-azacytidine, a methyltransferase inhibitor [29]. In addition, also deacetylase inhibitors (e.g. trichostatin A and valproic acid) have demonstrated a capacity to prevent transgene silencing *in vitro* and *in vivo* [30, 31]. However, in a first trial experiment, neither *in vitro* treatment of hMSC-NT3-EGFP cells, nor *in vivo* treatment of cell-transplanted rats, with 5-aza-2-deoxycytidine (a less toxic analogue of 5-azacytidine [32]), trichostatin A or valproic acid could reverse the observed transgene silencing *in vivo* in spinal cord (data not shown). Next, further analysis towards a putative explanation for the observed transgene silencing indicated: (i) that transgene silencing did not occur in subcutaneously transplanted hMSC-NT3-EGFP cells (Figure 6), and (ii) that silenced transgene expression was reactivated upon re-establishment of hMSC-NT3-EGFP cultures out of cell-transplanted spinal cords (Figure7). Based on these observations, we conclude that the observed transgene silencing in our hMSC-NT3-EGFP cell transplants is not due to epigenetic changes, but is most likely due to a natural site-dependent adaptation of transgene expression *in* *vivo*. Therefore, in order to use MSC or fibroblast cell transplants as cellular minipumps, genetic modification needs to result in sufficiently high transgene expression able to persist even when cellular activity might be reduced in spinal cord [19, 33].

In the third part of this study, because we observed a slowly progressing tumour growth following transplantation of all our hMSC cell transplants, we investigated whether omitting immune suppressive therapy is sufficient to prevent further tumour growth and to eradicate malignant xenogeneic cell transplants. Despite some tumorigenic characteristics of our hMSC cell transplants (high degree of Ki67-positivity, multiple mitoses, central necrosis, neovascularisation,... (data not shown)), total regression was seen over a 2-week period after cessation of systemic immune suppression. In addition, despite the extent of the resulting inflammatory infiltration in the spinal cord, we did not observe any influence on functional behaviour of the animals (data not shown). Of note, the observed tumour formation was not totally unexpected given the fact that prolonged *in vitro* stem cell culture might favour growth of MSC clones which potentially can become tumorigenic in vivo as reported by several investigators [34–37]. This *in vitro* culture-based selection of potentially tumorigenic MSC clones might indeed be a serious problem when aiming transplantation of gene-marked autologous or allogeneic MSC populations under immune suppression, where the immune system will not be able to control tumorigenicity. However, this does not mean that all hMSC cultures will behave this way. Additional experiments that were performed with non-modified early passage hMSC cultures indicated excellent cell survival with no signs of tumour formation upon transplantation into healthy spinal cord (data not shown). The absence of in vivo tumour formation for early passage MSC cultures has also been reported by other investigators [34, 37]. Therefore, based on these and other published reports/reviews about mesenchymal stromal cells [38] and supported by our findings, we conclude that new techniques need to be developed in order to safely culture and genetically modify hMSC cell transplants for research, and eventually clinical applications.

## CONCLUSION

In this study, we demonstrate that genetically modified hMSC lines can survive in healthy rat spinal cord over at least 3 weeks by using adequate immune suppression and can serve as vehicles for transgene expression. However, before genetically modified hMSC can potentially be used in a clinical setting to treat spinal cord injury, more research on standardization of hMSC culture and genetic modification needs to be done in order to prevent tumour formation and transgene silencing *in vivo*.

# METHODS

## Human Bone Marrow-Derived Stromal Cells (hMSC)

Four cryopreserved human bone marrow-derived stromal cell cultures used in a previously published study [39] were thawed and cultured in complete culture medium (CCM) consisting of Iscove's modified Dulbecco's medium supplemented with 2 mM L-glutamine (IMDM; Cambrex), 100 U/ml penicillin (Invitrogen), 100 mg/ml streptomycin (Invitrogen), 1.25 mg/ml amphotericin B (Invitrogen) and 10% fetal calf serum (FCS; Hyclone). In one out of four cultures, outgrowths of one or more cells lead to the establishment of an immortal bone marrow-derived stromal cell line. This line was (i) analyzed by immunostaining and found to be CD166+, CD44+, CD29+, CD73+, CD10+, HLA ABC+ and CD34-, CD45-, MHCII-, CD31-, CD13-, and (ii) analyzed by *in vitro* adipogenic and osteogenic differentiation studies and found to have lost their *in vitro* differentiation potential. This parental line, designated as "hMSC", was further cultured in CCM in T75 culture flasks (corning) at 37°C in a humidified atmosphere supplemented with 5% $CO_2$. For splitting, cells were harvested once a week using Trypsin/EDTA (Invitrogen) treatment and replated at a concentration of $6 \times 10^3$ cells/ml in 20 ml CCM in a new T75 culture flask.

## DNA Plasmids

The following DNA plasmids were used in this study for genetic modification of cultured hMSC: (i) the commercially available pEGFP-N1 plasmid encoding the enhanced green fluorescent protein (EGFP) (Clontech), the pCMV-NT3 plasmid encoding the rat neurothrophin-3 (NT3) protein, and (iii) the pIRES2-NT3-EGFP plasmid encoding rat NT3 and EGFP. The pCMV-NT3 plasmid was constructed by replacing the IRES-EGFP sequence from the pIRES2-EGFP plasmid (Clontech) with the rat NT3 cDNA (pAd-EF-NT3 plasmid, kindly provided by Prof. HD Shine, Baylor College of Medicine, Houston, TX, USA) [40]. The pIRES2-NT3-EGFP was constructed by inserting the NT3 cDNA into the multiple cloning site of the pIRES2-EGFP plasmid. After selection of successfully ligated pCMV-NT3 and pIRES2-NT3-EGFP clones via restriction digest mapping, several clones were confirmed by sequence analysis. All plasmids were propagated in *E. coli* supercompetent cells (Stratagene) and purified using plasmid midiprep columns (Qiagen). Directly before use in electroporation experiments, the plasmids were purified again using a PCR purification kit (Qiagen) and resuspended in DNA/RNA-free $H_2O$ at a concentration of 0.3 µg/µl.

## Stable Genetic Modification of hMSC

Stable genetic modification of hMSC following plasmid DNA electroporation was performed by culture under combined antibiotics selection, fluorescence activated cell sorting (FACS) and/or single cell cloning [41]. Briefly, hMSC were harvested, washed twice with CCM, and resuspended at $5 \times 10^6$ cells/ml in OptiMem medium (Invitrogen) supplemented with 10% FCS. Next, 500 µl cell suspension was mixed with 10 µg plasmid DNA in a 4 mm electroporation cuvette (Thermo Electron) and electroporation at 260 V and 1050 µF was carried out using a mammalian cell electroporation device (EquiBio). Following electroporation, cells were directly resuspended in CCM and cultured for 48 hours. Next, medium was changed to CCM supplemented with 250 µg/ml neomycin-analogue G418 (Sigma) for 3–4 weeks of selection. Next, in order to establish polyclonal "hMSC-EGFP" and "hMSC-NT3-EGFP" lines, cells highly positive for EGFP (highest 5%) were sorted twice using a FACS-Vantage cell sorter (Becton Dickinson). For establishment of a clonal "hMSC-NT3" line, single clones were grown and screened for NT3 expression by ELISA and RT-PCR (see further). All three lines were further cultured in CCM supplemented with 250 µg/ml G418 in T75 culture flasks at 37°C in a humidified atmosphere supplemented with 5% $CO_2$. For splitting, cell were harvested once a week using trypsin/EDTA treatment and replated at a concentration of $6 \times 100^3$ cells/ml in 20 ml CCM in a new T75 culture flask.

## Flow Cytometry

Flow cytometric analysis was used for routine (weekly) and pre-transplant measurement of EGFP transgene expression and cell viability of harvested genetically modified hMSC populations. For this, a sample ($0.5 \times 10^6$ cells) of harvested cells was resuspended in 1 ml CCM and analysed for EGFP expression on a FACS-scan analytical flow cytometer (Becton Dickinson). Cell viability was measured after addition of 1 µl/ml propidiumiodide (1 mg/ml stock solution, Sigma) to the cell suspension directly before flow cytometric analysis. In some experiments, in order to discriminate hMSC within fibroblast/neural cell cultures derived from dissected spinal cord (see below), cell samples were stained with a phycoerythrin (PE)-labelled monoclonal anti-human CD73 (Becton Dickinson) and a PE-Cy5 labelled monoclonal anti-human CD29 (Becton Dickinson) antibody. For this, cell samples ($1 \times 10^6$ cells/staining) were washed twice with Phosphate Buffered Saline (PBS) supplemented with 1% FCS, and resuspended in 100 µl PBS + 1% FCS. Next, 1 µg of each antibody was added for 15 min., followed by a washing step with PBS + 1% FCS. Finally, cells were resuspended in 1 ml PBS + 1% FCS and analyzed on a FACS-scan analytical flow cytometer.

## ELISA

Secretion of the NT3 neurotrophic factor by hMSC-NT3 and hMSC-NT3-EGFP cells *in vitro* was determined using the NT3 Emax ImmunoAssay Systems (Promega), according to manufacturers' instructions.

## Preparation of Cell Transplants

Following harvesting of hMSC-EGFP, hMSC-NT3 and hMSC-NT3-EGFP cell populations via trypsin/EDTA treatment, cells were washed twice with (PBS) supplemented with 5% FCS. Next, cells (mean viability of cell populations: 75–90%) were resuspended in PBS + 5% FCS at a concentration of $100 \times 10^6$ cells/ml for intraspinal cell transplantation. Cell preparations were kept at room temperature until injection. For subcutaneous transplantation, cells were resuspended in a cooled (1–4°C) Matrigel solution (Becton Dickinson, dilution: 1/2 Matrigel + 1/2 PBS+5%FCS) at a concentration of $100 \times 10^6$ cells/ml. Cell preparations were kept on ice until injection.

## Animals

Female Wistar rats (n = 75, Charles River Laboratories), starting weight 175–200 grams, were divided ad random into different experimental groups. For all experiments, rats were kept in normal day-night cycle (12/12) and got food and water ad libitum. All experimental procedures were approved by the "Ethical Committee for Animal Experiments" of the Antwerp University (approval no. 2004/68).

## Intraspinal Cell Transplantations

All surgical interventions were done under proper sterile conditions. One hour before general induction, rats were premedicated by subcutaneous (SC) injection of buprenorphine (0.1 mg/kg), five minutes before skin incision cefazoline (25 mg/kg) was injected subcutaneously and repeated after 6 hours in order to prevent post-operative infections. Animals were then anaesthetized in an induction chamber with a mixture of $O_2$ and $N_2O$ (0.5 l/min/1.0 l/min) and 4.0% isoflurane (Forene, Abbott). During surgery, anaesthesia was maintained by using the same mixture of $O_2$ and $N_2O$ (0.5 l/min/1.0 l/min) and 0.75% isoflurane (Forene) by a face mask. After skin incision and spreading of the paraspinal muscles a laminectomy was performed on Th10–Th11, using a 10× Zeiss OpMi1 operation microscope. Afterwards an automatic micro-injector pump (kdScientific) with 10 µl Hamilton Microliter™ Syringe was positioned above the exposed dura. A 33-Gauge Hamilton needle, attached to the syringe was stereotactically placed through the intact dura on midline position on a depth of 1.0 mm. After 2 minutes of pressure equilibration 5 µl cell suspension (containing $5 \times 10^5$ cells) was injected over 5 minutes (1 µl/min). Again a waiting period of 2 minutes with the needle still in position in the spinal cord was used for pressure equilibration and prevention of backflow of cell suspension. Muscle and skin was closed with respectively Vicryl (Ethicon) and staplers after proper desinfection of the operation field. Postoperative application of 10 ml glucose 5% solution prevented possible dehydration. During surgery, body temperature was kept on 37% by using a heating path with feedback control by an intrarectal placed sensor. Operated animals recovered by a 5 minutes period of 100% $O_2$ and were placed individually in plastic cages till day 1 postoperative with ad libitum water and food. Skin staplers were removed 6 days post-operative. Daily follow-up was done by measuring weight and assessment of general health status. During the entire experiment, immune mediated rejection of cell transplants was prevented by daily subcutaneous injections of 10 mg/kg cyclosporine A.

## Subcutaneous Cell Transplantations

Subcutaneous injections were done in the interscapular region. After shaving the region of interest, 300 µl of fluid matrigel-cell suspension was injected after which the matrigel immediately coagulated to a palpable subcutaneous tumour. During the entire experiment, immune mediated rejection of cell transplants was prevented by daily subcutaneous injections of 10 mg/kg cyclosporine A.

## Spinal Cord Dissection for Molecular and Cellular Analysis

At different time points post-transplantation, animals were re-anaesthetised as described above. After exposing the cell injection site, an expansion of the previous laminectomy of Th10 was performed to a total of three thoracal levels (e.g. Th 9–10–11). The spinal cord was cut 5 mm cranial and 5 mm caudal from the injection site. This 10 mm section was then extracted and further processed. (i) For standard and real-time PCR/RT-PCR analysis on whole spinal cord sections, the extracted spinal cord sections were preserved in RNAlater (Ambion), as described by the manufacturers' instructions, until further processing (see below). (ii) For gene expression analysis following Laser Capture Micro-dissection (LCM), extracted spinal cord sections were immediately frozen into liquid nitrogen and stored at -80°C until further processing (see below). (iii) For cellular analysis of transplanted cell populations, extracted spinal cord sections were washed in PBS and incubated for 90 min in a 0.1% collagenase (Sigma) solution dissolved in PBS. Next, dissociated cells were plated in CCM in T75 culture flasks. When cultures reached confluence, human cells were stained with antibodies against human

CD73 and CD29 and analysed for EGFP expression using a FACS-scan analytical flow cytometer (see above).

## Spinal Cord Dissection for Histological Analysis

Rats were deeply anaesthetized by intraperitoneal injection of sodium pentobarbital (90 mg/kg) and then perfused transcardially with 150 ml heparinised (1 U/ml) NaCl 0.9%, followed by 300 ml of cold buffered paraformaldehyde 4% (pH 7.4) over a period of 25 min. The spinal cord was than excised from low lumbal towards high cervical level, including the injection site, and immersed for an additional 2 hours in the same fixative.

## Matrigel Dissection

Subcutaneously placed matrigels were removed under gas anaesthesia and processed for molecular or histological analysis as described above.

## Standard and Real-Time PCR/RT-PCR Analysis on Cells and Whole Spinal Cord Sections

Genetically modified hMSC cell cultures and dissected spinal cord sections were processed for simultaneous DNA and RNA isolation using an AllPrep DNA/RNA Mini Kit (Qiagen), as described by the manufacturers' instructions. Following spectrophotometric quantification, isolated DNA was directly used for further PCR analysis. Isolated RNA was first reverse-transcribed into cDNA for 1 hour using an Omniscript RT kit (Qiagen), as described by manufacturers' instructions. Next, both DNA and cDNA was analysed for the presence of GAPDH, EGFP and NT3 sequences by standard PCR on a Thermocycler Px2 machine (Thermo Cycler) or by real-time PCR on an iCycler Thermal Cycler machine (Biorad). The following primer pairs were used for specific amplification: for GAPDH forward 5'-ACC ACA GTC CAT GCC ATC AC-3' and reverse 5'-TCC ACC ACC CTG TTG CTG TA-3' (this primer pair recognizes both rat and human GAPDH sequences), for EGFP forward 5'-AGA ACG GCA TCA AGG TGA AC-3' and reverse 5'-TGC TCA GGT AGT GGT TGT CG-3', and for NT3 forward 5'-GAT CCA GGC GGA TAT CTT GA-3' and reverse 5'-AAT CAT CGG CTG GAA TTC TG-3' (combination of this primer pair recognizes rat NT3 sequences, not human NT3 sequences; used for results described in Figure 1B) or reverse 5'-CTT ATC ATC GTC ATC CTT GTA GTC-3' (use of the latter primer recognizes a FLAG sequence on transgenic rat NT3 sequences, not on endogenous rat NT3 sequences; used for results described in Figure 1C and Figure 3). Standard PCR reactions were setup using a Taq PCR Core Kit (Qiagen) and real-time PCR reactions were setup in duplicate using a SYBR Green ER qPCR Supermix (Invitrogen), both according to the manufacturers› instructions. Reactions were carried out as follows: after an initial denaturation step at 94°C for 4 min., amplifications consisted of 30 (standard PCR) or 60 (real-time PCR) cycles of denaturation at 95°C for 1 min., annealing at 56°C for 1 min. and extension at 72°C for 1 min. PCR products obtained after standard PCR were analysed in a 1% agarose gel stained with 0.5 µg/ml ethidiumbromide and visualised by UV light. For analysis of real-time PCR data, expression levels of EGFP and NT3 were calculated versus expression of GAPDH.

## Gene Expression Analysis Following Laser Capture Micro-Dissection (LCM)

Frozen spinal cords were embedded in Neg-50™ (Richard-Allan Scientific) and 8 µm thick sections were cut using a cryomicrotome (Microm HM 500). These sections were then caught on a PEN-membrane, stretched on a metal frame, and instantly fixed with a 70% ethanol solution to inhibit RNA degradation. Next, sections were stained with cresyl-violet (LCM Staining Kit, Ambion) and dried in xylene. During the cutting process, some sections were stained with hematoxylin-eosin to verify the presence of the injected cells. The frames were sandwiched with an RNase-free slide and mounted on the LCM microscope (SL µCut, MMI). Areas with cells of interest in a range from 0.5 $mm^2$ to 1.5 $mm^2$ were selected with a software interface at 400× magnification and cut with a UV-laser. RNA was extracted using the Picopure™ RNA Isolation kit (Arcturus), and the RNA quality was assessed using the Agilent 2100 BioAnalyzer (Agilent Technologies). The overall RNA quality was very good and the calculated yields ranged from 3 – 10 ng of total RNA. RNA was reverse transcribed using the Sensiscript RT Kit (Qiagen). Real-Time PCR was carried out in triplicate on an ABI Prism 7700 Sequence Detection System (Applied Biosystems) using a custom Taqman™ Primers-Probes set for the EGFP gene and a Primers-Probe set for GAPDH as housekeeping gene. Real-time PCR amplification was carried out using following scheme: 2 initial steps of 2 min at 50°C and 10 min at 95°C, then 55 amplification loops with a denaturation of 1 sec at 95°C and an annealing and extention phase of 1 min at 60°C in a reaction volume of 25 µL.

## Histological Analysis

Fixed spinal cord segments and matrigels were dehydrated in sucrose gradients (5%, 10% and 20%), frozen in liquid nitrogen and stored at -80°C until further processing. Consecutive 10 µm-thick longitudinal cryosections were cut using a Microm HM5000 cryostat and stained with hematoxylin-eosin (HE) to locate the transplantation site. Further immunohistochemical analysis were done using either a mouse anti-human mitochondrial antigen (hMA) monoclonal antibody (Chemicon, MAB1273, 1/50 dilution) for human cell identification, a goat anti-EGFP polyclonal antibody (Abcam, AB6673, 1/1200 dilution) to detect EGFP transgene expression by hMSC cell transplants, and a mouse anti-rat CD68 monocyte/macrophage monoclonal antibody (Chemicon, MAB1435, 1/150 dilution) to detect macrophage infiltration into the transplantation site. In brief, slides were rinsed with a commercial washing buffer (DAKO S3006) and endogeneous peroxidase sites were blocked following 30 min incubation with methanol + 1% hydrogen peroxide. Next, slides were washed with water and washing buffer, followed by incubation with normal serum for 1 hour at room temperature (species dependent on the secondary antibody; normal goat serum (Dako, X0907) for staining with MAB1435 and normal rabbit serum (Dako, X0902) for staining with AB6673 and MAB1273). Subsequently slides were incubated overnight at 4°C with the primary antibody. Following this, slides were rinsed with washing buffer and incubated for 1 hour at room temperature either with peroxidase-coupled rabbit anti-goat Ig antibody (Rockland, 605-4302, 1/200 dilution) to detect EGFP, with peroxidase-coupled goat anti-mouse Ig antibody (Rockland, 610-1319, 1/200 dilution) to detect CD68, or with biotin-coupled rabbit anti-mouse Ig (Abcam, ab67271, 1/200 dilution) to detect hMA. For the latter, an extra incubation step with a HRP-coupled streptavidin based detection complex (StreptABComplex, Dako) was required and performed as described in the manufacturers' instructions. Visualisation for all slides was carried out after staining with DAB (Diaminobenzidine, Dako), according to manufacturers instructions, and nuclei were counterstained with hematoxylin carazzi. Fluorescence imaging (standard 2000 ms exposure time) and bright-field immuno-histochemical analysis was done using an Olympus Bx41 microscope equipped with an Olympus DP50 camera. Olympus DP Software was used for image collection and Photoshop for image processing.

## ACKNOWLEDGEMENTS

This work was supported by research grants 7.0004.03N, WO.012.02.N and G.0132.07 of the Fund for Scientific Research-Flanders (FWO-Vlaanderen, Belgium), by research grants 1081 BOF-KP 2005, 20872 BOF-KP 2006 and 1730 BOF-NOI 2006 from the Antwerp University, and by the Fund for Cell Therapy from the Antwerp University Hospital. Mark Ronsyn holds a PhD fellowship of FWO-Vlaanderen. Gie Spaepen holds a PhD fellowship of the Institute for Science and Technology (IWT). Peter Ponsaerts and Viggo Van Tendeloo are postdoctoral fellows of FWO-Vlaanderen. We also acknowledge helpful assistance from Dirk Van Bockstaele and Marc Lenjou (Laboratory of

Experimental Hematology) with flow cytometry and cell sorting, from August Van Laer (Laboratory of Experimental Surgery) with animal handling and surgical procedures, and from Frank Rylant and Gunther Vrolix (Laboratory of Pathology) with histological analysis.

## AUTHORS' CONTRIBUTIONS

MWR carried out cell culture, cell transplantations, histological analysis, data collection, data interpretation and drafted the manuscript. JD carried out plasmid DNA construction, cell culture and molecular analysis. GS carried out laser capture micro-dissection experiments. SC assisted in planning and evaluation of histological analysis. KV assisted in planning and evaluation of real-time PCR analysis. PD'H assisted in planning and evaluation of laser capture micro-dissection experiments. VFIVT acquired funding and assisted in evaluation of cell culture experiments. EVM assisted in evaluation of histological analysis. DY assisted in evaluation of animal experiments. ZNB acquired funding and assisted in study design and evaluation of cell culture experiments. PGJ acquired funding and assisted in study design and evaluation of animal experiments. PP carried out study design, data collection, data interpretation and drafted the manuscript. All authors have read and approved the final manuscript.

## REFERENCES

1. Bracken MB, Shepard MJ, Collins WF, Holford TR, Young W, Baskin DS, Eisenberg HM, Flamm E, Leo-Summers L, Maroon J, et al: A randomized, controlled trial of methylprednisolone or naloxone in the treatment of acute spinal-cord injury. Results of the Second National Acute Spinal Cord Injury Study. N Engl J Med. 1990, 322: 1405-1411.
2. Ramer MS, Harper GP, Bradburry EJ: Progress in spinal cord research – A refined strategy for the International Spinal Research Trust. Spinal Cord. 2000, 38: 449-472. 10.1038/sj.sc.3101055.
3. Enzmann GU, Benton RL, Talbott JF, Cao Q, Whittemore SR: Functional considerations of stem cell transplantation therapy for spinal cord repair. J Neurotrauma. 2006, 23: 479-495. 10.1089/neu.2006.23.479.
4. Okano H: Stem cell biology of the central nervous system. J Neurosci Res. 2002, 69: 698-707. 10.1002/jnr.10343.
5. Goh EL, Ma D, Ming GL, Song H: Adult neural stem cells and repair of the adult central nervous system. J Hematother Stem Cell Res. 2003, 12: 671-679. 10.1089/15258160360732696.
6. McDonald JW, Sadowsky C: Spinal-cord injury. Lancet. 2002, 359: 417-

425. 10.1016/S0140-6736(02)07603-1.
7.  Hall ED, Springer JE: Neuroprotection and acute spinal cord injury: a reappraisal. NeuroRx. 2004, 1: 80-100. 10.1602/neurorx.1.1.80.
8.  Conte V, Royo NC, Shimizu S, Saatman KE, Watson DJ, Graham DI, Stocchetti N, McIntosh TK: Neurotrophic Factors. Pathophysiology and Therapeutic Applications in Traumatic Brain Injury. Eur J Trauma. 2003, 29: 335-355. 10.1007/s00068-003-1335-z.
9.  Lu P, Jones LL, Tuszynski MH: BDNF-expressing marrow stromal cells support extensive axonal growth at sites of spinal cord injury. Exp Neurol. 2005, 191: 344-360. 10.1016/j.expneurol.2004.09.018.
10. Longhi L, Watson DJ, Saatman KE, Thompson HJ, Zhang C, Fujimoto S, Royo N, Castelbuono D, Raghupathi R, Trojanowski JQ, Lee VM, Wolfe JH, Stocchetti N, McIntosh TK: Ex vivo gene therapy using targeted engraftment of NGF-expressing human NT2N neurons attenuates cognitive deficits following traumatic brain injury in mice. J Neurotrauma. 2004, 21: 1723-1736.
11. Schnell L, Schneider R, Kolbeck R, Barde YA, Schwab ME: Neurotrophin-3 enhances sprouting of corticospinal tract during development and after adult spinal cord lesion. Nature. 1994, 367: 170-173. 10.1038/367170a0.
12. Chen Q, Zhou L, Shine HD: Expression of neurotrophin-3 promotes axonal plasticity in the acute but not chronic injured spinal cord. J Neurotrauma. 2006, 23: 1254-1260. 10.1089/neu.2006.23.1254.
13. Coumans JV, Lin TT, Dai HN, MacArthur L, McAtee M, Nash C, Bregman BS: Axonal regeneration and functional recovery after complete spinal cord transection in rats by delayed treatment with transplants and neurotrophins. J Neurosci. 2001, 21: 9334-9344.
14. Giehl KM, Tetzlaff W: BDNF and NT-3, but not NGF, prevent axotomy-induced death of rat corticospinal neurons in vivo. Eur J Neurosci. 1996, 8: 1167-1175. 10.1111/j.1460-9568.1996.tb01284.x.
15. Ramer MS, Priestley JV, McMahon SB: Functional regeneration of sensory axons into the adult spinal cord. Nature. 2000, 403: 312-316. 10.1038/35002084.
16. Cheng H, Fraidakis M, Blomback B, Lapchak P, Hoffer B, Olson L: Characterization of a fibrin glue-GDNF slow-release preparation. Cell Transplant. 1998, 7: 53-61. 10.1016/S0963-6897(97)00122-X.
17. Piantino J, Burdick JA, Goldberg D, Langer R, Benowitz LI: An injectable, biodegradable hydrogel for trophic factor delivery enhances

axonal rewiring and improves performance after spinal cord injury. Exp Neurol. 2006, 201: 359-367. 10.1016/j.expneurol.2006.04.020.

18. Cao Q, Xu XM, Devries WH, Enzmann GU, Ping P, Tsoulfas P, Wood PM, Bunge MB, Whittemore SR: Functional recovery in traumatic spinal cord injury after transplantation of multineurotrophin-expressing glial-restricted precursor cells. J Neurosci. 2005, 25: 6947-6957. 10.1523/JNEUROSCI.1065-05.2005.

19. Himes BT, Liu Y, Solowska JM, Snyder EY, Fischer I, Tessler A: Transplants of cells genetically modified to express neurotrophin-3 rescue axotomized Clarke's nucleus neurons after spinal cord hemisection in adult rats. J Neurosci Res. 2001, 65: 549-564. 10.1002/jnr.1185.

20. Shumsky JS, Tobias CA, Tumolo M, Long WD, Giszter SF, Murray M: Delayed transplantation of fibroblasts genetically modified to secrete BDNF and NT-3 into a spinal cord injury site is associated with limited recovery of function. Exp Neurol. 2003, 184: 114-130. 10.1016/S0014-4886(03)00398-4.

21. Ryan JM, Barry FP, Murphy JM, Mahon BP: Mesenchymal stem cells avoid allogeneic rejection. J Inflamm (Lond). 2005, 2: 8-10.1186/1476-9255-2-8.

22. Beyth S, Borovsky Z, Mevorach D, Liebergall M, Gazit Z, Aslan H, Galun E, Rachmilewitz J: Human mesenchymal stem cells alter antigen-presenting cell maturation and induce T-cell unresponsiveness. Blood. 2005, 105: 2214-2219. 10.1182/blood-2004-07-2921.

23. Mansilla E, Marin GH, Sturla F, Drago HE, Gil MA, Salas E, Gardiner MC, Piccinelli G, Bossi S, Salas E, Petrelli L, Iorio G, Ramos CA, Soratti C: Human mesenchymal stem cells are tolerized by mice and improve skin and spinal cord injuries. Transplant Proc. 2005, 37: 292-294. 10.1016/j.transproceed.2005.01.070.

24. Wang Y, Chen X, Armstrong MA, Li G: Survival of bone marrow-derived mesenchymal stem cells in a xenotransplantation model. J Orthop Res. 2007, 25 (7): 926-932. 10.1002/jor.20385.

25. Feng YQ, Desprat R, Fu H, Olivier E, Lin CM, Lobell A, Gowda SN, Aladjem MI, Bouhassira EE: DNA methylation supports intrinsic epigenetic memory in mammalian cells. PLoS Genet. 2006, 2: e65-10.1371/journal.pgen.0020065. DOI:10.1371/journal.pgen.0020065

26. Eliopoulos N, Stagg J, Lejeune L, Pommey S, Galipeau J: Allogeneic marrow stromal cells are immune rejected by MHC class I- and class II-mimatched recipient mice. Blood. 2005, 106: 4057-4065. 10.1182/blood-2005-03-1004.

27. Swanger SA, Neuhuber B, Himes BT, Bakshi A, Fischer I: Analysis of allogeneic and syngeneic bone marrow stromal cell graft survival in the spinal cord. Cell Transplant. 2005, 14: 775-786.
28. Newell-Price J, Clark AJ, King P: DNA methylation and silencing of gene expression. Trends Endocrinol Metab. 2000, 11: 142-148. 10.1016/S1043-2760(00)00248-4.
29. Di Ianni M, Terenzi A, Perruccio K, Ciurnelli R, Lucheroni F, Benedetti R, Martelli MF, Tabilio A: 5-azacytidine prevents transgene methylation in vivo. Gene Ther. 1999, 6: 703-707. 10.1038/sj.gt.3300848.
30. Krishnan M, Park JM, Cao F, Wang D, Paulmurugan R, Tseng JR, Gonzalgo ML, Gambhir SS, Wu JC: Effects of epigenetic modulation on reporter gene expression: implications for stem cell imaging. FASEB J. 2006, 20: 106-108.
31. Kim YH, Lee DS, Kang JH, Lee YL, Chung JK, Roh JK, Kim SU, Lee MC: Reversing the silencing of reporter sodium/iodide symporter transgene for stem cell tracking. J Nucl Med. 2005, 46: 305-311.
32. Momparler RL, Momparler LF, Samson J: Comparison of the antileukemic activity of 5-AZA-2'-deoxycytidine, 1-beta-D-arabinofuranosylcytosine and 5-azacytidine against L1210 Leukemia. Leuk Res. 1984, 8: 1043-1049. 10.1016/0145-2126(84)90059-6.
33. Lu P, Jones LL, Tuszynski MH: Axon regeneration through scars and into sites of chronic spinal cord injury. Exp Neurol. 2007, 203: 8-21. 10.1016/j.expneurol.2006.07.030.
34. Rubio D, Garcia-Castro J, Martin MC, de la Fuente R, Cigudosa JC, Lloyd AC, Bernad A: Spontaneous human adult stem cell transformation. Cancer Res. 2005, 65: 3035-3039.
35. Wang Y, Huso DL, Harrington J, Kellner J, Jeong DK, Turney J, McNiece IK: Outgrowth of transformed cell population derived from normal human BM mesenchymal stem cell culture. Cytotherapy. 2005, 7: 509-519. 10.1080/14653240500363216.
36. Liu C, Chen Z, Chen Z, Zhang T, Lu Y: Multiple tumor types may originate from bone marrow-derived cells. Neoplasia. 2006, 8 (9): 719-724. 10.1593/neo.06253.
37. Tolar J, Nauta AJ, Osborn MJ, Panoskaltsis Mortari A, McElmurry RT, Bell S, Xia L, Zhou N, Riddle M, Schroeder TM, Westendorf JJ, McIvor RS, Hogendoorn PC, Szuhai K, Oseth L, Hirsch B, Yant SR, Kay MA, Peister A, Prockop DJ, Fibbe WE, Blazar BR: Sarcoma derived from cultured mesenchymal stem cells. Stem Cells. 2007, 25: 371-379. 10.1634/stemcells.2005-0620.

38. Keating A: Mesenchymal stromal cells. Curr Opin Hematol. 2006, 13: 419-425. 10.1097/01.moh.0000245697.54887.6f.
39. Smits E, Ponsaerts P, Lenjou M, Nijs G, Van Bockstaele DR, Berneman ZN, Van Tendeloo VF: RNA-based gene transfer for adult stem cells and T cells. Leukemia. 2004, 18: 1898-1902. 10.1038/sj.leu.2403463.
40. Zhou L, Baumgartner BJ, Hill-Felberg SJ, McGowen LR, Shine HD: Neurotrophin-3 expressed in situ induces axonal plasticity in the adult injured spinal cord. J Neurosci. 2003, 23 (4): 1424-1431.
41. Van Tendeloo VF, Ponsaerts P, Van Broeckhoven C, Berneman ZN, Van Bockstaele DR: Efficient generation of stably electrotransfected human hematopoietic cell lines without drug selection by consecutive FACsorting. Cytometry. 2000, 41: 31-35. 10.1002/1097-0320(20000901)41:1<31::AID-CYTO4>3.0.CO;2-W.

# Chapter 3

## HLA ENGINEERING OF HUMAN PLURIPOTENT STEM CELLS

Laura Riolobos[1], Roli K Hirata[1], Cameron J Turtle[1,2], Pei-Rong Wang[1,5], German G Gornalusse[1], Maja Zavajlevski[1], Stanley R Riddell[1,2,3] and David W Russell[1,4]

[1]Department of Medicine, University of Washington, Seattle, Washington, USA
[2]Program in Immunology, Clinical Research Division, Fred Hutchinson Cancer Research Center, Seattle, Washington, USA
[3]Institute of Advanced Study, Technical University Munich, Munich, Germany
[4]Department of Biochemistry, University of Washington, Seattle, Washington, USA
[5]Department of Pediatrics, Second Hospital of Shandong University, Jinan, China

## ABSTRACT

The clinical use of human pluripotent stem cells and their derivatives is limited by the rejection of transplanted cells due to differences in their human leukocyte antigen (HLA) genes. This has led to the proposed use of histocompatible, patient-specific stem cells; however, the preparation of many different stem cell lines for clinical use is a daunting task. Here, we develop two distinct genetic engineering approaches that address this problem. First, we use a combination of gene targeting and mitotic recombination to derive HLA-homozygous embryonic stem cell (ESC) subclones from an HLA-heterozygous parental line. A small bank of HLA-homozygous stem cells with common haplotypes would match a significant proportion of the population. Second, we derive HLA class I–negative cells by targeted disruption of both alleles of the Beta-2 Microglobulin (B2M) gene in ESCs. Mixed leukocyte reactions and peptide-specific HLA-restricted CD8$^+$ T cell responses were reduced in class I–negative cells that had undergone differentiation in embryoid bodies. These B2M$^{-/-}$ ESCs could act as universal donor cells in applications where the transplanted cells do not express HLA class II genes.

Both approaches used adeno-associated virus (AAV) vectors for efficient gene targeting in the absence of potentially genotoxic nucleases, and produced pluripotent, transgene-free cell lines.

## INTRODUCTION

If human pluripotent stem cells are to be used clinically, they must overcome the immunological barriers that limit the transplantation of allogeneic cells. A major immunologic barrier results from the cell surface expression of human leukocyte antigens (HLA), which are encoded by genes in the major histocompatibility complex on chromosome 6, and present self and foreign peptides to T cells. These polymorphic loci include the class I HLA-A, -B, and -C genes expressed on most nucleated cells in the body, and the class II HLA-DR, -DP, and -DQ genes expressed in specialized antigen-presenting cells such as dendritic cells and macrophages. Given that multiple alleles exist for each polymorphic HLA gene, the chance of any specific pair of HLA haplotypes being found in a potential transplant recipient is exceedingly small. Depending on the application, transplanted cells and organs can be rejected based on their HLA type, with hematopoietic stem cells requiring extensive matching of both class I and II alleles, and solid organs requiring less stringent matching of class I loci. Typically, prolonged treatment with immunosuppressive drugs is required to prevent the rejection of mismatched grafts, often with dangerous side effects.

One solution to the immunologic barrier imposed by HLA is to use autologous, induced pluripotent stem cells (iPSCs) derived from each patient. iPSCs resemble embryonic stem cells (ESCs) and can be derived from adult human somatic cells by introducing specific reprogramming factors.[1,2] Although this approach ensures histocompatibility, it will be difficult to translate into clinical practice, due to the high cost for each patient, the prolonged cell culture period needed for reprogramming and differentiation into a therapeutic cell type, and the extensive validation and regulatory approval required of the final product. In addition, when treating genetic diseases, the responsible mutations must also be corrected before the cells are returned to the patient.

An alternative solution to this problem is to bank multiple stem cell lines with different HLA types, which allows therapeutic cell products derived from these lines to be prepared ahead of time. However, this would require large number of cell lines. The US bone marrow registry has >4,000,000 donors but accurately matches only 50–60% of the population at HLA-A and -B loci.[3] One study estimated that 150 ESC lines derived from donors in the United Kingdom would produce a cell bank that matches <20% of the population at HLA-A,

-B, and -DR loci.[4] The use of HLA-homozygous cell lines would decrease the number required for matching. For example, 50 iPSC lines derived from HLA-homozygous individuals with common haplotypes could match ~73% of the relatively non-diverse Japanese population at HLA-A, -B, and -DRB1 loci, although it may still be difficult to identify donors homozygous for rare haplotypes.[5] Any solution that requires the banking of multiple independent cell lines must also deal with the inherent variability of different pluripotent stem cell clones,[6,7] and in the case of iPSCs, genetic and epigenetic variations may also occur during reprogramming that could influence the behavior of individual clones.[8] This interclonal variation means that differentiation protocols must be optimized for each independent cell line, and that patients treated with distinct stem clones could have very different clinical outcomes.

Thus, there is a real need for developing pluripotent stem cell lines that are compatible with multiple allogeneic recipients, so that the number of cell lines required for clinical use can be reduced to a manageable level. Here, we develop two genetic engineering approaches that address this problem. First, we describe a method for deriving HLA-homozygous subclones from HLA-heterozygous ESC lines. A single HLA-homozygous line can be compatible with multiple recipients because only one haplotype requires matching. In the second approach, we develop HLA-negative stem cells that do not express any class I proteins on their cell surface by targeted disruption of the *B2M* gene. These $B2M^{-/-}$ ESCs could act as universal donor cells in applications where the transplanted cells do not express HLA class II genes.

# RESULTS

## Creating HLA-Homozygous Cells

Figure 1a shows our two-step strategy for deriving HLA-homozygous cells from a heterozygous parental cell line. First, adeno-associated virus (AAV) gene targeting vectors are used to insert a *HyTK* fusion gene centromeric to the HLA locus on the short arm of chromosome 6, and targeted cells are selected with hygromycin. Then ganciclovir is used to select for loss of the *TK* component of the *HyTK* gene and isolate mitotic recombinants resulting from crossovers centromeric to HLA. Studies in mouse ESCs have shown that these recombination events typically produce homozygosity extending from the crossover point to the telomere,[9] which includes the entire HLA locus. The crossover also removes the *HyTK* transgene from the cells.

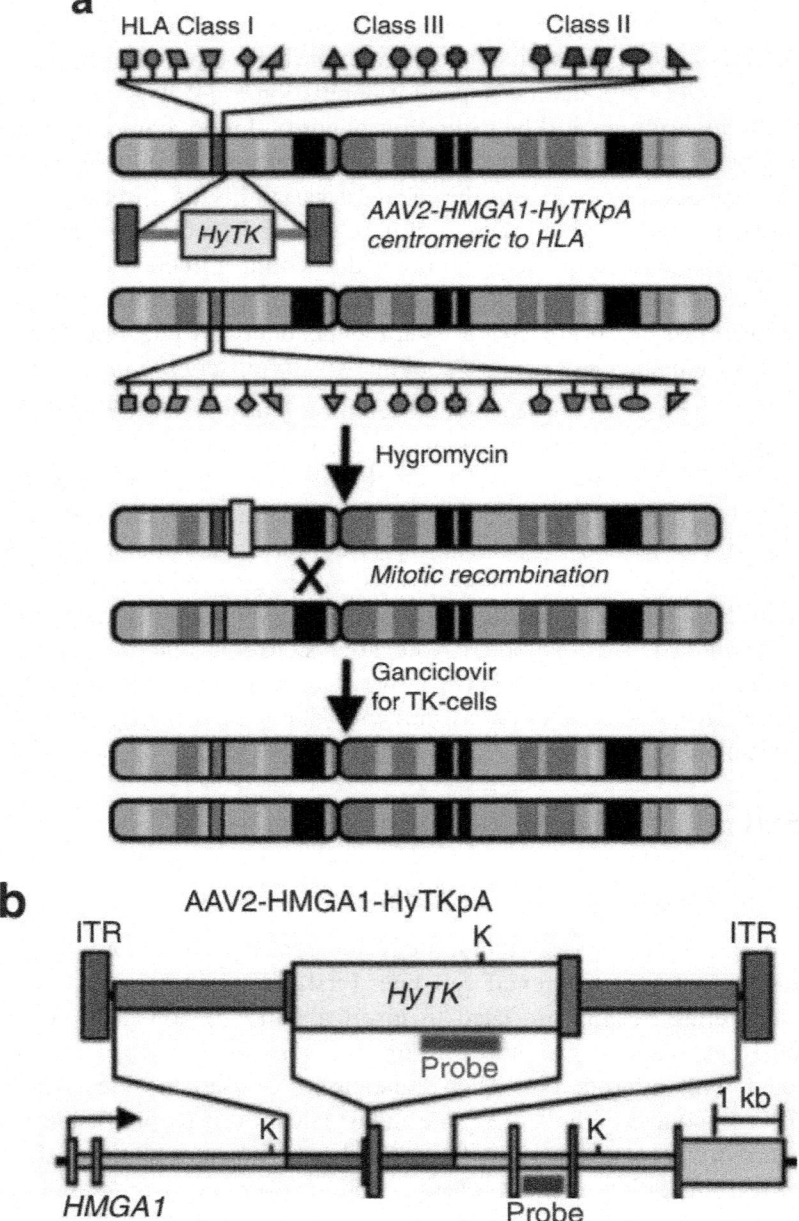

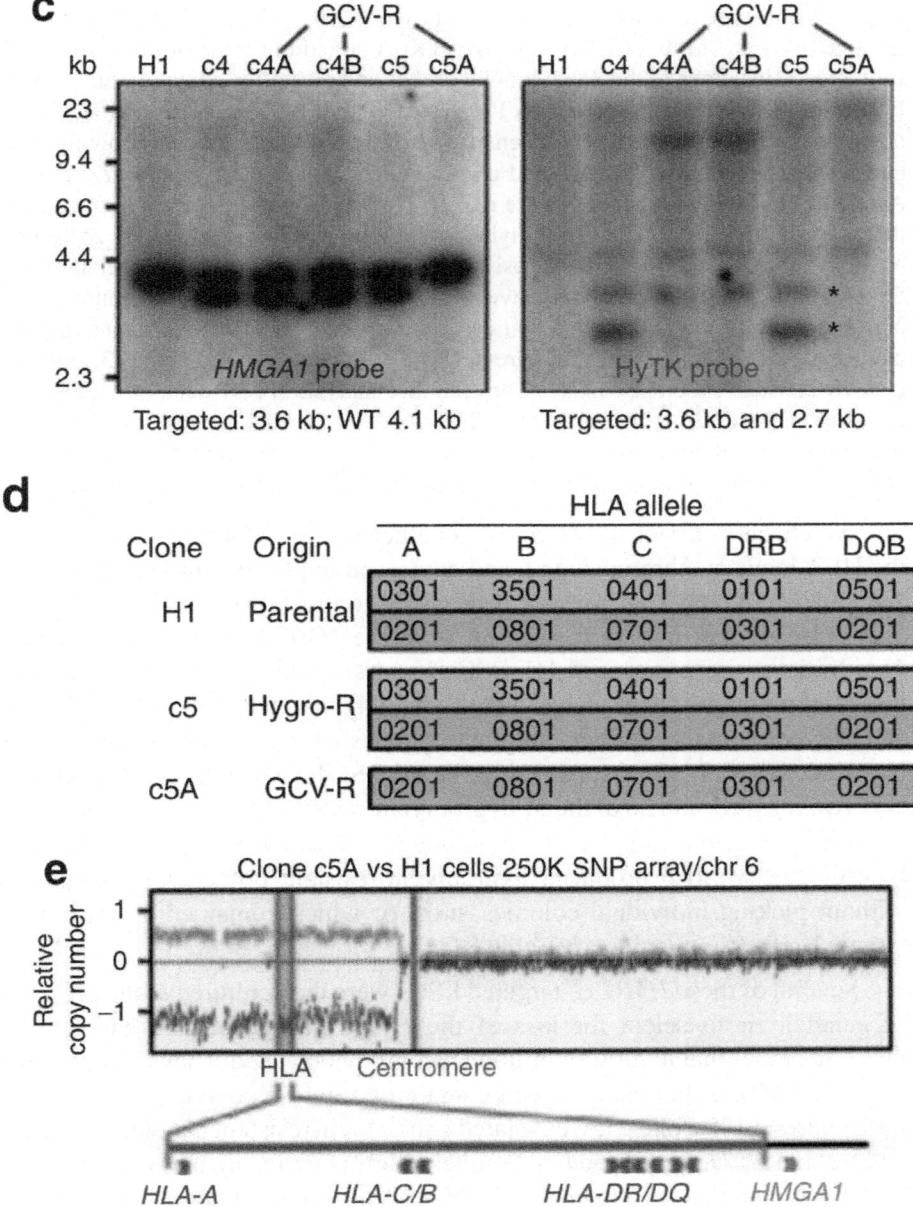

**Figure 1.** Derivation of HLA-homozygous ESCs. (a) Diagram showing the strategy for obtaining HLA-homozygous clones by targeting the *HMGA1* gene centromeric to the HLA locus, then selecting for cells that had lost the *HyTK* gene by mitotic

recombination with ganciclovir. The red and blue chromosomes represent the two copies of chromosome 6 in each cell, with the clusters of HLA class I, II, and III genes indicated. (**b**) Map of the AAV2-HMGA1-HyTKpA targeting vector and *HMGA1* gene, showing the locations of the Southern blot probes and restriction enzyme sites (K, Kpn I). (**c**) Southern blot analysis of Kpn I–digested genomic DNA of parental H1 ESCs, two clones targeted at the *HMGA1* gene (c4 and c5), and three subclones obtained by ganciclovir selection (c4A, c4B, and c5A), probed with *HMGA1* and *HyTK* probes. Asterisks indicate the locations of the two *HyTK*-hybridizing fragments derived from the targeted allele. The faint *HyTK*-hybridizing band in clone c5A and H1 represents trace signal from the hygromycin-resistant mouse embryonic fibroblast feeder cells. Subclones c4A and c4B contain a novel *HyTK*-hybridizing fragment demonstrating a rearrangement of the *HyTK* gene that can account for ganciclovir resistance in these clones. (**d**) HLA typing results for parental H1 ESCs, gene-targeted clone c5, and ganciclovir-resistant subclone c5A. (**e**) Copy number analysis of SNP data. Loss of heterozygosity in clone c5A telomeric to the *PRIM2A* gene on chromosome 6 is shown by an increase in copy number of one allele and decrease of the other relative to the parental cell line (H1). GCV-R, ganciclovir-resistance; HLA, human leukocyte antigens.

We chose to target the *HMGA1* gene because it is ~1 Mb centromeric to the HLA locus on chromosome 6 and expressed in pluripotent stem cells. The AAV2-HMGA1-HyTKpA vector is designed to insert the *HyTK* gene at exon 3 of *HMGA1* and initiate translation from the *HMGA1* start codon (**Figure 1b**). Initially, we transduced H1 ESCs[10] and showed by Southern blots that the hygromycin-resistant clones were accurately targeted at the *HMGA1* locus and did not contain random integrants (clones c4 and c5 in Figure 1c). We transduced the H7,[10] BG01, BG02, and BG03 ESC lines[11] with AAV2-HMGA1-HyTKpA as well, and 26 of the 27 hygromycin-resistant colonies screened were accurately targeted (Supplementary Table S1). Because the targeting step was so efficient, we also produced hygromycin-resistant polyclonal populations without picking individual colonies, most of which contained a majority of targeted cells (Supplementary Table S1).

Several of these *HMGA1*-targeted ESCs were then cultured in the presence of ganciclovir to select for loss of the *HyTK* transgene. The ganciclovir-resistant clones that underwent mitotic recombination should have removed the targeted *HMGA1* allele and instead contain two wild-type alleles of *HMGA1*. In the case of H1 clone c5, we isolated ganciclovir-resistant subclone c5A that had lost the *HyTK* gene based on Southern blots (Figure 1c), and we confirmed by HLA typing that it contained a single HLA haplotype (Figure 1d). SNP chip analysis showed a loss of heterozygosity extending from the pericentromeric region to the telomere on chromosome 6 (Figure 1e), with the crossover occurring within the *DNA Primase* gene *PRIM2* (data not shown). Thus, H1 subclone c5A had undergone the expected mitotic recombination event and become HLA-homozygous. These cells produced a trilineage teratoma when grown in immunodeficient mice, confirming that they remained pluripotent after both the gene targeting and mitotic recombination steps (Supplementary Figure S1). Several other ganciclovir selection experiments performed on other *HMGA1*-targeted ESCs failed to produce HLA-homozygous cell lines based on Southern blots and HLA typing, but instead produced resistant clones with *HyTK* gene rearrangements, mutations, or deletions (Supplementary Table S1). So, although subclone c5A clearly demonstrates that HLA-homozygous lines can be derived by mitotic recombination events, ganciclovir-resistant clones produced by other mechanisms are more commonly recovered. The unexpectedly low frequency of mitotic recombination events led us to investigate the derivation of HLA class I–negative cells as an alternative strategy.

## Creating HLA Class I–Negative Cells

The polymorphic HLA-A, -B, and -C class I proteins are expressed on the surface of nucleated cells where they present peptide antigens to the immune system, and matching at these loci is important for successful allogeneic transplantation. The *Beta-2 Microglobulin* (*B2M*) gene encodes a common subunit essential for cell surface expression of all the HLA class I antigen heterodimers (the other subunits are the heavy chains for HLA-A, -B, -C, -E, -F, or -G). Thus, deleting both *B2M* alleles should render ES cells class I deficient and represents an alternative approach for reducing their immunogenicity. Figure 2a shows our strategy for inactivating both copies of *B2M* in ESCs, which consists of two sequential rounds of gene targeting, followed by Cre-mediated excision of the selectable marker cassettes. The two AAV targeting vectors were designed to insert either a *TKNeo* or *HyTK* fusion gene encoding thymidine kinase linked to G418 or hygromycin resistance genes respectively into the *B2M* initiation codon located in exon 1. Both vectors were packaged in serotype 3 capsids to improve ESC transduction frequencies.[12] H1 ESCs were transduced with the AAV3-B2M-ETKNpA vector and 30% of the G418-resistant cells were targeted at one *B2M* allele based on Southern blots (Supplementary Table S2). One of these clones was then transduced with the AAV3-B2M-EHyTKpA vector and 10% of hygromycin-resistant cells were targeted at *B2M*. None of the targeted clones analyzed contained random integrants.

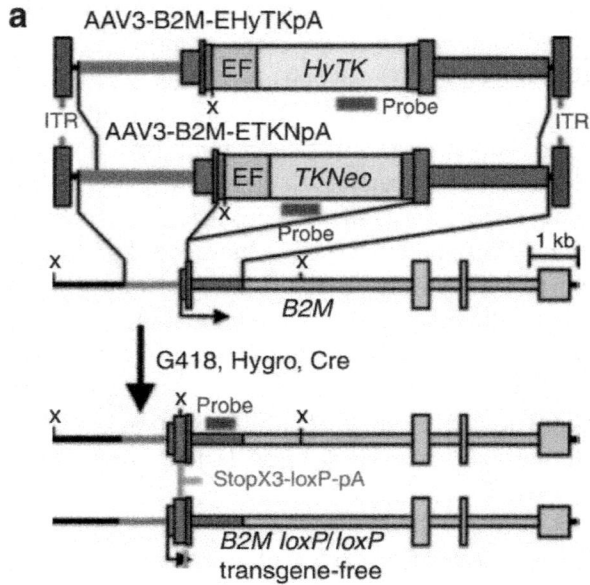

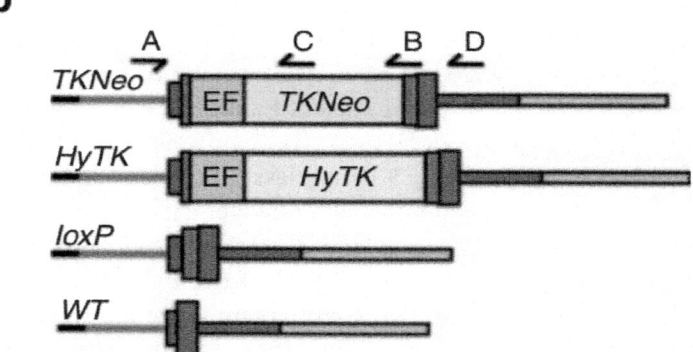

PCR of 45 Cre-treated clones

| Genotype | Primers & products (kb) | | | % of clones |
| --- | --- | --- | --- | --- |
| | A + B | A + C | A + D | |
| TKNeo/HyTK | 2.4 + 2.7 | 0.57 + 1.7 | None | 0 |
| TKNeo/loxP | 0.34 + 2.4 | 0.57 | 0.66 | 81 |
| loxP/HyTK | 0.34 + 2.7 | 1.7 | 0.66 | 5 |
| loxP/loxP | 0.34 | None | 0.66 | 14 |
| +/+ (WT) | None | None | 0.55 | 0 |

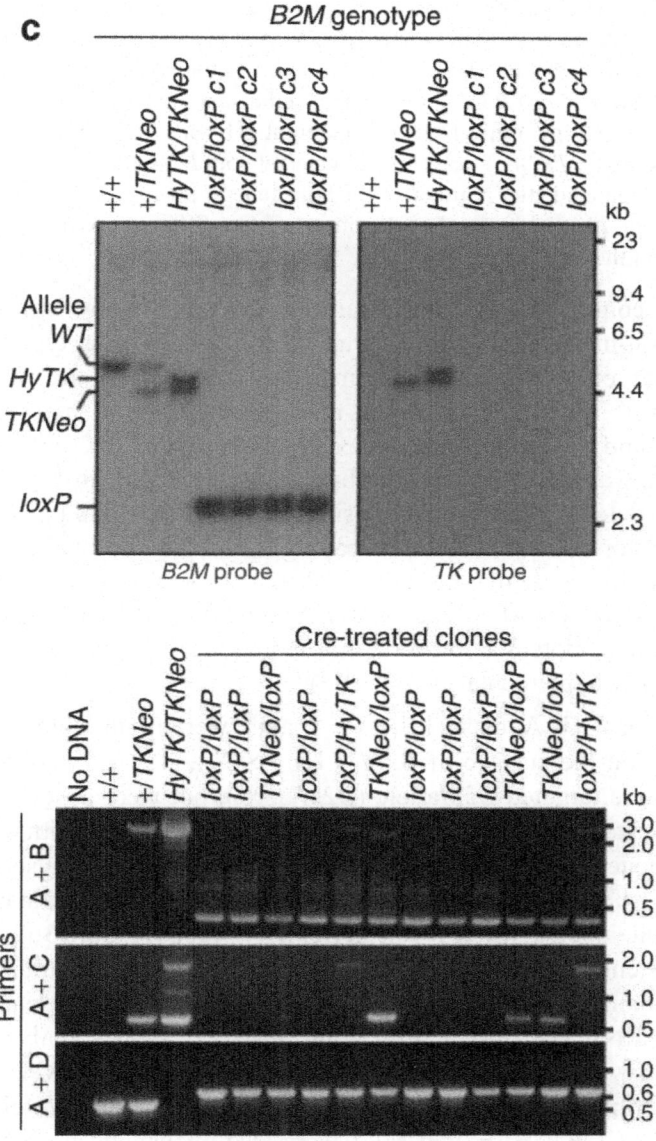

**Figure 2. *B2M* targeting and transgene removal.** (a) The structures of the AAV3-B2M-ETKNpA– and AAV3-EHyTKpA–targeting vectors, and the human *B2M* locus are shown before targeting, and after two rounds of targeting and transgene removal by Cre. The locations of Southern blot probes and enzymes (X, Xba I) are indicated. (b) Diagram of the wild-type, targeted, and transgene-deleted alleles of the *B2M* gene with the locations of PCR primers, and the results obtained from an analysis of 45 Cre-

treated clones. The gel below shows a representative PCR analysis of Cre-transduced clones with the different primer combinations, and the resulting allele designations are shown. DNA from the parental H1 cells (+/+), and clones targeted in one *(+/TKNeo)* or both *(HyTK/TKNeo) B2M* alleles were used as controls. (**c**) Southern blot analysis of Xba I–digested genomic DNAs of parental H1 cells (+/+), one clone targeted at one *B2M* allele (*+/TKNeo*), one clone targeted at both *B2M* alleles (*HyTK/TKNeo*), and four clones obtained after removal of the transgenes with Cre (*loxP/loxP* c1–4), probed with *B2M*- or *TK*-specific probes. Positions of the different possible *B2M* fragments are shown at the left.

Cre-recombinase was then transiently expressed by transducing cells with a non-integrating foamy virus vector, which is a type of retroviral vector that efficiently infects human ESCs.[13] Six of the 45 Cre-treated clones that were analyzed by allele-specific PCR had deleted both the *HyTK* and *TKNeo* transgenes (examples in Figure 2b), and deletion of the transgenes was confirmed by Southern blots (examples in Figure 2c). These transgene-deleted alleles contain a single residual loxP site in exon 1 of *B2M*, as well as stop codons in all three reading frames and a polyadenylation signal (Figure 2a) to ensure that no functional B2M protein can be synthesized. Two of these clones ($B2M^{loxP/loxP}$c1 and c2) were also shown to have trilineage developmental potential by teratoma assays as well as normal karyotypes (example in Supplementary Figure S2).

B2M and HLA class I heavy chains were both detectable by flow cytometry on the surface of $B2M^{+/+}$ H1 ESCs and at a lower level on cells with a single targeted *B2M*allele ($B2M^{+/TKNeo}$), but were absent on cells with both alleles targeted ($B2M^{TKNeo/HyTK}$) or targeted and treated with Cre to remove the transgenes ($B2M^{loxP/loxP}$) (Figure 3a). B2M protein was also undetectable in $B2M^{loxP/loxP}$ cells by western blot, whereas class I heavy chain proteins were still present, albeit at lower levels in $B2M^{loxP/loxP}$ cells (Figure 3b). These results are consistent with the known requirement for B2M in the transport of class I heavy chains to the cell surface.[14] Global gene expression analysis showed that *B2M* RNA was absent from both $B2M^{loxP/loxP}$ clones analyzed, but HLA class I heavy chain transcripts were present at wild-type levels (**Figure 3c**). The transcript levels of genes encoding other B2M-binding proteins, including CD1A-D, FCGRT, HFE, and MR1 were also unchanged in $B2M^{loxP/loxP}$ cells. Cluster analysis of the samples showed that the two $B2M^{-/-}$ cell lines did not cluster apart from the two parental H1 cell line duplicates (**Figure 3d**), demonstrating that the global gene expression pattern in these subclones had not diverged from the parental cell line. In addition, Pearson correlation coefficients for all crosswise comparisons were ≥0.98, suggesting that they had not undergone any major changes during their engineering, and that this approach could be used to generate cells for clinical use.

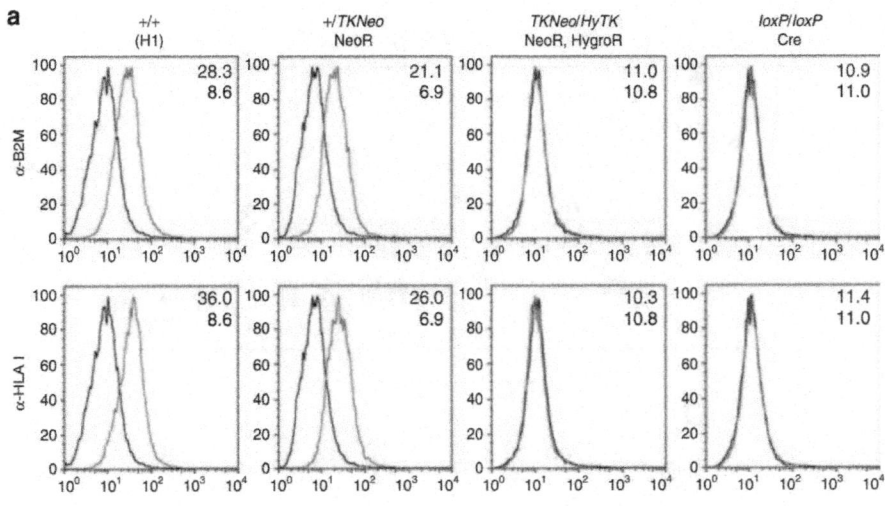
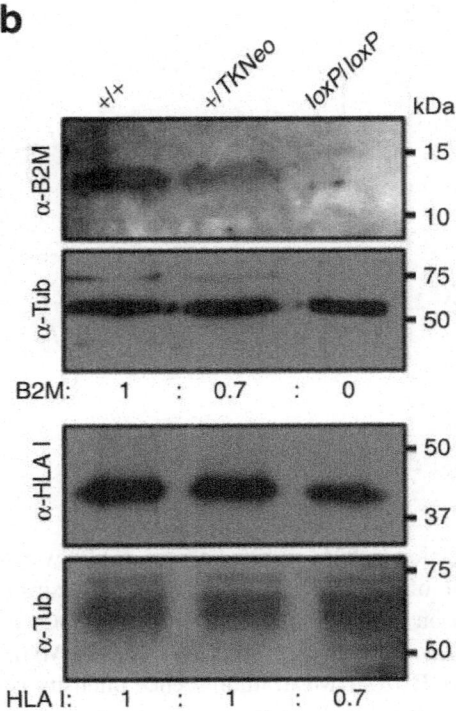

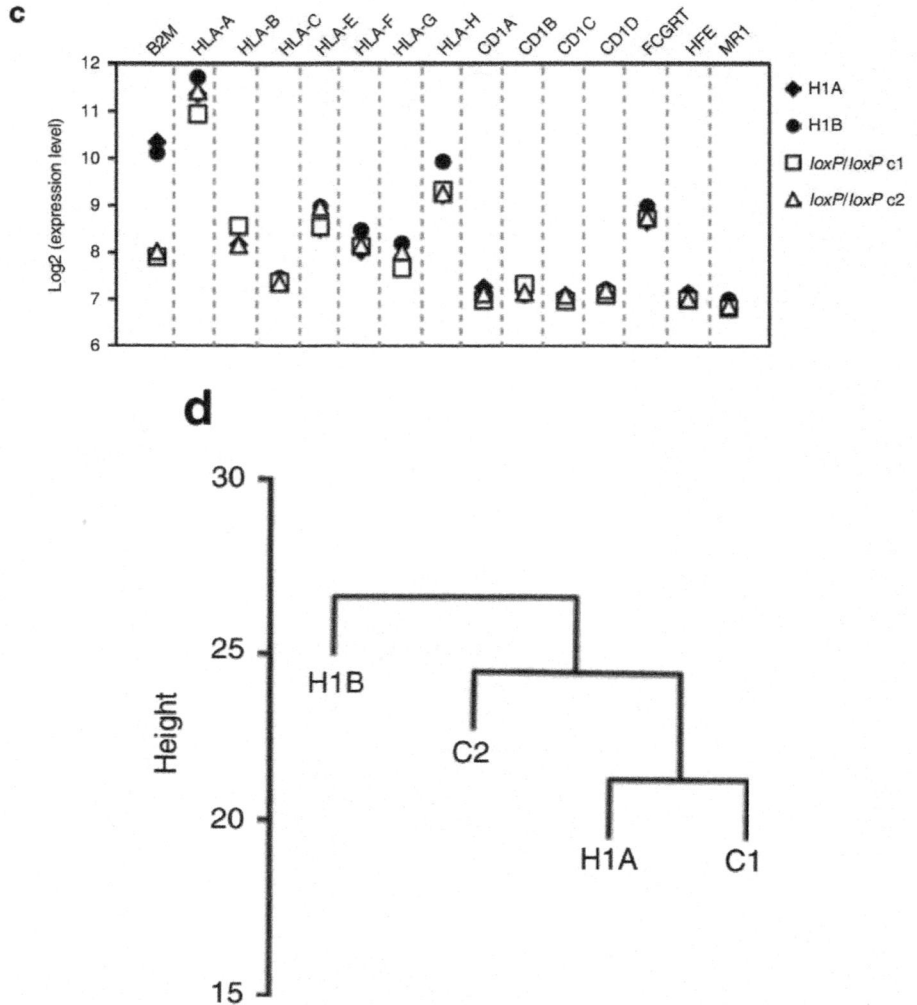

**Figure 3.** Gene expression in *B2M*-targeted clones. (a) Flow cytometry analysis showing surface expression of B2M and HLA class I heavy chains (antibody W6/32 binds HLA-A, -B, and -C) on undifferentiated parental H1 ESCs (+/+), ESCs targeted at one *B2M* allele *(+/TKNeo)*, both *B2M* alleles *(HyTK/TKNeo)*, or after Cre-mediated transgene removal *(loxP/loxP)*. Mean fluorescence intensity is indicated in each case for the specific antibody (blue numbers) and isotype controls (black numbers). (b) Western blot analysis of total cellular protein extracts from parental H1 ESCs (+/+), ESCs targeted at one *B2M* allele (+/TKNeo), or both *B2M* alleles after Cre-mediated transgene excision *(loxP/loxP)* performed with the indicated antibodies. Signal ratios for B2M or HLA class I heavy chains are shown after normalization to the wild-type

(+/+) sample. (**c**) Transcriptional array analysis. The relative mRNA expression levels for *B2M*, HLA class I heavy chains, and other B2M-binding protein genes (*CD1A-D, FCGRT,HFE*, and *MR1*) are shown for two *B2M$^{-/-}$* clones (*loxP/loxP* c1 and c2) and two independent cultures of parental H1 ESCs (+/+). Values are expressed as log2, after normalization and background subtraction. (**d**) Sample relation between two duplicate cultures of parental H1 cells (H1A and H1B) and two *B2M$^{-/-}$* clones (*loxP/loxP* c1 and c2) based on a global expression analysis of 18,174 genes with SD/mean >0.1. HLA, human leukocyte antigens.

## Human Immune Cell Responses to Class I–Deficient Cells

The lack of HLA class I surface expression on *B2M$^{loxP/loxP}$* cells should prevent their recognition by CD8$^+$ T cells. We first examined this with Mixed Leukocyte Reaction (MLR) assays in which allogeneic peripheral blood mononuclear cells (PBMCs) from normal blood donors were mixed with undifferentiated, irradiated *B2M$^{+/+}$* or*B2M$^{loxP/loxP}$* ESCs. In both cases, we did not observe proliferation responses as evidenced by the lack of $^3$H-thymidine uptake (**Figure 4a**, left panel). These findings confirm prior results showing a lack of MLR responses to undifferentiated ESCs,[15] and presumably reflect the absence of costimulatory molecules and/or the presence of inhibitors on pluripotent cells.[16] To elicit a more robust MLR response, we differentiated ESCs into embryoid bodies (EBs) for 15 days, which allows for the development of cells entering the endothelial/hematopoietic lineages that might be competent for antigen presentation.[17] At this timepoint, 12–15% of cells were CD34$^+$ by flow cytometry (**Figure 4b**), but HLA class II expression remained low (<1.2% of cells), allowing us to focus on class I–dependent responses. Wild-type*B2M$^{+/+}$* EB cells were capable of stimulating PBMC proliferation in an MLR assay, albeit at 14–28% the level of allogeneic PBMCs (**Figure 4a**, right panel), perhaps due to the smaller proportion of EB cells capable of antigen presentation. In contrast, the MLR assay using *B2M$^{loxP/loxP}$* EB cells was similar to that of unstimulated PBMCs. This suggested that class I–deficient cells produce a decreased allogeneic response, at least when HLA class II expression is limited. However, these MLR results should be interpreted cautiously, because human ESCs can also inhibit MLR responses of PBMCs against allogeneic dendritic cells, and ESC-derived EB cells did not induce significant PBMC proliferation in a prior study.[15]

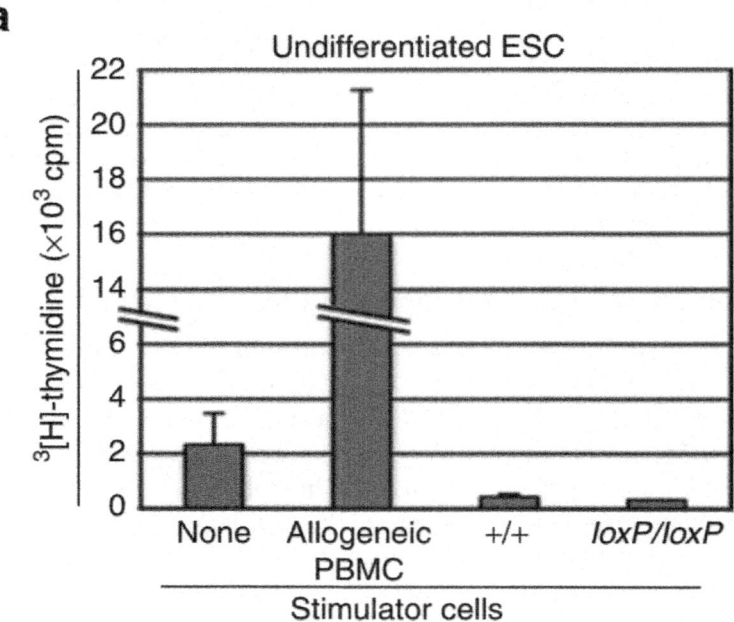
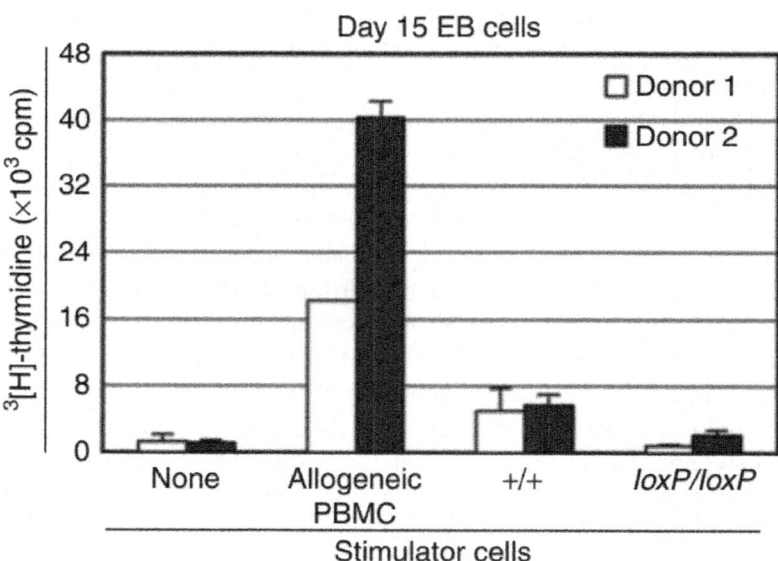

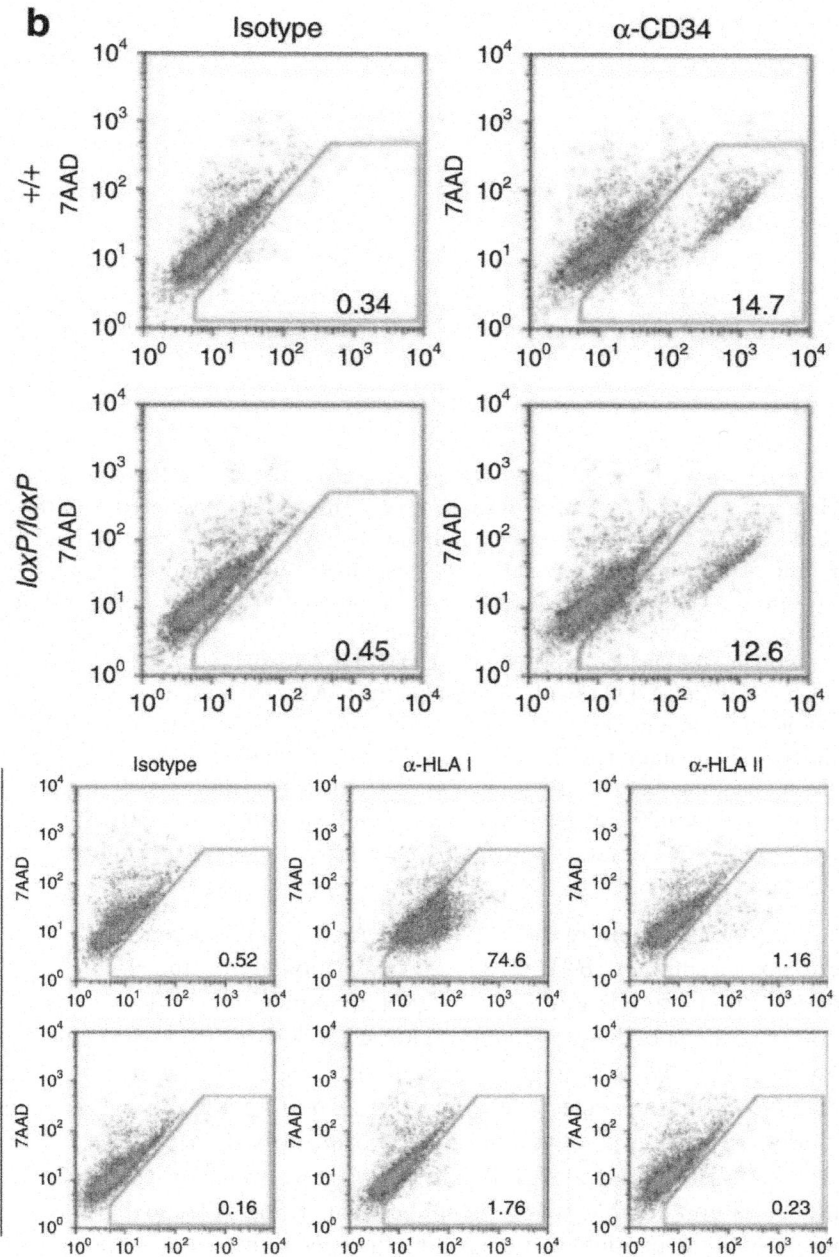

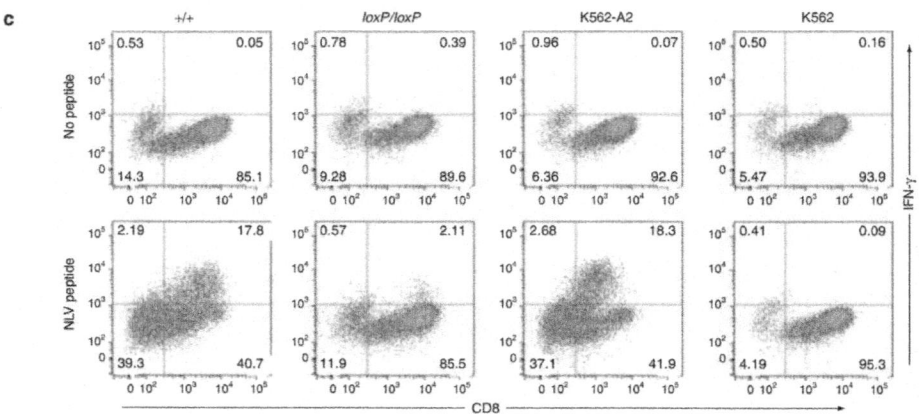

**Figure 4.** Immune responses to *B2M*-targeted cells. (**a**) MLR results showing [3][H]-thymidine incorporation by PBMC responder cells mixed with irradiated, undifferentiated $B2M^{+/+}$ or $B2M^{loxP/loxP}$ ESCs (left panel) or with day 15 EB-derived cells (right panel) at a ratio of 1:1. Two independent PBMC responder cell preparations were used when analyzing EB cells. Controls included unstimulated PBMCs, and PBMCs stimulated by allogeneic PBMCs. (**b**) Flow cytometry analysis of day 15 $B2M^{+/+}$ or $B2M^{loxP/loxP}$ EB cells showing the surface expression of CD34, HLA class I (HLA-A, -B, and -C), and HLA class II (HLA-DR), with isotype controls. All preparations were labeled with 7-amino-actinomycin D (7AAD) to improve gating and remove dead cells. (**c**) Intracellular cytokine staining of HLA-A*0201/NLV-CMVpp65-specific T cells stimulated with an equal number of $B2M^{+/+}$ or $B2M^{loxP/loxP}$ EB cells, HLA class I–deficient K562 cells, or HLA-A*0201–expressing K562-A2 cells, with or without prior pulsing with NLVPMVATV (NLV) peptide from the CMV pp65 protein. HLA, human leukocyte antigens.

To demonstrate more directly that HLA class I–restricted T cell responses were not elicited by $B2M^{loxP/loxP}$ cells, we looked for interferon-γ (IFN-γ) expression in an HLA-A*0201–restricted human CD8⁺ T cell clone that recognizes the specific peptide NLVPMVATV derived from cytomegalovirus. K562 cells served as a negative control lacking HLA class I antigens, and K562-A2 cells transfected with the *HLA-A*0201* gene and pulsed with the peptide served as a positive control. Exposure to wild-type $B2M^{+/+}$ EB cells (HLA-A*0201⁺) induced significant IFN-γ expression in these CD8⁺ T cells, which was comparable with that induced by K562-A2 cells, but only after loading with the peptide, demonstrating the specificity of the response (**Figure 4c**). In contrast, IFN-γ expression was significantly reduced in T cells exposed to $B2M^{loxP/loxP}$ EB cells (2.11% as compared with 17.8% for $B2M^{+/+}$ cells). The basis for this low-level residual IFN-γ expression is unclear, but possible explanations include peptide uptake and presentation by class I–expressing

T cells exposed to EB cell membranes, and/or the effects of free HLA class I heavy chains present in EB cells.

Class I–negative hematopoietic cells may be lysed by natural killer (NK) cells,[18] so this is a potential concern for our approach. We incubated human NK cells with either $B2M^{+/+}$ or $B2M^{-/-}$ EB cells and did not observe an increase in surface expression of CD107a, which is an indicator of NK cell degranulation and cytotoxicity (**Figure 5**). HLA class I–deficient K562 control cells did induce significant degranulation. Thus, there was no NK-mediated cytotoxicity against $B2M^{-/-}$ cells, at least at the EB stage of differentiation.

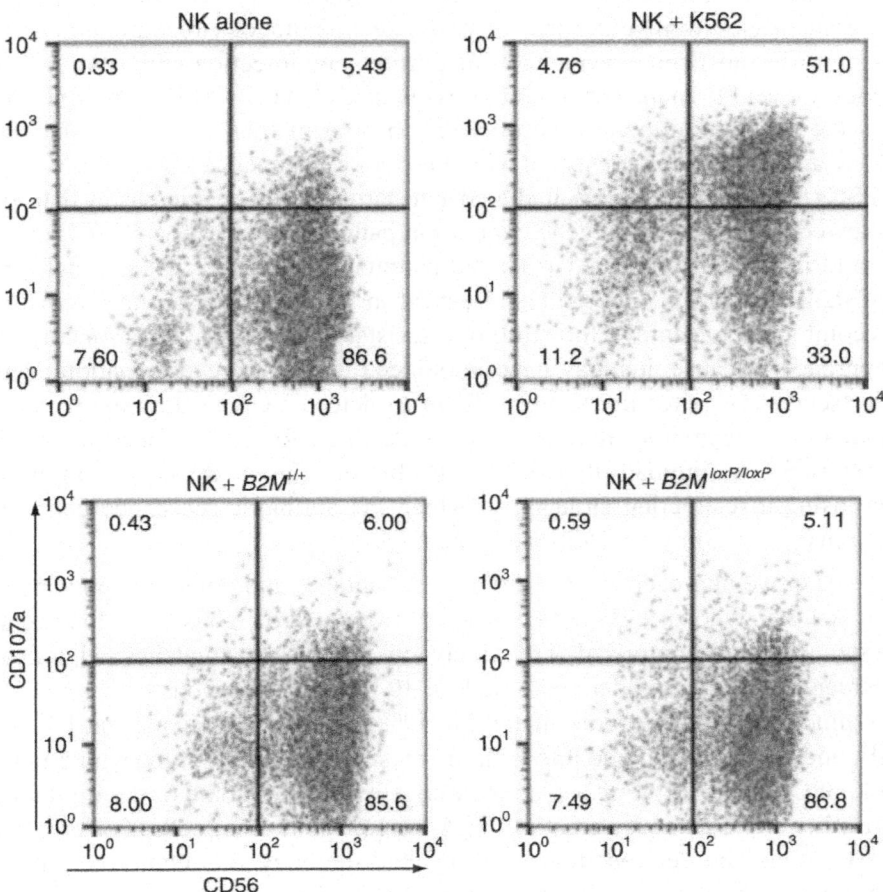

**Figure 5.** NK cells do not lyse $B2M^{-/-}$ EB cells. Flow cytometry analysis of CD107a expression on CD56+ NK effector cells when incubated with $B2M^{+/+}$ or $B2M^{loxP/loxP}$ day 15 EB cells at a ratio of 1:1. Controls included unstimulated NK cells, and NK cells incubated with class I–negative K562 cells. NK, natural killer.

## DISCUSSION

In this report, we describe two gene targeting approaches for engineering the HLA locus in human pluripotent stem cells to overcome immunogenicity and potentially enhance their therapeutic utility. AAV vectors were used in both cases, which allowed us to efficiently isolate targeted clones from a small number of antibiotic-resistant colonies. Over 95% of resistant clones were targeted by the *HMGA1* initiation codon trap vector, and 10–30% of resistant clones were targeted when inserting a functional promoter-transgene cassette at the *B2M* locus. Both approaches produced transgene-free cells. There was minimal genotoxicity from these manipulations, because site-specific endonucleases were not used to stimulate homologous recombination, the vector uses host cell enzymes for integration, and infection with AAV vectors does not lead to increased mutation frequencies.[19] This was largely confirmed by the normal karyotypes we obtained in both of the *B2M*$^{loxP/loxP}$ clones that were analyzed and is consistent with our prior cytogenetic studies of ESCs and iPSCs that underwent AAV-mediated gene targeting.[20] The microarray data also showed a consistency in gene expression patterns, because the *B2M*$^{-/-}$ clones could not be distinguished from the parental H1 cells except for their lack of *B2M* transcripts. One potential caveat in this regard was our use of Cre-recombinase to remove the antibiotic resistance genes in the *B2M* targeting approach, because this may have introduced unwanted recombination events at pseudo-loxP sites that were difficult to detect.[21] Given the importance of maintaining genomic integrity in cells destined for clinical use, it may be preferable to eliminate this step in the future. This could be accomplished by using a retargeting strategy, in which the antibiotic resistance genes are removed instead by an additional gene targeting step.

The two approaches have different advantages and potential applications. HLA-homozygous cells could be derived from heterozygous ESCs to generate a stem cell bank comprised of relatively few cell lines that matches a significant percentage of the population. A bank of 30 HLA-homozygous ESC lines with common haplotypes would match 38–59% of the US population at HLA-A, -B, and -DR, and a single line with the most common haplotype would match 9% of Caucasian Americans.[3] Our strategy overcomes some of the limitations of other approaches for creating HLA-homozygous stem cells. For example, ESCs could be derived directly from HLA-homozygous embryos, but this might require the ethically questionable use of HLA-typed sperm and egg donors specifically for ESC derivation. Parthenogenesis could be used to generate HLA-homozygous ESCs from heterozygous oocytes; however, this results in genome-wide homozygosity that could unmask recessive mutations. A more realistic alternative may be to derive a bank of iPSCs from HLA-homozygous individuals,[5] although iPSCs may not be identical to ESCs in their therapeutic potential.[22] If iPSCs are ever used clinically, our method would allow the derivation of histocompatible, HLA-homozygous iPSCs from the family members of patients that share one HLA haplotype, which could provide a source of matched, genetically normal cells for the treatment of genetic diseases.

One technical difficulty in our experiments was the low frequency of recovering HLA-homozygous cells by ganciclovir selection as compared to other types of *TK* mutations. This is consistent with recent results showing that mitotic recombination with loss of heterozygosity accounts for <1% of ganciclovir-resistant colonies in human pluripotent stem cells,[23] in contrast to what one would expect based on mouse ESC experiments.[9] Although this will hinder the routine application of our approach in regenerative medicine, there may be ways to improve the recovery of extended loss of heterozygosity events by inducing mitotic recombination or improving selection stringency by introducing a second *TK* gene telomeric to the HLA locus on chromosome 6.

The major advantage of *B2M* knockout cells is that they could serve as universal donors when transplanting cells that express HLA class I but not class II, including most solid organ cells. This would produce substantial savings in time and money, because a single cell line could be differentiated into therapeutic cell products, characterized, certified by regulatory agencies, and used in all recipients. Universal donor cells also avoid many of the potential problems of HLA-typed cell banks and patient-specific stem cells, including variability among different stem cell lines in gene expression patterns,[24] culture characteristics,[25] differentiation potential,[6,7] and genetic changes acquired during culture.[26] Unlike patient-derived iPSCs, universal donor cells can be prepared ahead of time and are ideal for treating genetic diseases, because they do not harbor disease-causing mutations. Additional benefits of HLA class I–deficient cells are that they will not present autoantigens that might otherwise prevent the successful treatment of autoimmune diseases, such as peptide antigens derived from GAD65 or insulin that can cause the destruction of β cells in diabetes,[27] and immune responses against therapeutic gene products can be avoided when treating genetic diseases, such as antidystrophin responses in muscular dystrophy.[28]

Significant evidence supports the concept that HLA class I–negative cells will function normally and avoid allogeneic rejection after transplantation. First, rare individuals who are class I–negative due to mutations in the *TAP1* or *TAP2* genes encoding the transporter associated with antigen processing are relatively healthy, with a spectrum of manifestations ranging from the absence of symptoms to chronic upper respiratory infections and skin inflammation,[29] showing that HLA class I–deficient human cells can form all essential organs. Second, $B2m^{-/-}$ mice have been extensively studied and they are relatively healthy except for a notable lack of CD4⁻CD8⁺ T cells.[30] Importantly, transplantation experiments have shown that organs or cells from $B2m^{-/-}$ mouse donors survive longer than allogeneic cells in $B2m^{+/+}$ recipients (sometimes persisting indefinitely), including liver cells,[31] kidneys,[32] hearts,[33] pancreatic islets,[34] and dendritic cells.[35] These mouse experiments suggest that $B2m^{-/-}$ human cells may also survive longer than allogeneic cells in many of the clinical settings being considered for pluripotent stem cells. In support of this, our data show that in comparison with wild-type cells, $B2M^{-/-}$ human cells induce less of an immune response in allogeneic human PBMCs and HLA-A*0201–restricted CD8⁺ T cells as determined by MLR and IFN-γ expression assays respectively. Although the lack of B2M could also affect the function of non-HLA, B2M-binding proteins such as HFE, which is associated with hepatic iron overload,[36] this would not be expected to impact transplant recipients with $B2M^{+/+}$ host cells.

The complete lack of HLA class I expression raises potential concerns. In mice, host NK cells have been shown to eliminate $B2m^{-/-}$ hematopoietic donor cells,[37] which occurs in the absence of specific inhibitory interactions between donor cell class I antigens and NK cell Ly49 receptors.[38] An analogous inhibition of human NK cells is mediated through interactions of HLA-C, -E, and -G with several NK cell receptors.[39,40] Although this is clearly an issue for hematopoietic cells, it is not known if this "missing self" response will be observed after transplanting non-hematopoietic cell types, because the same phenomena was not observed when solid organs from $B2m^{-/-}$ mice were transplanted in $B2m^{+/+}$ recipients.[31,32,33,34] We did not detect NK-mediated lysis in day 15 $B2M^{-/-}$ EB cells, so further differentiation down the hematopoietic lineage may be necessary to demonstrate the "missing self" phenomenon in ESC-derived cells. A second concern is the potential for tumor formation, because HLA expression is frequently lost or reduced in cancer cells.[41] However, neither HLA class I–deficient mice[30] nor humans[29] are tumor-prone. A third concern is viral infection of $B2M^{-/-}$ cells, which may escape class I–restricted immune responses. However, $B2M^{-/-}$ mice are able to survive infection with influenza[42] or Sendai virus,[43] and this would be mitigated when transplanting $B2M^{-/-}$ human cells to anatomically sequestered locations in individuals with fully functional immune systems.

Some of these concerns could be addressed by further engineering of $B2M^{-/-}$ stem cells, including the introduction of single chain fusion constructs encoding non-polymorphic HLA class I proteins fused to B2M.

Both HLA-G and HLA-E have been shown to interact with inhibitory NK cell receptors and prevent NK cell–mediated lysis.[39,40] The potential tumor risk could be addressed by the introduction of an inducible suicide gene that could be activated if aberrant cell proliferation occurred.[44] The additional disruption of HLA class II genes in $B2M^{-/-}$ cells could also be beneficial in some settings, because solid organ cells may be induced to express class II genes under various stress responses[45] and this would allow the derivation of universal donor hematopoietic cell products that might otherwise be rejected due to class II antigen presentation.

## MATERIALS AND METHODS

### Cell Culture

Undifferentiated human ESCs were cultured as described.[20] Differentiation of ESCs into embryoid bodies was performed as described.[46] Standard G-banding karyotypes were carried out in the Cytogenetics Laboratory of the Department of Pathology, University of Washington (Seattle, WA). Teratoma assays were performed as described.[20] All animal experiments (teratoma assays) were approved by the University of Washington Institutional Animal Care and Use Committee.

CD8$^+$ T cell clone 6G1-82, specific for the HLA-A*0201 presenting the peptide NLVPMVATV from the CMV pp65 protein,[47] was obtained and expanded as described.[48] K562 human chronic myeloid leukemia cells K562-A2 cells expressing HLA-A*0201 were previously described.[49] All cells were grown at 37 °C in a 5% $CO_2$ atmosphere.

### Viral Vectors

The AAV2-HMGA1-HyTKpA vector was described previously.[20] The AAV3-B2M-ETKNpA and AAV3-B2M-EHyTKpA vectors contain genomic DNA from the *B2M* locus (chromosome 15: nucleotides 45,002,685–45,004,747; and nucleotides 45,002,938–45,004,747, respectively, February 2009 assembly); with an EF1α (eukaryotic translation elongation factor 1α) promoter; *TKNeo* gene (herpes simplex virus thymidine kinase—neomycin phosphotransferase fusion gene) or *HyTK* gene (hygromycin phosphotransferase—herpes simplex virus thymidine kinase fusion gene) respectively, flanking loxP sites, stop codons in all three frames and a polyadenylation signal at the 3'end. The *TKNeo* and *HyTK* cassettes were inserted at the initiation codon in exon 1 of *B2M*. AAV vector stocks were produced and titered as described.[12] The non-integrating foamy virus vector NIFV-EokCreW is identical to the NIFV-

ECreW vector described previously,[50] but with an optimized Kozak sequence in the Cre gene.

## Gene Targeting and Ganciclovir Selection

AAV-mediated gene targeting experiments were performed as described previously. For *HMGA1* targeting, $5 \times 10^4$ H1 ESCs were plated per well of a 24-well plate on day 0, and on the next day cells were infected with the AAV2-HMGA1-HyTKpA vector at a multiplicity of infection of $2 \times 10^4$ genome-containing vector ps per cell. On day 4, hygromycin B (20 µg/ml) was added to the culture and selection was continued for 16 days. To isolate HLA-homozygous clones, *HMGA1*-targeted cells were selected with 3 µmol/l ganciclovir for 9 days.

For *B2M* targeting, $7.5 \times 10^5$ H1 ESCs were plated per 10 cm dish on day 0 and infected on day 1 with the AAV-B2M-ETKNpA vector at a multiplicity of infection of 20,000 genome-containing vector ps per cell. On day 4, the cells were cultured in 50 µg/ml G418 and selection was continued for 22–36 days. When targeting the second *B2M* allele, $2.5 \times 10^5$ ESCs targeted at the first allele with AAV-B2M-ETKNpA were plated per 6 cm dish on day 0 and infected on day 1 with the AAV-B2M-EHyTKpA vector at a multiplicity of infection of 2,000 genome-containing vector ps per cell. On day 4, the cells were cultured in 20 µg/ml hygromycin and selection was continued for 26–28 days. In each case, a portion of cells was also plated without selection to determine the percentage of transduced colony-forming units.

## Cre-Mediated Transgene Excision

$2.5 \times 10^4$ ESC targeted at both *B2M* alleles were plated per well of a 48-well plate on day 0. On day 1, the cells were infected with the non-integrating Foamy vector NIFV-EokCreW at a multiplicity of infection of 50,000 genome-containing vector ps per cell. On day 15, these cells were disaggregated into single cells with accutase (Stemgent, San Diego, CA) counted, and plated at densities of $10^6$ or $0.5 \times 10^6$ cells per 10 cm dish. On day 28, individual colonies were picked and expanded for further analysis.

## DNA Isolation and Analysis.

Genomic DNA was isolated by using the Puregene DNA purification system (Gentra Systems, Minneapolis, MN). Southern blot analysis, plasmid preparation, and restriction digestion were performed according to standard protocols. Radiolabeled probes were synthesized by random priming using Rediprime II (GE Healthcare, Piscataway, NJ). For PCR analysis, cells were lysed in 50 mmol/l Tris pH 8, 1 mmol/l EDTA, 0.2 mol/l sodium chloride, treated with 200 µg/ml Proteinase K (Invitrogen, Carlsbad, CA) at 55 °C for 1 hour, and heated at 95 °C for 5 minutes to inactivate Proteinase K. The lysate was clarified by centrifugation at 16,000$g$ for 10 minutes and the supernatant was used as template for the PCR. Primers used in the PCR were: (A) forward primer: CGCCGATGTACAGACAGCAAA; (B) reverse primer: TATCGACGGATCCCACACAA; (C) reverse primer: ACCGTCTATATAAACCCGCAGT; and (D) reverse primer: GCCAAAGGTCTCCCCTGCTCC. For the PCR, GoTaq polymerase (Promega, Madison, WI) was used, and the conditions were as follows: 150 ng of the template genomic DNA, 0.5 µmol/l each primer, 0.25 mmol/l dNTPs, and 2 mmol/l $MgCl_2$. The annealing temperatures were 56 °C (primers A + C and A + D) or 58 °C (primers A + B) and amplification was run for 35 cycles in a PTC-200 Thermo Cycler (Biorad, MJ Research, Hercules, CA). Short extension times (40 seconds) were used for primers A + D to amplify only the loxP or the wild-type alleles.

## Western Blots

Protein extracts were obtained by lysing cells in 50 mmol/l Tris pH 8, 150 mmol/l NaCl, 1 mmol/l EDTA, 1% (v/v) NP40, and Complete Protease Inhibitor Cocktail (Roche Diagnostics, Mannheim, Germany), and clarified by centrifugation at 16,000$g$ for 30 minutes at 4 °C. Samples were analyzed in precast SDS-PAGE gels (Biorad), under non-reducing conditions for the anti-HLA antibody or reducing conditions for the anti-B2M antibody; and electroblotted to nitrocellulose membranes (Biorad). After blocking with 5% milk in phosphate-buffered saline, the membranes were probed with the corresponding primary and secondary antibodies (see below), developed with the enhanced luminescence (ECL, Pierce, Thermo Scientific, Rockford, IL), and exposed to X-ray films. Anti-HLA-ABC antibody clone W6/32 was obtained from Sigma-Aldrich (St Louis, MO). Anti-B2M BBM.1 antibody was a gift from D Geraghty. Antitubulin antibody was from Abcam (Cambridge, MA). Secondary antibodies and antimouse horseradish peroxidase were obtained from Santa Cruz Biotechnology (Santa Cruz, CA).

## Flow Cytometry

Undifferentiated ESC suspensions were washed with flow buffer (phosphate-buffered saline containing 2% fetal bovine serum, 0.05% [v/v] sodium azide, and 2 mmol/l EDTA), then incubated with the appropriate antibodies for 30 minutes in the dark on ice. Embryoid body cells were obtained by trypsinization and preincubated with Fc blocking reagent (Miltenyi Biotec, Auburn, CA) for

10 minutes at 4 °C before staining with the antibodies. 7-amino-actinomycin D from Via-Probe (BD Biosciences, Bedford, MA) was added last to exclude dead cells from the analysis. Samples were analyzed on a FACscan or Calibur (BD Biosciences) instrument and the data were analyzed using FlowJo software (TreeStar, Ashland, OR). The antibodies used for flow cytometry were: anti-B2M PE clone B2M-01 (Santa Cruz Biotechnology), anti-HLA-ABC clone W6/32 FITC (eBioscience, San Diego, CA), anti-HLA-DR FITC (Miltenyi Biotec), anti-CD34 FITC (BD Pharmingen, San Diego, CA), anti-CD45 FITC (BD Pharmingen), anti-CD8 APC (BD Pharmingen), and anti-IFN-γ FITC (BD Pharmingen). Isotype control antibodies were obtained from Miltenyi Biotec or BD Biosciences.

## Mixed Lymphocyte Reactions

PBMCs were obtained from healthy donors at the Puget Sound Blood Center (Seattle, WA) by centrifugation in Ficoll (LSM, MP Biomedicals, Santa Ana, CA) gradients as previously described. MLRs were done following established protocols. Briefly, stimulator cells (hESCs, day 15 EB cells, or PBMCs for the allogeneic positive control) were irradiated with 4,000 rad, and $10^5$ cells were incubated with the responder PBMCs at a ratio of 1:1 in a 96-well round-bottom plate in RPMI-1640 supplemented with glutamine, 10% heat-inactivated human AB serum (GemCell, Sacramento, CA), and 50 μmol/l β-mercaptoethanol. After 6 days, these mixtures were pulsed with $^3$[H]-thymidine (Perkin Elmer, Santa Clara, CA) for 16–18 hours and harvested onto glass fiber filters (Filtermat A; Perkin Elmer) using a Filtermate Harvester. The amount of radioactivity incorporated was determined with an LS6500 counter (Beckman, Indianapolis IN).

## Intracellular Cytokine Staining

Cells derived from trypsinized embryoid bodies on day 15–18 of the differentiation protocol, K562 cells, or K562-A2 cells were pulsed with 5 μg/ml of peptide NLVPMVATV (>90% purity; GeneScript, Piscataway, NJ) at 37 °C for 2 hours. After the pulse, cells were washed four times then mixed with the responder CD8⁺ T cell clone 6G1-82 at a ratio 1:1, and incubated for 4 hours at 37 °C, with brefeldin A (Sigma-Aldrich) being added to the cells after 1 hour. The cells were then fixed and permeabilized with Fix and Perm kit (BD Biosciences), costained with FITC-conjugated anti-IFN-γ antibody (BD Biosciences), and APC-conjugated anti-CD8 antibody (BD Biosciences), and analyzed by flow cytometry.

## CD107a Expression

NK effector cells were isolated from blood samples by positive selection using Miltenyi's magnetic beads and the CliniMacs purification system. The sorted cells were >94% $CD56^+$ by flow cytometry. Effector (NK) and target cells (K562 or EB-derived cells) were incubated at a ratio of 1:1 in the presence of anti-CD107a-PeCy5 (BD Pharmingen) at 37 °C in 5% $CO_2$. After 1 hour, 6 μg/ml monensin (Golgi-Stop, BD Biosciences) was added to the culture and the cells were incubated for an additional 4 hours. The cells were then stained with anti-CD56 PE (Invitrogen) and the samples were analyzed by flow cytometry.

## Microarray Analysis

RNA was isolated from hESCs cultured without feeder cells and hybridized to the HumanHT 12 v4 BeadChip (Illumina, San Diego, CA) by the Fred Hutchinson Cancer Research Center Genomics Shared Resource. The gene expression levels of all RefSeq genes were quantile normalized. Cell comparisons were conducted using the Bioconductor package "limma" (http://bioconductor.org/) by calculating the 75th percentile of negative controls for each array and using this as a minimum signal intensity threshold value to be applied to each array. For probe sets considered detectable, microarray data normalized to the median and log2 ratios were calculated. Microarray data have been deposited under Gene Expression Omnibus accession number (GSE42289).

For the analysis of the of SNPs copy number, genomic DNA was isolated from parental H1 hESCs and GCV-resistant clone c5A, and hybridized to the Affymetrix GeneChip Human Mapping 250K Nsp Array at the University of Washington Center for Array Technology. Data were analyzed using the GeneChip Genotyping Analysis Software (GTYPE; Affymetrix, Santa Clara, CA) and Copy Number Analysis Tool.

## SUPPLEMENTARY MATERIAL

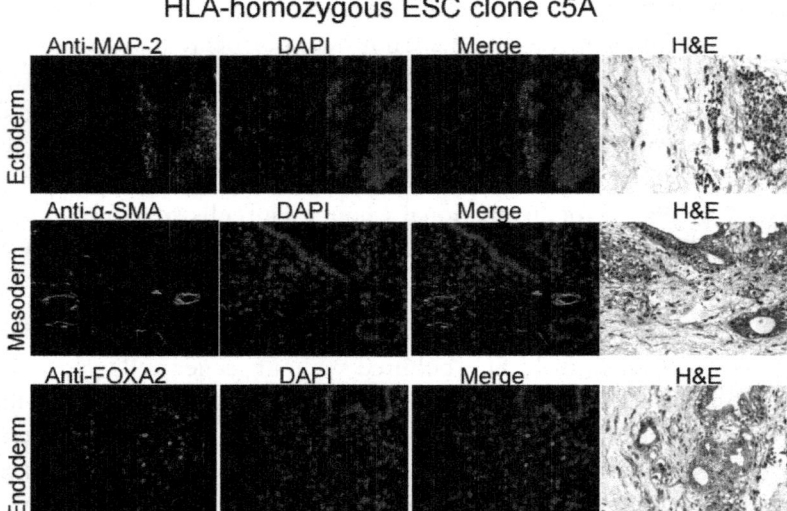

**Figure S1.** Trilineage teratoma formation in HLA-homozygous ESC clone c5A. Teratomas derived from clone c5A were stained with the indicated antibodies or DAPI and examined by fluorescence microscopy. Serial sections are shown stained with hematoxylin and eosin (H&E). MAP-2, Microtubule Associated Protein-2. α-SMA (alpha-Smooth Muscle Actin). FOXA2, Forkhead Box Protein A2. Scale bars = 100 microns.

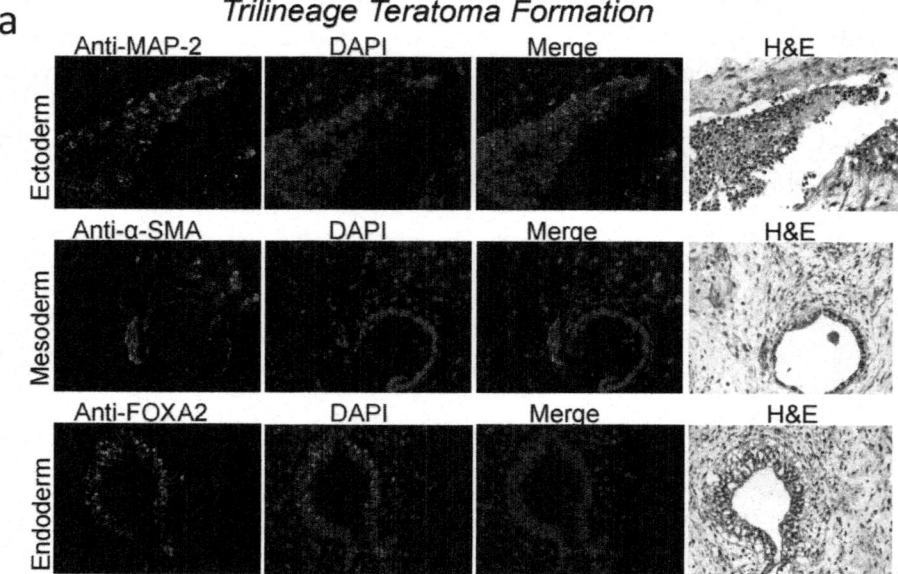

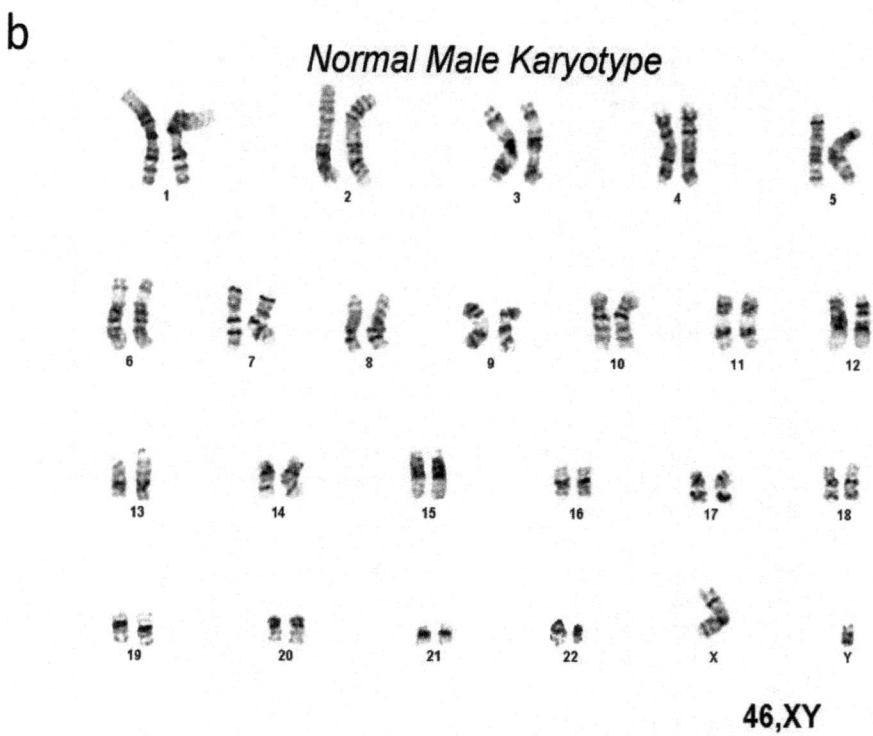

**Figure S2.** Trilineage teratoma formation and normal karyotype of $B2M^{loxP/loxP}$ cells. (a) Teratoma produced by Cre'd out $B2M^{loxP/loxP}$ HLA class I-negative clone c1 stained with the indicated antibodies or DAPI and examined by fluorescence microscopy. Serial sections are shown stained with hematoxylin and eosin (H&E). MAP-2, Microtubule Associated Protein-2. α-SMA (alpha-Smooth Muscle Actin). FOXA2, Forkhead Box Protein A2. Scale bars = 100 microns. (b) Karyotype of the same clone. Five out of five spreads examined were normal male cells.

**Table S1.** *HMGA1* targeting and GCV-resistant clone analysis

| Cell line | Hyg-R clones | Targeted clones | GCV-R subclones | *HMGA1* probe | HyTK probe | HLA type | Presumed basis for GCV resistance |
|---|---|---|---|---|---|---|---|
| H1 | 5 | 5 | c1A-L | WT | Missing | Hetero | Gene conversion |
|  |  |  | c2A | Targeted | Targeted | NA | HyTK mutation |
|  |  |  | c4A,B | Targeted | New band | NA | HyTK rearrangement |
|  |  |  | c4C-H | Targeted | Targeted | NA | HyTK point mutation |
|  |  |  | c5A | WT | Missing | Homo | Mitotic recombination |
| H7 | 5 | 4 | NA |  |  |  |  |
| BG01 | 3 | 3 | NA |  |  |  |  |
|  | pc1 | 100% | pc1A-N | WT | Missing | Hetero | Gene conversion |
|  |  |  | pc1O,P | Targeted | Targeted | NA | HyTK mutation |
|  | pc2 | 100% | pc2A-D | WT | New band | NA | HyTK rearrangement |
|  | pc3 | 58% | pc3A,B | WT | New band | NA | HyTK rearrangement |
|  | pc4 | 100% | pc4A,B | WT | Missing | Hetero | Gene conversion |
|  | pc5 | 100% | pc5A-C | WT | Missing | Hetero | Gene conversion |
|  | pc6 | 100% | NA |  |  |  |  |
| BG02 | pc1 | 100% | pc1A-J | WT | Missing | Hetero | Gene conversion |
|  | pc2 | 100% | pc1A | WT | New band | NA | HyTK rearrangement |
|  | pc3 | 38% | NA |  |  |  |  |
|  | pc4 | 100% | NA |  |  |  |  |
|  | pc5 | 100% | NA |  |  |  |  |
|  | pc6 | 100% | pc1A | WT | Missing | Hetero | Gene conversion |
|  | pc7 | 100% | NA |  |  |  |  |
| BG03 | 14 | 14 | c1A,B | Targeted | Targeted | NA | HyTK mutation |

**Table S2.** *B2M* targeting experiments

| Cell line | Allele 1 vector | MOI | Total CFU | # G418$^R$ CFU | % G418$^R$ CFU | G418$^R$ CFU/ infected cell | # G418$^R$ CFU screened | # G418$^R$ CFU targeted | % G418$^R$ CFU targeted | % Total CFU targeted | Targeted CFU/ infected cell |
|---|---|---|---|---|---|---|---|---|---|---|---|
| H1 | AAV2-B2M-ETKNpA | 20,000 | 1575 | 291 | 18.5 | 1.9E-04 | 30 | 9 | 30.0 | 6.1 | 6.35E-05 |
|  |  |  |  | 345 | 21.9 | 2.3E-04 |  |  |  |  |  |
|  |  |  |  | 336 | 21.3 | 2.2E-04 |  |  |  |  |  |
|  |  |  |  | 298 | 18.9 | 2.0E-04 |  |  |  |  |  |
| H1 | none | 0 | 1575 | 0 | <0.06 | 0 |  |  |  |  |  |

| Cell line | Allele 2 vector | MOI | Total CFU | Hyg$^R$ CFU | % Hyg$^R$ CFU | Hyg$^R$ CFU/ infected cell | # Hyg$^R$ CFU screened | # Hyg$^R$ CFU targeted | % Hyg$^R$ CFU targeted | % Total CFU targeted | Targeted CFU/ infected cell |
|---|---|---|---|---|---|---|---|---|---|---|---|
| C23-2 | AAV3-B2M-EHyTKpA | 2,000 | 450 | 7 | 1.6 | 1.4E-05 | 60 | 6 | 10.0 | 0.26 | 2.34E-06 |
|  |  |  |  | 8 | 1.8 | 1.6E-05 |  |  |  |  |  |
|  |  |  |  | 10 | 2.2 | 2.0E-05 |  |  |  |  |  |
|  |  |  |  | 11 | 2.4 | 2.2E-05 |  |  |  |  |  |
|  |  |  |  | 16 | 3.6 | 3.2E-05 |  |  |  |  |  |
|  |  |  |  | 10 | 2.2 | 2.0E-05 |  |  |  |  |  |
|  |  |  |  | 8 | 1.8 | 1.6E-05 |  |  |  |  |  |
|  |  |  |  | 14 | 3.1 | 2.8E-05 |  |  |  |  |  |
|  |  |  |  | 19 | 4.2 | 3.8E-05 |  |  |  |  |  |
|  |  |  |  | 14 | 3.1 | 2.8E-05 |  |  |  |  |  |
| C23-2 | none | 0 | 450 | 0 | <0.2 | 0 |  |  |  |  |  |

Abbreviations: CFU, colony-forming unit; Hyg, Hygromycin; MOI, multiplicity of infection.

## ACKNOWLEDGMENTS

We thank Daniel Geraghty (Fred Hutchinson Cancer Research Center, Seattle, WA) for providing the anti-B2M BBM-1 antibody, Li B Li for help with array data analysis, and Raisa Stolitenko for technical assistance. This work was supported by US National Institutes of Health grants DK55759, HL53750, and GM086497 to D.W.R. and AI053193 to S.R.R. D.W.R. is on the Advisory Board of Horizon Discovery. The other authors declare no conflict of interest.

## REFERENCES

1. Takahashi, K, Tanabe, K, Ohnuki, M, Narita, M, Ichisaka, T, Tomoda, K et al. (2007). Induction of pluripotent stem cells from adult human fibroblasts by defined factors. *Cell* **131**: 861–872.
2. Yu, J, Vodyanik, MA, Smuga-Otto, K, Antosiewicz-Bourget, J, Frane, JL, Tian, S et al. (2007). Induced pluripotent stem cell lines derived from human somatic cells. *Science* **318**: 1917–1920.
3. Faden, RR, Dawson, L, Bateman-House, AS, Agnew, DM, Bok, H, Brock, DW et al. (2003). Public stem cell banks: considerations of justice in stem cell research and therapy. *Hastings Cent Rep* **33**: 13–27.
4. Taylor, CJ, Bolton, EM, Pocock, S, Sharples, LD, Pedersen, RA and Bradley, JA (2005). Banking on human embryonic stem cells: estimating the number of donor cell lines needed for HLA matching. *Lancet* **366**: 2019–2025.
5. Okita, K, Matsumura, Y, Sato, Y, Okada, A, Morizane, A, Okamoto, S et al. (2011). A more efficient method to generate integration-free human iPS cells.*Nat Methods* **8**: 409–412.
6. Chang, KH, Nelson, AM, Fields, PA, Hesson, JL, Ulyanova, T, Cao, H et al. (2008). Diverse hematopoietic potentials of five human embryonic stem cell lines. *Exp Cell Res* **314**: 2930–2940.
7. Pekkanen-Mattila, M, Kerkelä, E, Tanskanen, JM, Pietilä, M, Pelto-Huikko, M, Hyttinen, J et al. (2009). Substantial variation in the cardiac differentiation of human embryonic stem cell lines derived and propagated under the same conditions–a comparison of multiple cell lines. *Ann Med* **41**: 360–370.
8. Mayshar, Y, Ben-David, U, Lavon, N, Biancotti, JC, Yakir, B, Clark, AT et al. (2010). Identification and classification of chromosomal aberrations

in human induced pluripotent stem cells. *Cell Stem Cell* **7**: 521–531.
9. Lefebvre, L, Dionne, N, Karaskova, J, Squire, JA and Nagy, A (2001). Selection for transgene homozygosity in embryonic stem cells results in extensive loss of heterozygosity. *Nat Genet* **27**: 257–258.
10. Thomson, JA, Itskovitz-Eldor, J, Shapiro, SS, Waknitz, MA, Swiergiel, JJ, Marshall, VS *et al.* (1998). Embryonic stem cell lines derived from human blastocysts. *Science* **282**: 1145–1147.
11. Mitalipova, M, Calhoun, J, Shin, S, Wininger, D, Schulz, T, Noggle, S *et al.* (2003). Human embryonic stem cell lines derived from discarded embryos.*Stem Cells* **21**: 521–526.
12. Khan, IF, Hirata, RK and Russell, DW (2011). AAV-mediated gene targeting methods for human cells. *Nat Protoc* **6**: 482–501.
13. Gharwan, H, Hirata, RK, Wang, P, Richard, RE, Wang, L, Olson, E *et al.* (2007). Transduction of human embryonic stem cells by foamy virus vectors.*Mol Ther* **15**: 1827–1833.
14. Arce-Gomez, B, Jones, EA, Barnstable, CJ, Solomon, E and Bodmer, WF (1978). The genetic control of HLA-A and B antigens in somatic cell hybrids: requirement for beta2 microglobulin. *Tissue Antigens* **11**: 96–112.
15. Li, L, Baroja, ML, Majumdar, A, Chadwick, K, Rouleau, A, Gallacher, L *et al.* (2004). Human embryonic stem cells possess immune-privileged properties.*Stem Cells* **22**: 448–456.
16. Drukker, M, Katz, G, Urbach, A, Schuldiner, M, Markel, G, Itskovitz-Eldor, J *et al.* (2002). Characterization of the expression of MHC proteins in human embryonic stem cells. *Proc Natl Acad Sci USA* **99**: 9864–9869.
17. Zhan, X, Dravid, G, Ye, Z, Hammond, H, Shamblott, M, Gearhart, J *et al.* (2004). Functional antigen-presenting leucocytes derived from human embryonic stem cells *in vitro*. *Lancet* **364**: 163–171.
18. Ljunggren, HG and Kärre, K (1990). In search of the 'missing self': MHC molecules and NK cell recognition. *Immunol Today* **11**: 237–244.
19. Miller, DG, Petek, LM and Russell, DW (2004). Adeno-associated virus vectors integrate at chromosome breakage sites. *Nat Genet* **36**: 767–773.
20. Khan, IF, Hirata, RK, Wang, PR, Li, Y, Kho, J, Nelson, A *et al.* (2010). Engineering of human pluripotent stem cells by AAV-mediated gene targeting. *Mol Ther* **18**: 1192–1199.
21. Thyagarajan, B, Guimarães, MJ, Groth, AC and Calos, MP (2000). Mammalian genomes contain active recombinase recognition sites. *Gene* **244**: 47–54.

22. Kim, K, Doi, A, Wen, B, Ng, K, Zhao, R, Cahan, P *et al*. (2010). Epigenetic memory in induced pluripotent stem cells. *Nature* **467**: 285–290.
23. Li, LB, Chang, KH, Wang, PR, Hirata, RK, Papayannopoulou, T and Russell, DW (2012). Trisomy correction in down syndrome induced pluripotent stem cells. *Cell Stem Cell* **11**: 615–619.
24. Rosler, ES, Fisk, GJ, Ares, X, Irving, J, Miura, T, Rao, MS *et al*. (2004). Long-term culture of human embryonic stem cells in feeder-free conditions. *Dev Dyn* **229**: 259–274.
25. Ware, CB, Nelson, AM and Blau, CA (2006). A comparison of NIH-approved human ESC lines. *Stem Cells* **24**: 2677–2684.
26. Närvä, E, Autio, R, Rahkonen, N, Kong, L, Harrison, N, Kitsberg, D *et al*. (2010). High-resolution DNA analysis of human embryonic stem cell lines reveals culture-induced copy number changes and loss of heterozygosity. *Nat Biotechnol* **28**: 371–377.
27. Di Lorenzo, TP, Peakman, M and Roep, BO (2007). Translational mini-review series on type 1 diabetes: Systematic analysis of T cell epitopes in autoimmune diabetes. *Clin Exp Immunol* **148**: 1–16.
28. Mendell, JR, Campbell, K, Rodino-Klapac, L, Sahenk, Z, Shilling, C, Lewis, S *et al*. (2010). Dystrophin immunity in Duchenne's muscular dystrophy. *N Engl J Med* **363**: 1429–1437.
29. Zimmer, J, Andrès, E, Donato, L, Hanau, D, Hentges, F and de la Salle, H (2005). Clinical and immunological aspects of HLA class I deficiency. *QJM* **98**: 719–727.
30. Zijlstra, M, Bix, M, Simister, NE, Loring, JM, Raulet, DH and Jaenisch, R (2010). Beta 2-microglobulin deficient mice lack CD4-8+ cytolytic T cells. 1990. *J Immunol* **184**: 4587–4591.
31. Li, X and Faustman, D (1993). Use of donor beta 2-microglobulin-deficient transgenic mouse liver cells for isografts, allografts, and xenografts. *Transplantation* **55**: 940–946.
32. Coffman, T, Geier, S, Ibrahim, S, Griffiths, R, Spurney, R, Smithies, O *et al*. (1993). Improved renal function in mouse kidney allografts lacking MHC class I antigens. *J Immunol* **151**: 425–435.
33. Qian, S, Fu, F, Li, Y, Lu, L, Rao, AS, Starzl, TE *et al*. (1996). Impact of donor MHC class I or class II antigen deficiency on first- and second-set rejection of mouse heart or liver allografts. *Immunology* **88**: 124–129.
34. Prange, S, Zucker, P, Jevnikar, AM and Singh, B (2001). Transplanted MHC class I-deficient nonobese diabetic mouse islets are protected from autoimmune injury in diabetic nonobese recipients. *Transplantation* **71**:

982–985.
35. Matsunaga, Y, Fukuma, D, Hirata, S, Fukushima, S, Haruta, M, Ikeda, T et al. (2008). Activation of antigen-specific cytotoxic T lymphocytes by beta 2-microglobulin or TAP1 gene disruption and the introduction of recipient-matched MHC class I gene in allogeneic embryonic stem cell-derived dendritic cells. *J Immunol* **181**: 6635–6643.
36. Feder, JN, Gnirke, A, Thomas, W, Tsuchihashi, Z, Ruddy, DA, Basava, A et al. (1996). A novel MHC class I-like gene is mutated in patients with hereditary haemochromatosis. *Nat Genet* **13**: 399–408.
37. Bix, M, Liao, NS, Zijlstra, M, Loring, J, Jaenisch, R and Raulet, D (1991). Rejection of class I MHC-deficient haemopoietic cells by irradiated MHC-matched mice. *Nature* **349**: 329–331.
38. Karlhofer, FM, Ribaudo, RK and Yokoyama, WM (1992). MHC class I alloantigen specificity of Ly-49+ IL-2-activated natural killer cells. *Nature* **358**: 66–70.
39. Pazmany, L, Mandelboim, O, Valés-Gómez, M, Davis, DM, Reyburn, HT and Strominger, JL (1996). Protection from natural killer cell-mediated lysis by HLA-G expression on target cells. *Science* **274**: 792–795.
40. Lee, N, Llano, M, Carretero, M, Ishitani, A, Navarro, F, López-Botet, M et al. (1998). HLA-E is a major ligand for the natural killer inhibitory receptor CD94/NKG2A. *Proc Natl Acad Sci USA* **95**: 5199–5204.
41. Hicklin, DJ, Wang, Z, Arienti, F, Rivoltini, L, Parmiani, G and Ferrone, S (1998). beta2-Microglobulin mutations, HLA class I antigen loss, and tumor progression in melanoma. *J Clin Invest* **101**: 2720–2729.
42. Eichelberger, M, Allan, W, Zijlstra, M, Jaenisch, R and Doherty, PC (1991). Clearance of influenza virus respiratory infection in mice lacking class I major histocompatibility complex-restricted CD8+ T cells. *J Exp Med* **174**: 875–880.
43. Hou, S, Doherty, PC, Zijlstra, M, Jaenisch, R and Katz, JM (1992). Delayed clearance of Sendai virus in mice lacking class I MHC-restricted CD8+ T cells. *J Immunol* **149**: 1319–1325.
44. Di Stasi, A, Tey, SK, Dotti, G, Fujita, Y, Kennedy-Nasser, A, Martinez, C et al. (2011). Inducible apoptosis as a safety switch for adoptive cell therapy. *N Engl J Med* **365**: 1673–1683.
45. Yousem, SA, Curley, JM, Dauber, J, Paradis, I, Rabinowich, H, Zeevi, A et al. (1990). HLA-class II antigen expression in human heart-lung allografts.*Transplantation* **49**: 991–995.
46. Silk, KM, Tseng, SY, Nishimoto, KP, Lebkowski, J, Reddy, A and

Fairchild, PJ (2011). Differentiation of dendritic cells from human embryonic stem cells.*Methods Mol Biol* **767**: 449–461.

47. Wills, MR, Carmichael, AJ, Mynard, K, Jin, X, Weekes, MP, Plachter, B *et al.* (1996). The human cytotoxic T-lymphocyte (CTL) response to cytomegalovirus is dominated by structural protein pp65: frequency, specificity, and T-cell receptor usage of pp65-specific CTL. *J Virol* **70**: 7569–7579.

48. Manley, TJ, Luy, L, Jones, T, Boeckh, M, Mutimer, H and Riddell, SR (2004). Immune evasion proteins of human cytomegalovirus do not prevent a diverse CD8+ cytotoxic T-cell response in natural infection. *Blood* **104**: 1075–1082.

49. Terakura, S, Yamamoto, TN, Gardner, RA, Turtle, CJ, Jensen, MC and Riddell, SR (2012). Generation of CD19-chimeric antigen receptor modified CD8+ T cells derived from virus-specific central memory T cells. *Blood* **119**: 72–82.

50. Deyle, DR, Khan, IF, Ren, G, Wang, PR, Kho, J, Schwarze, U *et al.* (2012). Normal collagen and bone production by gene-targeted human osteogenesis imperfecta iPSCs. *Mol Ther* **20**: 204–213.

# Chapter 4

## USE OF RECOMBINATION-MEDIATED GENETIC ENGINEERING FOR CONSTRUCTION OF RESCUE HUMAN CYTOMEGALOVIRUS BACTERIAL ARTIFICIAL CHROMOSOME CLONES

Kalpana Dulal, Benjamin Silver, and Hua Zhu

Department of Microbiology and Molecular Genetics, UMDNJ-NJ Medical School, 225 Warren Street, Newark, New Jersey 07101-1709, USA

## ABSTRACT

Bacterial artificial chromosome (BAC) technology has contributed immensely to manipulation of larger genomes in many organisms including large DNA viruses like human cytomegalovirus (HCMV). The HCMV BAC clone propagated and maintained inside E. coli allows for accurate recombinant virus generation. Using this system, we have generated a panel of HCMV deletion mutants and their rescue clones. In this paper, we describe the construction of HCMV BAC mutants using a homologous recombination system. A gene capture method, or gap repair cloning, to seize large fragments of DNA from the virus BAC in order to generate rescue viruses, is described in detail. Construction of rescue clones using gap repair cloning is highly efficient and provides a novel use of the homologous recombination-based method in E. coli for molecular cloning, known colloquially as recombineering, when rescuing large BAC deletions. This method of excising large fragments of DNA provides important prospects for in vitro homologous recombination for genetic cloning.

## INTRODUCTION

Human cytomegalovirus (HCMV), also known as human herpesvirus-5, belongs to the betaherpesvirinae subfamily of the herpesviridae family. The virus is ubiquitously found in all geographic locations and among different socioeconomic groups with seroprevalence up to 90% [1]. Though a large percentage of the human population is infected with this virus, it does not cause significant clinical illness in immunocompetent individuals; however, high mortality and morbidity rate is found in immunocompromised individuals such as AIDS patients, transplant recipients, newborns, and developing fetuses [1]. Like other herpesviruses, this virus also retains the ability to remain latent in its host for life [2]. The virus can infect many different cell types including epithelial cells, endothelial cells, monocytes, macrophages, fibroblasts, smooth muscle cells, and neurons [1]. The virus is transmitted among individuals via bodily fluids, prenatal exposure, and sexual contact [2].

The HCMV genome is the largest and most complex among human herpesviruses, with 200–235 kb nucleotide pairs encoding for more than 200 proteins [3]. The virus genome encodes 40 herpesvirus core proteins and the rest of the genes encode betaherpesvirus virus-specific protein or HCMV-specific proteins. At present, the function of all of the viral genes has not been fully understood. One method for studying gene function is to delete a specific gene and study the mutant virus phenotype in the host. Due to the large genome size of HCMV, making recombinant virus is a challenging and cumbersome task; as a result, the functional study of many viral genes is hindered.

Several approaches have been used for the construction of recombinant HCMV. One of the earlier methods was to transfect human permissive cell lines with virus genome along with a plasmid carrying a marker gene with long flanking homology to the viral region to be modified into a host cell and then selecting for recombinants [4]. Some of the drawbacks in this method were:

1. the primary permissive cells are difficult to transfect and the probability of getting both the viral genome and the plasmid in the same cell was low;
2. efficiency of homologous recombination was low and long homology sequences (over 500 bp) are required;
3. replication cycle of the virus was slow and selecting a recombinant virus took a few days, and
4. the recombinant viruses and parental viruses were mixed and therefore isolation of the recombinant virus required extensive plaque purification.

However, the development of a cosmid system did lead to a better method for construction of recombinants [5–7]. Manipulation of the viral genome could be done in one of the cosmids and cotransfection of this mutated cosmid along with other wild-type cosmids in a permissive cell would give rise to recombinant virus upon homologous recombination [5, 7]. Although the cosmid method was an improvement, it was far from perfect as HCMV has a large genome and a total of eight cosmids were required for recombinant generation [5]. This does not seem to be a problem intuitively, but the resultant virus from this method of mutagenesis would be the product of several recombination events that are not only difficult to control, but verification of the viral genome can only be done after growth and isolation of the mutant virus, leading to a great deal of lost time and effort if all of the recombination events did not occur properly.

Overcoming the aforementioned difficulties, the development of the bacterial artificial chromosome clone (BAC) of the virus was a breakthrough in the HCMV mutagenesis field. The HCMV BAC clone can be maintained stably and propagated inside a bacterial cell, which allows for easy manipulation of the viral genome. Any required mutation can easily be achieved inside the E. coli cell and this mutation can be verified before making recombinant virus. Recombinant viruses can be made rapidly and straightforwardly when utilizing BAC genetic methods of manipulation. The first CMV BAC was made by using the murine cytomegalovirus (MCMV) Smith strain by Messerle et al. in 1997 [6]. Shortly after the construction of the MCMV BAC, Borst et al. described the construction of first HCMV BAC clone using the AD169 strain of the virus [8]. Recombinant viruses were made inside a CBTS E. coli strain, which carries a recA amber allele and a temperature-sensitive amber suppressor, so that the cells are recA positive at 30°C and recA negative at temperatures higher than 37°C [8]. The CBTS E. coli carrying the virus BAC were transformed with a shuttle plasmid containing virus sequence with desired modification and upon homologous recombination, the altered virus sequence replaced the wild-type sequence from the BAC, resulting in recombinant BAC clones. The recombinant BAC was then verified, isolated, and transfected into mammalian cells in order to generate recombinant virus. Since the development of this novel method for construction of recombinant viruses, detailed functional study of virus genome has been done using the BAC mutagenesis approach [3, 9–11].

However, the methods for BAC mutagenesis we use today were not at first easily attained as the large size of BACs posed a serious obstacle for their exact manipulation, as is required for viral research. Techniques were developed that utilized the power of homologous recombination in order to create recombinant viruses with the ability to safeguard every step in the process. We can now replace, and therefore delete large DNA fragments by inserting antibiotic resistance markers for proper and accurate selection of recombinant BAC clones and insert a luciferase reporter gene in order to measure expression of the recombinant virus. These new systems circumvented the problems associated with conventional genetic engineering as there was no longer a size restriction as seen when using restriction enzymes or other previous methods [12, 13]. Various modifications have been made to increase the efficacy of the homologous recombination-based method for the construction of recombinant viruses. This method is known as recombineering, or, less colloquially, the recombination-mediated genetic engineering method in E. coli that uses the recombination proteins derived from $\lambda$ phage [13]. Discussed below are the methods commonly used to make recombinant clones via this method.

## USES OF RECOMBINATION-MEDIATED GENETICALLY ENGINEERED BACS IN VIRAL RESEARCH

### Construction of Deletion Mutants

Several different approaches have been utilized to make deletion clones of HCMV using the BAC clone. Use of site-directed mutagenesis in E. coli using positive and negative selection markers is a common method of making recombinants. The virus BAC is maintained inside a modified DH10B strain of E. coli such as DY380 or SW102, which are recA negative and harbor repressed prophage recombination systems. The recombination system consists of exo, bet, and gam genes and is under the control of temperature-sensitive cIrepressor protein [12, 13]. A generalized summary of the straightforward procedure of using this recombination system in E. coli to generate deletion mutant BACs is given in Figure 1. When using a positive selection marker to replace a region of interest (ROI) by homologous recombination (Figure 1), the marker is usually a PCR-amplified antibiotic resistant gene, kanamycin-resistant gene in this case (a), containing short flanking homology to the region of interest to be deleted that was conferred by primer design (b). The recombination system of the E. coli harboring target DNA (d) can be induced, only when required, by inactivating the repressor protein by incubating the bacteria at 42°C for 15 minutes during electrocompetent cell preparation (e). The kanamycin cassette is transformed into electrocompetent and recombination-activated E. coli, in this case DY380 with its recombination system induced, harboring the wild-type virus BAC (f). The recombinants produced by homologous recombination event (g) are selected in the presence of kanamycin, which allows for the growth of only the clones containing the kanamycin resistant gene. Advantage of this method is that two genes can be deleted sequentially using two different drug resistance markers [14].

The second approach, which makes use of a positive and counterselection protocol, utilizes the SW105 (or SW102 or SW106) strain of E. coli, which is derived from the DY380 strain [15]. The SW105 strain contains the same prophage recombination system as DY380, but the major difference is that the SW105 contains a galactose operon in which the galactokinase (galK) gene is defective, so when SW105 bacteria are incubated on minimal media with galactose as the sole source of carbon, the bacteria cannot grow. However, when the galKgene is provided in trans, it will complement the defective galK gene, thus allowing the bacteria to grow in minimal media with galactose. When making deletion mutants using this approach, instead of a cassette conferring drug resistance, a PCR-amplified galK gene is used to replace the gene of interest by homologous recombination. The bacteria containing the correct deletion clones are selected on minimal media with galactose, where positive growth indicates the presence of the inserted galK gene. Colonies are further verified by growing them on MacConkey agar, where the clones containing the galK gene produce dark pink colonies [15].

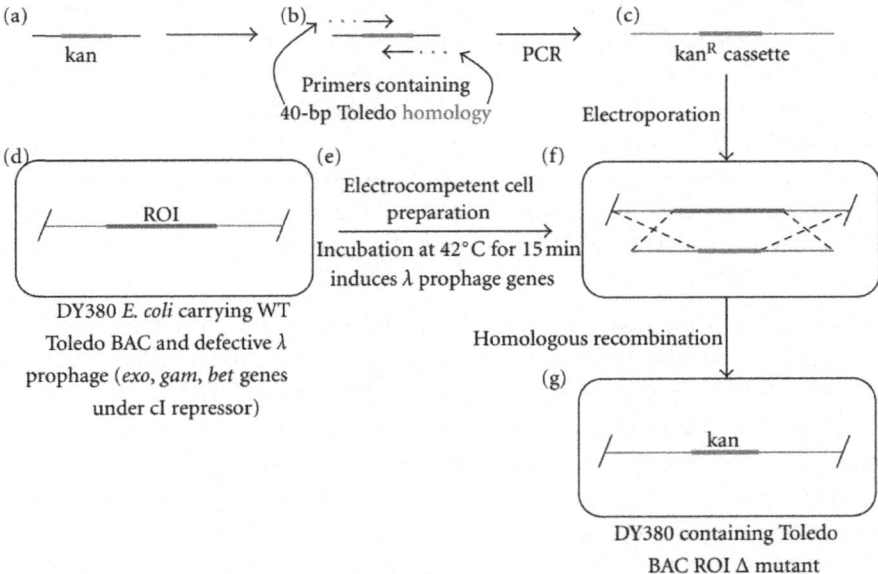

**Figure 1:** Toledo ROI Deletion Mutant BAC Clone Generation. A kanamycin resistant gene (a) is amplified by primers containing 40-bp homology to the sequences flanking the region of interest (ROI) in Toledo BAC (b) to create the kanamycin resistant cassette (c). In order to generate an ROI deletion mutant BAC clone (ROI Δ), DY380 E. coli carrying WT Toledo BAC (d) must undergo electrocompetent cell preparation and induction for recombination gene expression (e). The kanamycin resistant cassette is then electroporated into the electrocompetent DY380 harboring WT Toledo BAC (f). Upon homologous recombination, the ROI will be replaced by the kanamycin resistant gene, generating the Toledo ROI Δ mutant BAC clone (g).

## Construction of Rescue Clones

When making rescue or mutant clones by a positive selection method that takes advantage of homologous recombination in DY380, first the gene of interest is deleted by recombination with the antibiotic resistant cassette methods discussed above. Next, the gene is cloned from WT BAC and mutated (if making a point, frameshift, or nonsense mutation) either by PCR in which the primers confer the mutation, or via site-directed mutagenesis. The mutated gene is cloned into a plasmid either upstream or downstream of a drug resistance gene for positive selection (Note: antibiotic-resistant gene must be different than the gene used to construct the deletion mutant). The gene antibiotic-resistant cassette is then amplified by PCR using primers containing homology arms flanking the BAC genetic locus. The homologous sequences can be as short as 40 bp. The cassette is transformed into bacteria that contain the gene deletion virus BAC clone, and the resulting recombinant clones are selected using the new drug resistance gene marker. The clones can be further verified by confirming their sensitivity to the previous drug used to make the deletion. The drawback of this method is that the virus genome will contain a drug resistance gene. To improve the shortcomings of this method, the drug resistance gene can be flanked with loxP sites, which can be removed by addition of cre during transfection of the recombinant BAC DNA into mammalian cells. This will leave only 34 bp loxP site in the mutant BAC clone and virus [16].

When making rescue clones in SW105 using the galK counterselection marker method, the WT gene or mutated gene is PCR amplified along with the homology arms flanking the BAC genetic locus. This PCR cassette is electroporated into electrocompetent and recombination-activated SW105 harboring the deletion mutant BAC, and by homologous recombination, the galK gene will be replaced by the WT or mutated gene from the cassette. The recombinants are selected in minimal media containing 2-DOG and glycerol. The bacteria that contain galK genes will be unable to grow due to the toxic metabolites released from 2-DOG metabolism, whereas the clones in which the galK gene has been replaced by the wild-type or the mutated sequence will survive using glycerol as the source of carbon [15]. This method has advantages over the first one in that the rescue clones do not contain any foreign sequences in the viral BAC clone, hence in the virus. Even though this method is more advantageous over the previous one, making rescue clones in this manner is still a difficult process.

As simple as the previously cited methods sound, they are equally unsuitable when making rescue clones of large deletion mutants such as a 6-kb deletion, due to the requirement of PCR amplification of the region. In order to bypass the size restriction of PCR, a slightly different approach to making rescue clones of a large fragment deletion mutant (15-kb) was used by Wang et al. [17]. Since the 15-kb region in WT BAC DNA was flanked by two EcoRI restriction enzyme sites, the BAC DNA was digested by EcoRI and separated on the gel. From the gel, an 18-kb size product, which contains the 15-kb region fragment with 1.5-kb homology arms on either side, was extracted and transformed into the DY380 E. coli containing the deletion mutant BAC. It was possible to use this method because the deleted region was flanked by restriction enzyme sites in such a way that it provided long homology arms at both ends of the region of interest. The applicability of this method depends entirely upon the presence of restriction enzyme sites flanking the gene/region of interest. When working with large DNA like HCMV BAC, it is likely not to have suitable enzyme sites flanking the gene/region of interest. Furthermore, even if restriction enzyme sites flank the region appropriately, there could be additional restriction sites in other locations of the viral genome, resulting in the digested band of interest in the gel containing multiple different species of DNA fragments.

To resolve the issues associated with each of the methods described above, a gene capture method was developed and used for the first time to rescue a large 15-kb deletion mutant HCMV BAC [18]. This approach utilizes the principle of gap repair cloning in yeast with modification. Gap repair cloning is a commonly used method to repair double-strand breaks in yeast. This method has been described as the most efficient and error proof method of cloning [19]. It was shown that the plasmid was integrated into the chromosome if a crossover event was followed by the gap repair, whereas nonintegrated plasmid with repaired double-stranded break was the result of no crossover event [19]. As small as 30-bp homology arms have been shown to be sufficient for efficient gap repair cloning [20]. A modification of the method was to use oligonucleotide linkers as homology arms so that unique restriction enzyme sites can be added to the vector plasmid along with the oligonucleotide arms [21].

When making rescue clones by this gene capture method using DY380 containing the wild-type virus BAC (Figure 2), first the region of interest (ROI) is deleted by the above-mentioned method in DY380, generating an ROI deletion mutant. Next, a linearized plasmid (a) containing a positive selection marker is amplified by PCR with primers containing homologous sequences flanking the ROI in the BAC (b). In order to facilitate screening of rescue clones, a gene providing resistance to an antibiotic different than the antibiotic markers in both the BAC and rescue plasmid, flanked with loxP sites, is inserted within the captured ROI by homologous recombination. The PCR-amplified cassette (c) is transformed into electrocompetent and recombination-activated DY380 harboring the virus BAC (d). By homologous recombination, the ROI will be captured into the plasmid due to gap repair cloning (e). These recombinant clones are selected using the drug resistance marker gene on the plasmid (Note: the plasmid antibiotic resistance should be separate than the antibiotic resistance of the BAC). This capture can be further verified by amplifying an open reading frame within the ROI by PCR and also by digesting the plasmid with restriction enzymes to check for the presence of the captured region.

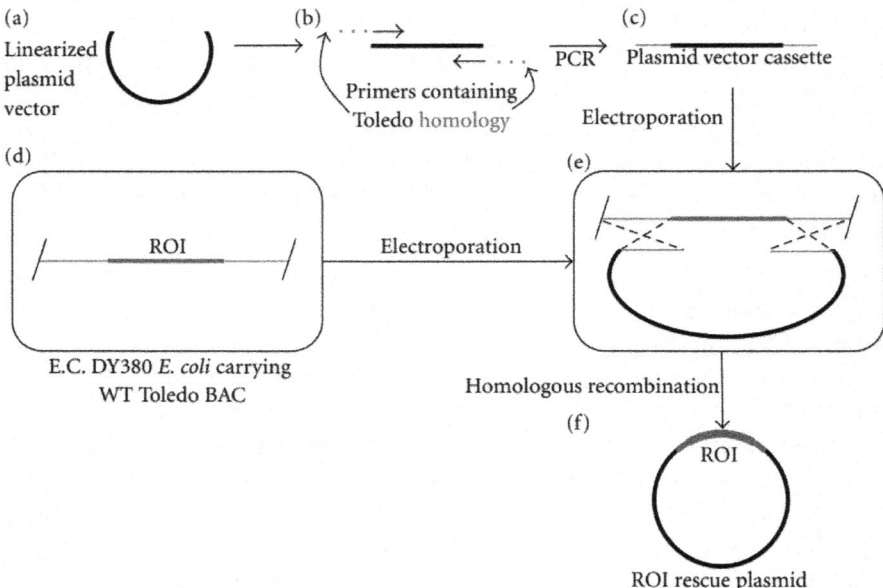

**Figure 2:** Gene Capture Method. (a) Linearized plasmid vector (b) is amplified by primers containing homology arms to the sequences flanking the region of interest (ROI) in Toledo to create the plasmid vector cassette (c). Electrocompetent and recombination-activated DY380 harboring WT Toledo BAC (d) are used for electroporation with the plasmid vector cassette (e). Upon homologous recombination and circularization, the ROI is captured, creating the ROI rescue plasmid (f).

Figure 3 provides the method for generating the rescued BAC clone after ROI capture. First, the plasmid containing the ROI (a) is digested by specific restriction enzymes with recognition sites flanking the region of interest (b).

This rescue cassette (c) is then electroporated into electrocompetent and recombination-activated DY380 (d) containing the ROI deletion BAC (e). By homologous recombination, the rescue cassette will be inserted into the ROI deletion BAC, replacing the gene conferring kanamycin resistance, and thereby generating a rescued BAC clone (f). Since the ROI contains an antibiotic resistance marker different than that in the BAC, the clones are screened for resistance to that antibiotic in LB agar. The rescue clones can be confirmed further by verifying sensitivity to the previous antibiotic resistance (kanamycin) used to make the deletion.

It is important to note the subtle yet significant differences between the insertion of a cassette into the BAC DNA for creation of a deletion of mutant and the excision, or capture, of a DNA fragment from the BAC vector into a plasmid for rescue virus generation. As shown in Figure 4, directly after PCR amplification, the homology arms of the plasmid vector cassette are oriented differently than the cassette for deletion mutant construction.

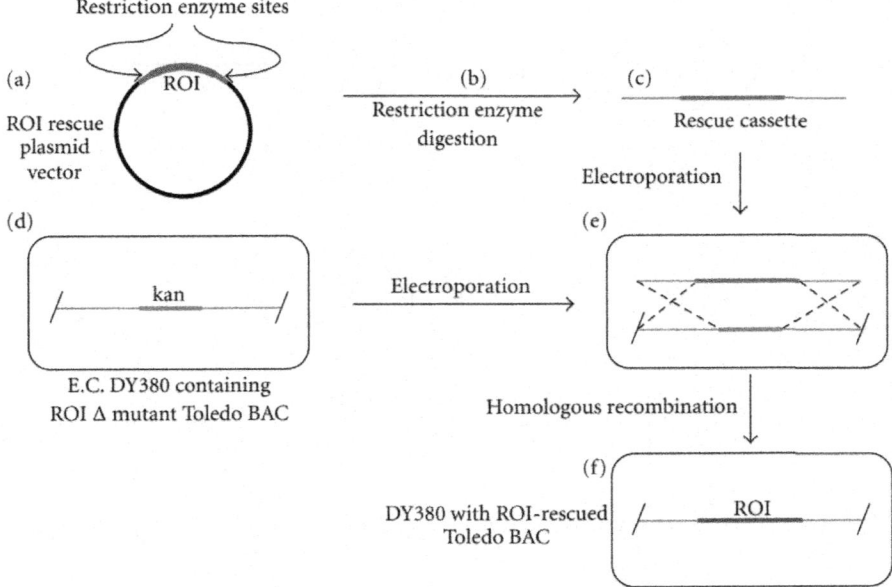

**Figure 3:** Toledo ROI Rescue BAC Clone Construction. Since the ROI rescue plasmid contains restriction enzyme sites (a), the plasmid is digested by restriction endonucleases (b), leading to isolation of the rescue cassette (c). Electrocompetent and recombination-activated DY380 harboring the ROI Δ mutant Toledo BAC (d) are used for electroporation with the rescue cassette (e). Upon homologous recombination, the ROI is inserted into the ROI Δ mutant Toledo BAC, generating the ROI-Rescued Toledo BAC (f).

The forward primer sequence that generates the positive marker cassette with homology for the ROI flanking sequences in the BAC is the same sequence oriented in the same direction as the coding sequence upstream of the ROI. However, the reverse primer used in this positive marker cassette generation needs to be the reverse complement of the downstream coding sequence as to allow for proper insertion required for deletion mutant generation (a). The slight but compulsory variance between insertion and capture lies in the fact that during PCR amplification of the plasmid vector cassette, the template is linear, whereas when it is electroporated into E. coli for gene capture, the plasmid vector cassette becomes circular and the homology arms invert (b). This inversion necessitates the forward primer to be the reverse complement sequence of the upstream homology region coding sequence, and the reverse primer to be identical to the coding sequence downstream of the ROI in sequence and in orientation. This inversion needs to be taken into careful consideration when designing suitable primers for the gene capture method to work properly.

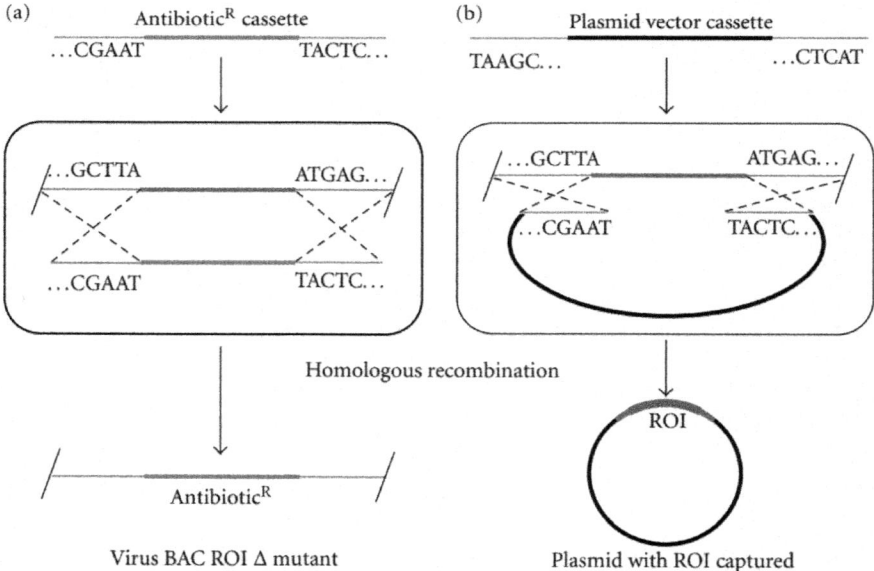

**Figure 4:** Variation in Gene Capture Method Recombineering. (a) Orientation of homology arms in generation of deletion mutant BAC clone. (b) Variant orientation of homology arms in linearized plasmid cassette for generation of plasmid with ROI captured.

## SPECIFIC RECOMBINEERING METHODS IN CONSTRUCTION OF RESCUE BAC CLONE

### Construction of 6-kb Region IV Deletion Mutant

We deleted a 6-kb region (UL132–151, referred to as region IV hereafter) from wild-type (WT) Toledo BAC by replacing it with a positive antibiotic selection marker. A kanamycin cassette with added homology arms flanking the region IV of Toledo BAC DNA was PCR amplified from pGEM-oriV/kan1 plasmid using Hotstar Taq polymerase (Qiagen Company, CA). Forward and reverse primers for kanamycin cassette were designed with 20-bp sequences from the kanamycin resistance gene and 40 bases homologous nucleotide sequences flanking region IV (Table 1). The amplified kanamycin cassette was treated overnight with DpnI restriction endonuclease (New England Biolab, MA) to digest any template plasmid and then purified using Qiagen's PCR purification kit (Qiagen, CA). The kanamycin cassette was used to replace region IV in the Toledo BAC. 300 ng of the purified cassette was transformed into electrocompetent DY380 cells carrying WT Toledo BAC. During electrocompetent DY380 cell preparation, the prophage recombination system in the cells was induced by incubating the bacteria in a 42°C water bath for 15 minutes with vigorous shaking. The presence of recombination proteins in the E. coli strain prevents degradation of the electroporated linear DNA cassette and allows for homologous recombination [13]. The transformed cells were grown at 32°C on LB agar with kanamycin (30 $\mu$g/mL). Colonies were restreaked onto LB agar with ampicillin (100 $\mu$g/mL) and LB agar with kanamycin plates, respectively, to confirm the antibiotic resistance of the colonies and since the kanamycin cassette was amplified from pGEM-oriV/kan1 plasmid containing an ampicillin resistance gene, the colonies were screened for ampicillin sensitivity in order to prevent selection of clones containing the pGEM-oriV/kan1 plasmid. The colonies that were sensitive to ampicillin and resistant to kanamycin were used for minipreparation. Miniprep of the BAC DNA was done as described by Zhang et al. [14]. After confirmation of the deletion mutant BAC, a maxiprep of the bacteria using NucleoBond BAC Maxi kit (BD Biosciences, CA) extracted the Toledo Region IV deletion (IV-$\Delta$) mutant BAC DNA from the E. coli on a large scale. Deletion was once more verified by PCR and integrity of the region IV-$\Delta$ BAC DNA was analyzed by restriction endonuclease digestion with EcoRI (New England Biolab, MA) and compared with the digestion pattern of Toledo BAC DNA.

**Table 1:** Table of primers used for PCR. ToledoUL132KanF and ToledoUL151KanR: primers to amplify $kan^R$ cassette for insertion of positive antibiotic selection marker and simultaneous deletion mutant construction. ZeoInsertF and ZeoInsertR: amplification of $zeo^R$ cassette for insertion of positive antibiotic selection marker insertion within region IV between UL 130 and UL 149. pUC19-IVCap_F and pUC19-IVCap_R: amplification of plasmid cassette for region IV capture for rescue clone construction.

| Primer | Sequence (5′→ 3′) | Function of primers |
|---|---|---|
| ToledoUL132KanF | CGCGGACATAGCAAGAAATCCA | Amplify $kan^R$ cassette for antibiotic selection and region IV deletion mutant construction |
|  | CGTCGCCACATCTCGAGAGCTCT |  |
|  | TGTTGGCTAGTGCGTA |  |
| ToledoUL151KanR | CGACCAGCGCTTTGTGCGCT |  |
|  | GCCTGTGCGTGTCGTCCCTCTGC |  |
|  | CAGTGTTACAACCAA |  |
| ZeoInsertF | GTCCGGCAGGATAGCGGTTAAG | Amplify $zeo^R$ cassette for antibiotic selection within region IV between UL 130 and UL 149 |
|  | GATTCGGTGCTAAGGCCGCATG |  |
|  | GCCGCGGATGGATCC |  |
| ZeoInsertR | TATCTGCGTGGGTCTAATCATGG |  |
|  | GTGTCACCGTGATCGCGGCCGC |  |
|  | ACTAGTGATAGATCT |  |
| pUC19-IVCap_F | CATTCAGGCGCGCCGGTAGTGT | Amplify plasmid cassette for region IV capture into puc19 plasmid for rescue clone generation |
|  | GTACAAAGGGAGGCGTGCTCAC |  |
|  | GGCCCGCAACCCGGGTACCGAG |  |
|  | CTCGAAT |  |
| pUC19-IVCap_R | TACTCAGGCCGGCCATCAAAAC |  |
|  | GCGAGCCCATATCGCCGCCATC |  |
|  | ATTGTAATCAGATGTGTGAAATT |  |
|  | GTTATCC |  |

## Construction of Rescue Clone through Gene Capture

Region IV was cloned into a linearized rescue vector (pUC19) with flanking homology arms, as previously described by gene capture, and the captured fragment was then put back into the region IV-Δ BAC from the rescue vector. In order to facilitate screening of rescue clones, a zeocin resistance gene flanked with loxP sites was inserted between ORF UL149 and UL130 region, which is included in the region IV ROI, by homologous recombination. The zeocin resistance gene, along with the flanking loxP sites, was amplified from pGEM-lox-zeo plasmid by PCR. The primers contained 40-bp homology flanking the BAC DNA locus where the gene was inserted (Table 1). The PCR product was treated with DpnI, as mentioned before, to remove the template plasmid and then gel purified using Qiagen's gel purification kit (Qiagen Company, CA). ~300 ng of the purified PCR product was electroporated into recombination-activated electrocompetent DY380 cells harboring WT Toledo BAC. The colonies were grown on LB agar with zeocin (50 μg/mL) and the insertion of the zeocin resistance gene was confirmed by PCR (Figure 5).

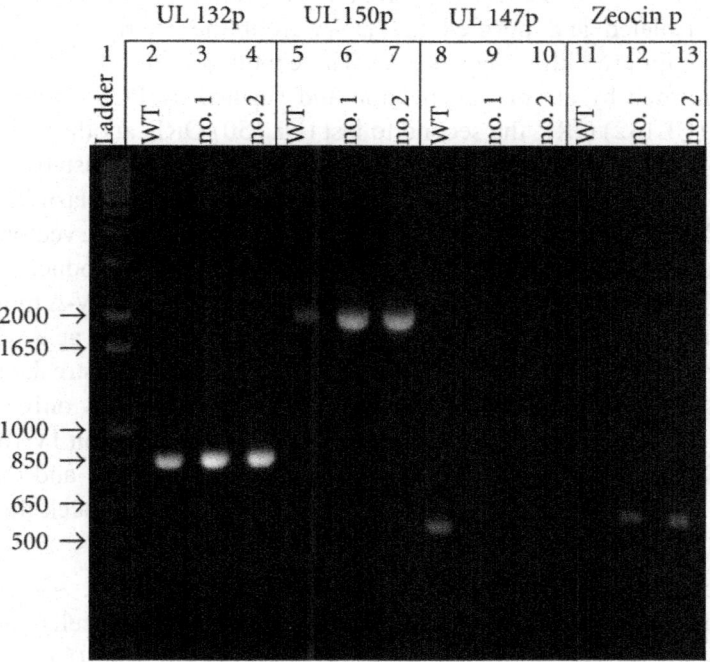

**Figure 5**: 1% Agarose gel electrophoresis of targeted amplification reactions in WT Toledo BAC (WT) and two pUC-19-IV capture plasmids. Lane 1: 1 kb Plus DNA Ladder. Lanes 2–4: amplification of UL 132 (~850 bases) with UL 132 F and R primers. Lanes 5–7: amplification of UL 150 (~1.9 kb) with UL 150 F and R primers. Lanes 8–10: amplification of UL 147 (negative control, not within region IV) (~500 bases)

with UL 147 F and R primers. Lanes 11–13: amplification of zeocin marker (positive control for capture only) (~550 bases) with Zeocin F and R primers.

Region IV was captured into a rescue vector, pUC19, along with an extra 150 bp sequences at each end that were used as homology arms. To capture the region IV DNA fragment, the rescue plasmid was first linearized by BamHI digestion and amplified using primers containing homology flanking region IV (Table 1). Since the large size of the region being captured did not possess adequate restriction enzyme sites for digestion, unique restriction sites for AscI (upstream) and FseI (downstream) were added to the ends of the region via these primers. The PCR product with the homology arms was purified and 300 ng of the purified PCR cassette was electroporated into DY380 carrying WT Toledo BAC with zeocin inserted into region IV. The transformed cells were grown overnight at 32°C on LB agar with zeocin and ampicillin. The colonies grown were picked from the LB agar plate and cultured overnight in 5 mL of LB with zeocin and ampicillin. Miniprep of pUC19-IV capture plasmid isolated and purified the DNA from the overnight culture using Qiagen's miniprep kit (Qiagen, CA). The captured region IV DNA fragment was confirmed by antibiotic selection and verified by PCR amplification of the first (UL132) ORF, the second to last (UL150) ORF, another ORF outside of region IV as a negative control (UL147), and zeocin resistance gene was verified as a positive control for capture (Figure 5). The plasmid was then digested with AscI and FseI restriction enzymes to separate the vector sequence from the region IV ROI. 500 ng of the purified digestion product (the rescue cassette) was transformed into DY380 carrying the Toledo IV-Δ mutant BAC by electroporation and the transformants were cultured on LB agar with zeocin and hygromycin (50 μg/mL). The resulting colonies were restreaked onto LB agar with either kanamycin or zeocin. The colonies that grew only on zeocin, and not on kanamycin plates, were used for miniprep. Region IV rescue (IV-R) BAC DNA was used for PCR verification along with WT and IV-Δ BACs (Figure 6). The IV-R BAC was also digested by EcoRI to check the integrity of the BAC DNA.

In order to produce mutant and rescue Toledo viruses, ~3–5 μg of the respective BAC DNA was electroporated into MRC-5 cells along with pCDNA-71 (pp71 expressing plasmid) and pGS403Cre (Cre expressing plasmid). The pp71 protein is a viral transactivator and is used to enhance viral growth [22]. The Cre protein is a recombinase that is used to remove the loxP flanking BAC vector and zeocin gene from the viral genome [16]. The electroporation conditions used were 260 V, 975 μF and time constant of ≤50 ms.

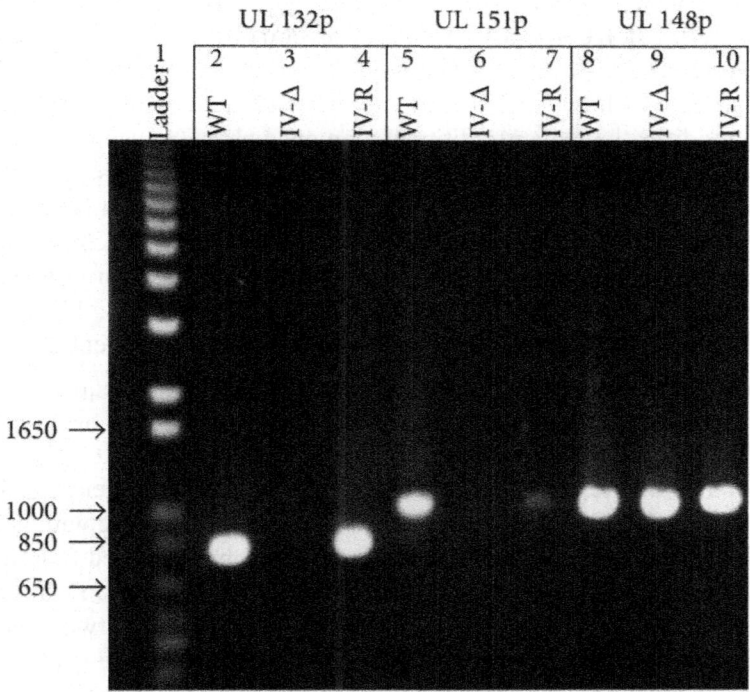

**Figure 6:** 1% Agarose gel electrophoresis of ORF targeted amplification reactions in WT Toledo BAC (WT), Region IV Deletion Mutant BAC (IV-Δ), and Region IV Rescued BAC (IV-R). Lane 1: 1 kb Plus DNA ladder. Lanes 2–4: amplification of UL 132 (~850 bases) with UL 132 F and R primers. Lanes 5–7: amplification of UL 151 (~1100 kb) with UL 151 F and R primers. Lanes 8–10: amplification of UL 148 (~1100 bases) with UL 148 F and R primers (positive control outside of region for deletion).

MRC-5 cells electroporated with the IV-R BAC were grown in a 10-cm cell culture dish and split when cells reached 100% confluency, with wild-type BAC used as a control. Six-to-eight days following transfection, plaques with similar growth kinetics and characteristics were observed from both WT and rescue BAC DNA, indicating that the recombinant, deletion-rescued BAC DNA was as infectious as the wild-type BAC.

## DISCUSSION

BAC cloning technique of viral genome has been a useful tool, especially for mutagenesis studies of large DNA viruses such as HCMV, in which the BAC DNA has been used to mutate or delete the individual genes to understand their functions in viral replications. Its greatest usefulness was marked by studies in which viral gene functions were screened at global scale for both MCMV and HCMV [6, 9, 10]. HCMV BAC DNAs of several clinical and attenuated strains have been generated so that the genomic DNA can be propagated and sustained inside an E. coli host [3, 23]. The cloning technique using BAC DNA has several modifications over traditional cloning such as (1) using E. coli strains containing recombination proteins to enhance homologous recombination efficiency [12, 24–27]; (2) using cre/lox system so that extra DNA segments can be removed when making viruses using BAC DNA [15]; (3) inserting luciferase gene in the viral genome so that viral growth curve experiments can be more reproducible and less time consuming [12, 14, 17, 22]. These modifications have made BAC cloning easier and more efficient than before.

So far, most mutagenesis studies using the BAC system require generation of a rescue clone of the mutation to assure that no mutations occur anywhere other than the target site. The procedure of making rescue BAC DNA requires an original DNA fragment to be inserted back into the viral genome. This can be achieved by PCR amplification of the DNA fragment along with homology arms flanking the mutated region and insertion of the amplified fragment back into the mutant genome by homologous recombination. If the DNA fragment were larger than the PCR limit, difficulties would arise in two aspects: (1) it is not easy to amplify; and (2) even if it is possible to be synthesized by currently available specific polymerases, the errors in DNA sequences are still unavoidable. To solve this problem, we have utilized the gap repair cloning method in BACs and modified it for the purpose of generating rescue BAC clones in a method called gene capture. Although the gap repair method of BACs has been utilized in yeast [19–21] and E. coli [13, 24–27], it has not been optimized for viral BACs in order to generate rescue viruses. It was first accomplished in HCMV when we presented the rescue of a 15-kb deletion BAC clone by the gene capture method, previously unattainable with the then current methods of PCR-based amplification of the region of interest for generation of rescue virus BAC clones [18]. It has here been presented as an optimized gene capture method for rescuing the region IV deletion mutant HCMV Toledo BAC clone with more practical methods that can be applied across many fields. By capturing the region IV DNA fragment into the rescue vector and putting it back into the mutant BAC, we successfully rescued the region IV deletion in HCMV genome. This gene capture method has three advantages: (1) no PCR was needed so that possible mutations induced by PCR can be avoided; (2) cloning of large DNA segments can be accomplished in only one step with high efficiency; and (3) different selection markers can be used during each step to ensure a higher cloning efficiency.

An additional advantage of the gene capture method is that it provides flexibility for the selection of the rescue vector. Any commercial or lab constructed plasmid can be used as a vector for capture. However, smaller plasmids are desirable, as they need to be PCR amplified with homologous regions for the capture method to proceed. As a proof of principle, we used a commonly available pUC19 plasmid as a rescue vector and successfully captured region IV from Toledo BAC. While choosing the vector plasmid, unique restriction sites flanking the capture gene are desirable. Those sites are later required to separate the captured region from the plasmid backbone for rescue cassette isolation and rescue procedure. If there are no unique restriction sites in the plasmid, they can be added while amplifying the plasmid via PCR primers, as was done in this case.

Previously, when generating a 15-kb rescue virus, we used homology arms of more than 1-kb on each side [18]. After this study, it has been shown that the size of the homology arms can be much smaller, which will reduce the total size of the rescue plasmid. In the case of this region IV deletion rescue, we used homology arms of 150-bp on each side of the rescue cassette, a significant improvement over the 1.5-kb arms used previously.

In summary, this enhanced gene capture method allows for easy cloning of large DNA fragments, previously a dubious practice when done by PCR, which can be used to rescue large deletions from any genome. By using region IV, a 6-kb DNA piece from HCMV genome as an example, we have demonstrated that the gene capture method is efficient and easier to perform than ever before for the construction of rescue clones from large deletion fragment mutants.

## FUTURE DIRECTIONS

The potential for this method is undoubted, as it is not limited to just HCMV and can be used for mutagenesis study of other large DNA viruses as well as other organisms. Furthermore, the prospects of using the homologous recombination system in vitro for methods like gene capture has numerous applications, the extent of its reach has yet to be calculated in genetic and molecular cloning.

## REFERENCES

1. E. Mocarski, Cytomegaloviruses and Their Replication, in Fields Virology, Lippincott Williams & Wilkins, Philadelphia, Pa, USA, 2007.
2. J. Sinclair and P. Sissons, "Latency and reactivation of human cytomegalovirus," Journal of General Virology, vol. 87, part 7, pp. 1763–1779, 2006.
3. E. Murphy, D. Yu, J. Grimwood et al., "Coding potential of laboratory and

clinical strains of human cytomegalovirus," Proceedings of the National Academy of Sciences of the United States of America, vol. 100, no. 25, pp. 14976–14981, 2003.
4. R. R. Spaete and E. S. Mocarski, "Insertion and deletion mutagenesis of the human cytomegalovirus genome," Proceedings of the National Academy of Sciences of the United States of America, vol. 84, no. 20, pp. 7213–7217, 1987.
5. G. Kemble, G. Duke, R. Winter, and R. Spaete, "Defined large-scale alteration of the human cytomegalovirus genome constructed by cotransfection of overlapping cosmids," Journal of Virology, vol. 70, no. 3, pp. 2044–2048, 1996.
6. M. Messerle, I. Crnkovic, W. Hammerschmidt, H. Ziegler, and U. H. Koszinowski, "Cloning and mutagenesis of a herpesvirus genome as an infectious bacterial artificial chromosome," Proceedings of the National Academy of Sciences of the United States of America, vol. 94, no. 26, pp. 14759–14763, 1997. ·
7. M. van Zijl, W. Quint, J. Briaire, T. De Rover, A. Gielkens, and A. Berns, "Regeneration of herpesviruses from molecularly cloned subgenomic fragments," Journal of Virology, vol. 62, no. 6, pp. 2191–2195, 1988.
8. E. M. Borst, G. Hahn, U. H. Koszinowski, and M. Messerle, "Cloning of the human cytomegalovirus (HCMV) genome as an infectious bacterial artificial chromosome in Escherichia coli: a new approach for construction of HCMV mutants," Journal of Virology, vol. 73, no. 10, pp. 8320–8329, 1999.
9. W. Dunn, C. Chou, H. Li et al., "Functional profiling of a human cytomegalovirus genome," Proceedings of the National Academy of Sciences of the United States of America, vol. 100, no. 2, pp. 14223–14228, 2003.
10. D. Yu, M. C. Silva, and T. Shenk, "Functional map of human cytomegalovirus AD169 defined by global mutational analysis," Proceedings of the National Academy of Sciences of the United States of America, vol. 100, no. 21, pp. 12396–12401, 2003. ·
11. A. Dolan, C. Cunningham, R. D. Hector et al., "Genetic content of wild-type human cytomegalovirus,"Journal of General Virology, vol. 85, part 5, pp. 1301–1312, 2004.
12. D. Yu, H. M. Ellis, E. C. Lee, N. A. Jenkins, N. G. Copeland, and D. L. Court, "An efficient recombination system for chromosome engineering in Escherichia coli," Proceedings of the National Academy of Sciences of the United States of America, vol. 97, no. 11, pp. 5978–5983, 2000.

13. Y. Zhang, J. P. P. Muyrers, G. Testa, and A. F. Stewart, "DNA cloning by homologous recombination in Escherichia coli," Nature Biotechnology, vol. 18, no. 12, pp. 1314–1317, 2000.
14. Z. Zhang, Y. Huang, and H. Zhu, "A highly efficient protocol of generating and analyzing VZV ORF deletion mutants based on a newly developed luciferase VZV BAC system," Journal of Virological Methods, vol. 148, no. 1-2, pp. 197–204, 2008.
15. S. Warming, N. Costantino, D. L. Court, N. A. Jenkins, and N. G. Copeland, "Simple and highly efficient BAC recombineering using galK selection," Nucleic Acids Research, vol. 33, no. 4, p. e36, 2005.
16. P. K. Stricklett, R. D. Nelson, and D. E. Kohan, "The Cre/loxP system and gene targeting in the kidney," American Journal of Physiology, vol. 276, no. 5, part 2, pp. F651–F657, 1999.
17. W. Wang, S. L. Taylor, S. A. Leisenfelder et al., "Human cytomegalovirus genes in the 15-kilobase region are required for viral replication in implanted human tissues in SCID mice," Journal of Virology, vol. 79, no. 4, pp. 2115–2123, 2005.
18. K. Dulal, Z. Zhang, and H. Zhu, "Development of a gene capture method to rescue a large deletion mutant of human cytomegalovirus," Journal of Virological Methods, vol. 157, no. 2, pp. 180–187, 2009.
19. T. L. Orr-Weaver and J. W. Szostak, "Yeast recombination: the association between double-strand gap repair and crossing-over," Proceedings of the National Academy of Sciences of the United States of America, vol. 80, no. 14, pp. 4417–4421, 1983.
20. S. B. Hua, M. Qiu, E. Chan, L. Zhu, and Y. Luo, "Minimum length of sequence homology required for in vivo cloning by homologous recombination in yeast," Plasmid, vol. 38, no. 2, pp. 91–96, 1997.
21. A. A. Kitazono, "Improved gap-repair cloning method that uses oligonucleotides to target cognate sequences," Yeast, vol. 26, no. 9, pp. 497–505, 2009.
22. C. J. Baldick Jr., A. Marchini, C. E. Patterson, and T. Shenk, "Human cytomegalovirus tegument protein pp71 (ppUL82) enhances the infectivity of viral DNA and accelerates the infectious cycle," Journal of Virology, vol. 71, no. 6, pp. 4400–4408, 1997.
23. A. Marchini, H. Liu, and H. Zhu, "Human cytomegalovirus with IE-2 (UL122) deleted fails to express early lytic genes," Journal of Virology, vol. 75, no. 4, pp. 1870–1878, 2001.
24. D. L. Court, J. A. Sawitzke, and L. C. Thomason, "Genetic engineering

using homologous recombination," Annual Review of Genetics, vol. 36, pp. 361–388, 2002.
25. L. Thomason, D. L. Court, M. Bubunenko et al., "Recombineering: genetic engineering in bacteria using homologous recombination," in Current Protocols in Molecular Biology, M. Ausubel, et al., Ed., chapter 1, unit 1.16, John Wiley & Sons, New York, NY, USA, 2007. View at Google Scholar
26. S. K. Sharan, L. C. Thomason, S. G. Kuznetsov, and D. L. Court, "Recombineering: a homologous recombination-based method of genetic engineering," Nature Protocols, vol. 4, no. 2, pp. 206–223, 2009.
27. K. Narayanan and Q. Chen, "Bacterial artificial chromosome mutagenesis using recombineering," Journal of Biomedicine and Biotechnology, vol. 2011, Article ID 971296, 10 pages, 2011.

# Chapter 5

# GENETICALLY ENGINEERED MOUSE MODELS FOR HUMAN LUNG CANCER

Kazushi Inoue[1,2,3], Elizabeth Fry[1,2], Dejan Maglic[1,2,3] and Sinan Zhu[1,3]

[1]The Department of Pathology, Wake Forest University Health Sciences, Medical Center Boulevard, Winston-Salem, NC, USA

[2]The Department of Cancer Biology, Wake Forest University Health Sciences, Medical Center Boulevard, Winston-Salem, NC, USA

[3]Graduate Program in Molecular Medicine, Wake Forest University Health Sciences, Medical Center Boulevard, Winston-Salem, NC, USA

## INTRODUCTION

Lung cancer is the leading cause of cancer deaths in the world, which is a cause for more solid tumor-related deaths than all other carcinomas combined. More than 170,000 new cases are diagnosed each year in the United States alone, of whom ~160,000 will eventually die, accounting for nearly 30% of all cancer deaths (Siegel *et al.*, 2012). The annual incidence for lung cancer per 100,000 population is highest among African Americans (76.1), followed by whites (69.7), American Indians/Alaska Natives (48.4), and Asian/Pacific Islanders (38.4). Hispanic people have much lower lung cancer incidence (37.3) than non-Hispanics (71.9) (CDC, 2010). These results identify the racial/ethnic populations and geographic regions that would benefit from enhanced efforts in lung cancer prevention, specifically by reducing cigarette smoking and exposure to environmental carcinogens.

Lung lobectomy provides the best chance for patients with early-stage disease to be cured. African American patients with early-stage lung cancer have lower five-year survival rates than whites, which has been attributed to lower rates of resection in former patients (Wisnivesky *et al.*, 2005). Several potential factors underlying racial differences in receiving surgical therapy include differences in pulmonary function, access to care, beliefs about tumor spread at the time of operation, and the possibility of cure without surgery. Of these, access to care is considered to be the most important factor underlying racial disparities. The most outstanding modifiable risk factor for lung cancer is cigarette smoking (Swierzewski III, 2011). Other risk factors include asbestos exposure, radon, occupational chemicals, radiation, and alcohol. People who smoke tend to drink more alcohols and consume more non-narcotic pain relievers than non-smokers, thus reducing the intoxicating effects of alcohol, promoting the progression from moderate to heavy drinking. Alcoholism is also associated with significant immune suppression - therefore, a history of drinking may increase a person's susceptibility to lung cancer.

Lung cancer has a high morbidity because it is difficult to detect early and is frequently resistant to available chemotherapy and radiotherapy. The overall 5-year survival rate for all types of lung cancer is around 15 % at most, and it is even worse in SCLC (~5 %) although SCLC is more sensitive to chemo/radiation therapy than NSCLC (Meuwissen & Berns, 2005; Schiller, 2001; Worden & Kalemkerian, 2000). Non-smokers who develop lung cancer may experience delays in diagnosis due to the fact that many early symptoms of lung cancer mimic those of non-specific respiratory infections (Menon, 2012). Thus, a physician may misdiagnose the malignant disease for asthma or other respiratory illnesses. Another reason for delayed diagnosis of lung cancer is that there is no sensitive and specific biomarker, such as prostate-specific antigen in prostate cancer (Brambilla *et al.*, 2003). Thus several biomarkers will have to be used together for early diagnosis of lung cancer at present, which include mutant Ras, mutant p53, and methylation of a variety of genes using bronchial biospies or bronchoalveolar lavage (Brambilla *et al.*, 2003).

Certain combinations of clinical signs and symptoms – e.g. endocrine, neurologic, immunologic, and hematologic - are associated with lung cancer as a manifestation of the secretion of cytokines/hormones by tumor cells or as an associated immunologic response (Yeung *et al.*, 2011). These paraneoplastic syndromes occur commonly in patients with SCLC. Since the syndromes can be the first clinical manifestation of malignant disease, increased awareness of these syndromes associated with lung cancer is critical to the earlier diagnosis of malignancies, thereby improving the overall prognosis of patients.

Lung cancer has been categorized into two major histopathological groups: non-small-cell lung cancer (NSCLC) (Moran, 2006) and small-cell lung cancer (SCLC) (Schiller, 2001), the latter of which show neuroendocrine features and thus are different from the former. Approximately 80 % of lung cancers are NSCLC, and they are subcategorized into adenocarcinomas (AdCA), squamous cell (SqCLC), bronchoalveolar, and large-cell carcinomas (LCLC) (Travis, 2002). SCLC and NSCLC show major differences in histopathologic characteristics that can be explained by the distinct patterns of genetic alterations found in both tumor types (Zochbauer-Muller *et al.*, 2002).

The *K-Ras* gene is mutated in 20~30 % of NSCLC while its mutation is rare in SCLC; *Rb* inactivation is found in ~90 % of SCLC while *p16*$^{INK4a}$ is inactivated by gene deletion and/or promoter hypermethylation in ~50 % of NSCLC (Fong *et al.*, 2003; Meuwissen & Berns, 2005). Responsiveness of tumor cells to chemotherapy and/or radiation therapy significantly varies between NSCLC and SCLC, and thus, has a dramatic effect on the prognosis of patients.

Progress in whole genome approaches to detect genetic alterations found in human lung cancer has resulted in the identification of a growing number of genes. Genome-wide association studies, whether they are based on single-nucleotide polymorphism array or in gene copy number assays, have identified mutations in lung cancer-related genes. Identification of these lung cancer-related genes will provide great potential as therapeutic targets for lung cancer intervention. Target validation should be done through intervention studies of specific genetic alterations in human lung cancer cell lines. Since *in vitro* cell culture studies cannot fully mimic more complex *in vivo* onset/development of lung carcinogenesis, developing endogenous lung cancer in mice that harbor specific mutations will undoubtedly provide a further insight into the mutation-specific effects on lung tumor initiation/development. Moreover, a high degree of pathophysiological similarity between mouse lung tumors and human lung carcinomas will make it possible to use these mouse models in pre-clinical tests for novel anticancer drug screening. Various intervention strategies against specific mutation can then be tested to evaluate both specificity and efficacy in mouse lung tumors at every developing stage. The number of genetically engineered mouse models for lung cancer is ever expanding. Continuous attempt to manipulate the mouse genome has enabled us to adjust compound mouse models of lung cancer in a way that they start to reproduce the more complex human lung cancer in a higher degree.

While susceptibility and incidence of spontaneous lung tumors vary among well-established mouse strains, endogenous mouse lung tumors share many similarities with human lung cancers. This was clearly demonstrated in early studies where defined chemical carcinogens were used to induce lung tumors in mice (Wakamatsu *et al.*, 2007). The incidence of spontaneous and induced lung tumors were very high (61%) in A/J and SWR strains, but very low (6%) in resistant strains such as C57BL/6 and DBA (Wakamatsu *et al.*, 2007). Contrary to human lung cancer with its complex molecular genetics and four distinct tumor types (adenocarcinoma, squamous cell carcinoma, large-cell carcinoma, and small-cell carcinoma) that easily metastasize, spontaneous and chemically-induced lung lesions in mice often result in pulmonary adenomas and more infrequent adenocarcinomas. Mouse lung adenocarcinomas are usually 5mm or more in diameter; however, they are categorized into carcinomas when nuclear atypia or signs of local invasion/metastasis is found in tumors less than 5mm. Mouse lung tumor development shows initial hyperplastic foci in bronchioles and alveoli, which then become benign adenomas and eventually adenocarcinomas (Shimkin et al., 1975). The tumor latency depends on mouse strain and carcinogen administration protocols. Most potent carcinogens are found in cigarettes, such as polycyclic aromatic hydrocarbons, tobacco-specific nitrosamine, and benzopyrene (BaP) (Pfeifer et al., 2002). It has been especially difficult to reproduce well-characterized pre-malignant lesions found in human airway epithelium in mice (Sato et al., 2007). Nevertheless, major histopathological features remain the same between the two species and molecular characterization of spontaneous and carcinogen-induced murine lung tumors revealed a high degree of similarity as compared to their human counterparts (Malkinson, 2001). A common early event is the occurrence of activating *K-ras* mutations in hyperplastic lesions. Besides overexpression of c-*Myc*, inactivation of well-known tumor suppressor genes, such as *p53, fhit, Apc, Rb, Mcc, p16*$^{Ink4a}$ and/or *Arf* occur in both mice and human lung cancers; only a small percentage of lung adenomas progress into AdCAs (Malkinson, 2001).

## THE FIRST GENERATION MOUSE MODELS FOR LUNG CANCER

The first generation transgenic models for lung cancer were created by ectopic transgene expression under control of lung-specific promoters. Thus transgenic expression was constitutive. Transgene expression was mainly found in specific subsets of lung epithelial cells. Lung *surfactant protein C(SPC)* promoter was used for constitutive gene expression in type II alveolar cells whereas *Clara Cell Secretory Protein (CCSP)* promoter was used to target the non-ciliated secretory (Clara) cells that exist on the airways. In early studies, *SV40 Tag* (Simian virus large T-antigen) that neutralizes the activity of both Rb and p53 was constitutively expressed under the control of *CCSP* (DeMayo et al., 1991;Sandmoller et al., 1994) or *SPC* promoters (Wikenheiser et al., 1992). Although each tumor originated from either Clara cells or type II alveolar cells, they both resulted in quite similar aggressive AdCAs without metastases (Wikenheiser et al., 1997). A similar strategy was used to express distinct oncogenes (such as *c-Raf* and *c-Myc* [Geick et al., 2001]) in the lung/bronchial epithelium, ending up with a milder phenotype, as both transgenic mice mainly developed adenomas, and a few progressed to AdCAs without any metastases.

Ehrhardt *et al.* (2001) created transgenic mouse models to study tumorigenesis of bronchiolo-alveolar AdCAs derived from alveolar type II pneumocytes. Transgenic lines expressing c-*Myc* under the control of the *SPC* promoter developed multifocal bronchiolo-alveolar hyperplasias, adenomas, AdCAs, whereas transgenic lines expressing a secretable form of the epidermal growth factor, TGFα, developed hyperplasias of the alveolar epithelium. Since the oncogenes c-Myc and TGFα are frequently overexpressed in human lung bronchiolo-alveolar carcinomas, these mouse lines will be useful as those for human lung bronchiolo-alveolar carcinomas (Ehrhardt *et al.*, 2001).

Sunday *et al.* created a transgenic model for primary pulmonary neuroendocrine cell hyperplasia/neoplasia using *v-Ha-ras* driven by the *neuroendocrine* (NE)-specific calcitonin promoter (named *rascal*). All rascal transgenic mouse lineages developed hyperplasias of NE and non-NE cells, but mostly non-NE cells developed lung carcinomas (Sunday *et al.*, 1999). Analyses of embryonic lung demonstrated *rascal* mRNA in undifferentiated epithelium, consistent with expression in a common pluripotent precursor cell. These observations indicate that *v-Ha-ras* can lead to both NE and non-NE hyperplasia/carcinoma *in vivo* (Sunday *et al.*, 1999).

A strong correlation exists between *p53* mutations and lung malignancies, and LOH for *p53* has been reported in 40% of NSCLC with specific primers (Mallakin *et* al., 2007). Preceding this study, Morris*et al.* (1998) established a transgenic mouse model with disrupted p53 function in the epithelial cells of the peripheral lung. A dominant-negative mutant form of *p53* was expressed from the human *SPC*promoter. The dominant-negative p53 (dnp53) expressed from the *SPC* promoter antagonized wild-type p53 functions in alveolar type II pneumocytes and some bronchiolar cells of the transgenic animals, and thereby promoted the development of carcinoma of the lung. This mouse model should prove useful to the study of lung carcinogenesis and to the identification of agents that contribute to neoplastic conversion in the lung. Another group later created *CCSP-dnp53* transgenic mice and reported significant increase in the incidence of spontaneous lung cancer in 18-month-old transgenic mice (Tchon-Wong *et al.*, 2002). In addition to the increased incidence of spontaneous lung tumor, these transgenic mice were more susceptible to the development of lung adenocarcinoma after exposure to BaP. The risk of lung tumors was 25.3 times greater in BaP-treated mice adjusted for transgene expression. These results suggest that p53 function is important for protecting mice from both spontaneous and BaP-induced lung cancers.

The receptor tyrosine kinase RON (recepteur d'origine nantais) is a member of the MET proto-oncogene family, which is expressed by a variety of epithelial-derived tumors and cancer cell lines and has been implicated in the pathogenesis of lung adenocarcinomas (Chen *et al.*, 2002). To determine the oncogenic potential of RON, transgenic mice were generated using the lung *SPC* promoter to express human wild-type RON in type II cell phenotypes (Chen *et al.*, 2002). The mice were born normal without morphological alterations in the lung, however, multiple adenomas appeared as a single mass in the lung around 2 months of age and gradually developed into multiple nodules throughout the lung. Most of the tumors were characterized as cuboidal epithelial cells with type II cell phenotypes which transformed from pre-malignant adenomas to adenocarcinomas. Interestingly, Ras expression was dramatically increased in the majority of tumors without mutation in the 'hot spots' of the *K-Ras* or*p53* genes suggesting that *SPC-RON* is a mouse lung tumor model with unique biological characteristics (Chen *et al.*, 2002).

Many prominent genetic lesions found in human lung cancer clearly link the inactivation of well-known tumor suppressor genes (Sekido *et al.*, 2003) to lung cancer development. Initial attempts to mimic some of these lesions implicated in lung cancer by using conventional knockout mice had limited success with respect to the onset of lung cancer. The main reason for this failure was that germ-line deletion of many essential tumor suppressor genes (such as the *retinoblastoma* gene (*Rb*) (Jacks *et al.*, 1992) lead to embryonal lethality. Non-essential tumor suppressor gene (for embryonic survival) knockout mice often had a very broad tumor spectrum of which lung tumors formed only a minor fraction. Thus, *p53*, *p16*$^{Ink4a}$ and *p19*$^{Arf}$ (Meuwissen & Berns, 2005) null allele mice seldom develop lung AdCAs. However, introducing similar mutations into endogenous *p53* alleles, such as those prominently found in Li–Fraumeni patients, generated *p53*$^{R270H/+}$ and *p53*$^{R172H/+}$ which had a different tumor spectrum compared with *p53*$^{+/-}$ mice (Olive *et al.*, 2004), although their mean survival times were identical. Interestingly these mice, but especially *p53*$^{R270H/+}$ and *p53*$^{R270H/-}$ mice, gave rise to more malignant lung AdCAs, and even their metastases, which never occurred in *p53*$^{-/-}$ mice. These results suggest that "humanized" *p53* mutations have a greater impact on lung tumor progression than complete *p53* loss (Olive *et al.*, 2004; Lang *et al.*, 2004).

Targeting genes deleted early in human lung tumorigenesis, such as the complete cluster at chromosome 3p21.3, showed that heterozygous deletion for this 370 kb region showed no obvious predisposition for lung cancer development albeit homozygous deletion caused embryonal lethality (Smith *et al.*, 2002). A more specific deletion of candidate tumor suppressor genes on chromosome 3 like *RassF1a*, *FHIT* and *VHL*, showed that 31% of *Rassf1a*$^{-/-}$ mice produced spontaneous mainly lymphomas but also lung adenomas (Tommasi *et al.*, 2005). Treatment of *Rassf1a*$^{-/-}$ mice with BP or urethane resulted in an even higher rate of lung tumors. No spontaneous lung tumors were observed in $Fhit^{-/-}$ or $Vhl^{+/-}$ mice, but 44% of $Fhit^{-/-};Vhl^{+/-}$ mice developed AdCAs by age 2 years. Again use of mutagens such as dimethylnitrosamine led to 100% adenoma and AdCA induction in $Fhit^{-/-};Vhl^{+/-}$ mice and even adenomas in 40% of $Fhit^{-/-}$ mice by age 20 months (Zanesi et al., 2005). This showed the usefulness of these knockout mice in recapitulating a pattern of early lung cancer development similar to human pattern.

## THE SECOND GENERATION MODELS

### K-ras$^{LA}$ and LSL K-ras Models

A different approach to address lung cancer onset was the use of knock-in alleles to activate oncogenes. One example of this is based on the somatic K-ras activation via an oncogenic $Kras^{G12D}$ knock-in allele ($Kras^{LA2}$), which is expressed only after a spontaneous recombination event (Johnson et al., 2001). In this way, sporadic $Kras^{G12D}$ expression occurred on an endogenous level, which in turn augments efficient development of lung AdCAs. However, these mice also developed other tumor lesions as K-Ras$^{G12D}$ expression was not limited to the lung epithelial tissues.

Dmp1 (Dmtf1) is a Myb-like protein with tumor suppressive activity that had been isolated in a yeast two-hybrid screen with cyclin D2 bait (Hirai and Sherr, 1996; Inoue and Sherr, 1998; for review,Inoue et al., 2007; Sugiyama et al., 2008a). The promoter is activated by oncogenic Ras-Raf signaling and induces cell-cycle arrest in an Arf, p53-dependent fashion (Inoue et al., 1999; Sreeramaneni et al., 2005). Both $Dmp1^{+/-}$ and $Dmp1^{-/-}$ mice are prone to spontaneous and carcinogen-induced tumor development, indicating that it is haplo-insufficient for tumor suppression, the mechanism of which have not been elucidated yet (Inoue et al., 2000, 2001, 2007). The survival of $K$-$ras^{LA}$ mice was shortened by approximately 15 weeks in both $Dmp1^{+/-}$ and $Dmp1^{-/-}$ backgrounds, the lung tumors of which showed significantly decreased frequency of p53 mutations compared to $Dmp1^{+/+}$. Approximately 40% of $K$-$ras^{LA}$ lung tumors from Dmp1 wild-type mice lost one allele of the Dmp1 gene, suggesting the primary involvement of Dmp1 in K-ras-induced tumorigenesis (Mallakin et al., 2007). Tumors from Dmp1-deficient mice showed more invasive and aggressive phenotypes than those from Dmp1 wild-type mice. Loss of heterozygosity (LOH) of the hDMP1 locus was detectable in approximately 35% of human lung carcinomas, which was found in mutually exclusive fashion with LOH of INK4a/ARF or that of p53. Thus, DMP1 is a novel tumor suppressor for both human and murine NSCLC (Mallakin et al., 2007; Sugiyama et al., 2008b).

Integration of gene expression data from a $Kras^{LA2}$ mouse model and *KRAS* mutated human lung tumors showed a significant overlap but also revealed a gene-expression signature for *K-ras* mutation in human lung cancer itself (Sweet-Cordero et al., 2005). By using $Kras^{LA2}$ knock-in mouse model and human lung cancer specimen, they compared gene expression patterns between these two species (Sweet-Cordero et al., 2005). They applied this method to the analysis of a model of $Kras^{LA2}$-mediated lung cancer and found a good relationship to human lung AdCA, thereby validating the usefulness of this transgenic model. Furthermore, integrating mouse and human data uncovered a gene-expression signature of *KRAS2* mutation in human lung cancer. They confirmed the importance of this signature by gene-expression analysis of shRNA-mediated inhibition of oncogenic $Kras^{LA2}$ (Sweet-Cordero et al., 2005). However, one problem of $Kras^{LA}$ mice is that they develop tumors other than lung cancer (Mallakin et al., 2007). To overcome this issue, Jackson et al. (2001) developed a new model of lung AdCA in mice having a conditionally activatable allele of oncogenic *K-ras* (*LSL $Kras^{G12D}$*). They show that the use of a recombinant adenovirus expressing Cre recombinase (AdenoCre) to induce $Kras^{G12D}$ expression in the lungs of mice allows control of the timing and multiplicity of tumor initiation. Through the ability to synchronize tumor initiation in these mice, they could characterize the stages of tumor progression. Of particular significance, this system led to the identification of a new cell type contributing to the development of pulmonary AdCA (Jackson et al., 2001). By using this Cre-lox system, the same group later created conditional knock-in mice with mutations in *K-ras* combined with one of mutant *p53* alleles (Jackson et al., 2005). *p53*-loss strongly promoted the progression of *Kras*-induced lung AdCAs, yielding a mouse model that precisely recapitulates advanced human lung AdCA. The influence of *p53*-loss on malignant progression was observed as early as 6 weeks after tumor initiation. They also found that the contact mutant p53R270H behaved in a dominant-negative fashion to promote *K-ras*-driven lung AdCAs. Of note, a subset of mice also developed sinonasal adenocarcinomas, suggesting specific expression of *K-ras* in this tissue. In contrast to the lung tumors, expression of the point-mutant *p53* alleles strongly promoted the development of sinonasal AdCAs compared with simple loss-of-function, suggesting a tissue-specific gain-of-function of mutant p53 (Jackson et al., 2005).

Since activating *K-ras* mutation models recapitulate the human lung tumor phenotypes well, closer analyses of early lung tumor initiating events were performed (Ji et al., 2006). A combination of both*CCSP-Cre* recombinase and *LSL $Kras^{G12D}$* alleles (Jackson et al., 2005) resulted in a progressive phenotype of cellular atypia, adenoma and finally AdCA. The activation of *K-ras* mutant allele in CC10-positive cells resulted in a progressive phenotype characterized by cellular atypia, adenoma and ultimately AdCA. Surprisingly, *Kras* activation in the bronchiolar epithelium was associated with a robust inflammatory response characterized by an abundant infiltration of alveolar macrophages and neutrophils. These mice displayed early mortality in the setting of this pulmonary inflammatory response. Bronchoalveolar lavage fluid from these mutant mice contained the MIP-2, KC, MCP-1 and LIX chemokines that increased significantly with age. Thus, *Kras* activation in the lung induces inflammatory chemokines and provides an excellent means to study the complex interactions between inflammatory cells, chemokines, and tumor progression (Ji *et al.*, 2006).

## Doxycycline (DOX)-Inducible/De-Inducible Lung Cancer Models

*In Kras*$^{LA}$ mice, oncogene can be induced, but it cannot be de-induced after lung carcinogenesis. To improve this mouse model, a better method of replicating gene expression patterns of target oncogenes had to be taken into account. Furthermore, a general knock-in or knockout procedure only poorly represents genetic events that occur during sporadic lung cancer since genes are already deleted already*in utero* (Jonkers & Berns, 2002). Conditional regulation of the temporal-spatial expression of oncogenes or inactivation of tumor suppressor genes in somatic tissues of choice can more accurately mimic the *in vivo* situation leading to the onset of sporadic cancer (Jonkers & Berns, 2002;Lewandoski, 2001). This is why the second generation of mouse models for lung cancer makes use of a conditional bitransgenic tet-inducible system (Lewandoski, 2001). Most often, the reverse tetracycline (tet)-controlled transactivator (*rt*TA) inducible system is used. The first transgene with the*rt*TA element behind a tissue-specific promoter causes the *rt*TA expression in a specific cell types, e.g. MMTV-*rt*TA, CCSP-*rt*TA. This transgene is then combined with a second transgene, consisting of a target gene behind a tet-responsive promoter (*tetO7*) vector, e.g. pTRE-Tight (2$^{nd}$ generation vector from Clontech). The presence of tet/dox ensures stable interaction of the *rt*TA element with the *tetO7*promoter, which, in turn, expresses the target gene upon exposure to tet or dox.

Therefore, on/off target gene expression is possible depending on administration or withdrawal of tet/dox (Gossen *et al.*, 1992). Both *SPC-rtTA* and *CCSP-rtTA* transgenes (Perl *et al.*, 2002) have been used for directing dox-responsive *rt*TA to either alveolar type II or Clara cells. Although both of these promoters have been used to create lung cancer models of mice, CCSP-*rt*TA has more widely been used than SPC-*rt*TA since the *CCSP* promoter is active in both Clara cells and alveolar type II cells while the *SPC* promoter is active only in alveolar type II cells (Floyd *et al.*, 2005). Several transgenic mice such as *CCSP-rtTA;tetO7-FGF-7* and *CCSP-rtTA;tetO7-Kras$^{G12D}$* have been successfully created to induce lung lesions in response to antibiotics (Tichelaar et al., 2000; Fisher et al., 2001). Induction of FGF-7 caused initial epithelial cell hyperplasia followed by adenomatous hyperplasia after dox application. All hyperplasia disappeared after withdrawal of dox (Tichelaar et al., 2000). However, mouse Kras$^{G12D}$ induction caused epithelial cell hyperplasia, adenomatous hyperplasia and, after 2 months dox application, multiple adenomas and AdCAs. Again, no lesion was detected after 1 month of dox withdrawal (Fisher et al., 2001). When the *CCSP-rtTA;tetO7 Kras$^{G12D}$* alleles were combined with conventional *p53* or *Ink4a/Arf*-null alleles, AdCAs with a more malignant phenotype appeared after 1 month dox treatment, thus showing a synergy of mutant *K-ras* and *p53* or *Ink4a/Arf* deficiencies. However, even in these compound *tet*-inducible mouse models, all lesions disappeared after dox withdrawal. This finding demonstrated the importance of mutant *K-ras* as a "driving" oncogene not only at tumor onset, but also during maintenance of AdCA in these mice (Fisher et al., 2001). Other models for early, benign lung tumor lesions have been created by using a bitransgenic *tet*-inducible human *Kras$^{G12C}$* allele that can be expressed in both Clara and/or alveolar type II cells (Tichelaar et al., 2000; Floyd et al., 2005). Expression of human Kras$^{G12C}$ caused multiple, small lung tumors over a 12-month time period. Although tumor multiplicity increased upon continued *K-ras* expression, most lung lesions were hyperplasias or well-differentiated adenomas (Floyd et al., 2005). This is in good contrast to the more severe phenotypes observed in other transgenic mouse models in which different mutant *K-ras* alleles were expressed in the lung. Expression of K-ras$^{G12C}$ was associated with a 2-fold increase in the activation of the Ras and Ral signaling pathways and increased phosphorylation of Ras downstream effectors, including Erk, p90 ribosomal S6 kinase, ribosomal S6 protein, p38 and MAPKAPK-2. In contrast, expression of K-ras$^{G12C}$ had no effect on the activation of the JNK and Akt signaling pathways explaining low tumor induction by human*Kras$^{G12C}$*. This observation was in strong contrast to the effects of the previously described mouse*Kras$^{G12D}$* models (Fisher et al., 2001).

## Cre/loxP or Flp/FRT Models

The *Cre/loxP* or *Flp/FRT* system (Jonkers & Berns, 2002; Lewandoski, 2001; Dutt et al., 2006) provided excellent tools for reproducing more complicated lung tumor genetics found in human lung cancers, by introducing somatic mutations in a limited number of differentiated cells of choice whereby other cells of the fully developed lung remained normal. In short, mutations of targeted regions, flanked by loxP (also known as being "floxed") or flippase recombination target (Frt) sequence sites, were introduced through deletion by their respective site-specific recombinases Cre or Flp. Thus, in the case of tumor suppressor genes, conditional hypomorphic mutations (i.e., lower than normal function of the protein) or null allele, several coding or non-coding exons are floxed and can, therefore, be deleted by its corresponding recombinase. Conversely, floxed transcription stops (Lox-Stop-Lox or LSL) in front of oncogene or knock-in alleles can control their respective conditional activation (Jackson et al., 2001) as in the case of *LSL KRas*$^{G12D}$ mice described in the previous section.

The determining factor of this conditional approach is the control of temporal-spatial Cre or FRT recombinase expression. For that purpose, several *Cre* transgenic lines have been generated, with or without *tet*-inducible promoters (Perl et al., 2002). Apart from this, Cre-mediated recombination can also be achieved through the administration of an engineered Adeno-Cre virus *via* nasal or tracheal inhalation (Meuwissen et al., 2001; Jackson et al., 2001). An advantage of the latter method is that a limited amount of adult lung cells can be targeted in a very concise, localized, and timely fashion. Efficacy of this method was tested with conditional alleles of *KRas*$^{G12D}$ and *KRas*$^{G12V}$ (Jackson et al., 2001; Guerra et al., 2003). Infection of adult lungs with Adeno-Cre virus rapidly resulted in the onset of adenomatous alveolar hyperplasia, followed by the development of adenomas and AdCAs at 3-4 months post-infection. Although a latency of 8 months was also observed (Guerra et al., 2003), no metastases could be found in any of the models. Most probably a single *K-ras* activation is not enough to allow the AdCAs to progress into a higher state of malignancy as would be required for fully metastasizing lesions. However, these straightforward experiments disclosed the important role of *K-ras* in human lung cancer onset and progression (Guerra et al., 2003). Another important aspect of this model was that lung tumor multiplicity could be controlled by the dose of *Adeno-Cre* virus infecting only a subset of lung epithelial cells. This, together with a controlled time-point of *Adeno-Cre* application, mimics sporadic character of human lung cancer development. However, one has to be careful to note that variability of the *Adeno-Cre* virus delivery and infection (especially with the intranasal method) might lead to inconsistent experimental results. Nevertheless this versatile method remains powerful in that it resembles human lung cancer events.

## SPECIFIC ONCOGENES IN MOUSE LUNG CANCER MODELS

### Kras Downstream Effectors and Lung Cancer – Roles of RAF

Since *Kras* mutations are very common (20-25%) in NSCLC, the understanding of the precise signaling cascade of the Kras pathway is very important (Ji *et al.*, 2007). One of the best characterized Ras pathways is Ras/Raf/MEK/ERK. In fact, *BRAF* gene mutations have been found in a variety of human cancers including NSCLC (Davies *et al.*, 2002; Ji *et al.*, 2007). Oncogenic mutations of *BRAF* render constitutively phosphorylation of the protein, resulting in continued ERK activation. Of all the *BRAF* mutations, *BRAF-V600E* is the most frequent. (Mercer *et al.*, 2003). Dankort *et al.* (2007) created BRaf(CA) (CA: constitutively active) mice to express normal BRaf prior to Cre-mediated recombination after which *BRaf(V600E)* was expressed at physiological levels. *BRaf(CA)* mice infected with an Adenovirus expressing Cre recombinase developed benign lung tumors that only rarely progressed to AdCA. The reason for this is the initial proliferation is halted by increased expression of senescence markers p53 and Ink4a/Arf. Consistent with the tumor suppressor function for Ink4a/Arf and p53, BRaf(V600E) expression combined with mutation of either locus led to lung cancer progression. Moreover, *BRaf(VE)*-induced lung tumors were prevented by pharmacological inhibition of MEK1/2.

In another study, Ji *et al* generated a lung-specific, *tet*-inducible, mice model in which the *CCSP-rtTA;tetO7-BRAFV600E* induced a development of lung AdCA with bronchioalveolar carcinoma type. The extracellular signal-regulated kinase (ERK)-1/2 (MAPK) pathway was highly activated by the expression of *BRAF(V600E)* mutant. Upon dox withdrawal, the deinduction of *BRAF*-mutant expression led to regression of lung tumors together with a marked decrease in phosphorylation of ERK1/2. Furthermore, the *in vivo* use of a specific MAPK/ERK kinase (MEK) inhibitor also induced lung tumor regression. All these results showed that both activated BRAF and KRAS signaling converge onto the same MAPK pathway, making this pathway a potential target for lung tumor intervention.

The significance of c-Raf was also investigated in *K-Ras*$^{G12V}$-driven NSCLCs. Ablation of c-Raf in *K-Ras*$^{+/G12V}$; *c-Raf*$^{lox/lox}$ mice induced dramatic increase of survival rate and life span due to the decrease of tumor burden. This result suggests the essential role of c-Raf in mediating oncogenic Ras signaling in NSCLCs (Blasco *et al*, 2011). Further investigation during *Kras*$^{G12D}$-driven lung tumorigenesis showed the MAPK antagonist Sprouty-2 (Spry-2) was upregulated. When *Spry-2* was knocked out in Cre/lox dependent *Spry-2$^{flox/flox}$;LSL Kras$^{G12D}$* mice, both tumor number and total tumor area were significantly increased. This clearly suggested a tumor suppressor activity for *Sprouty-2* during *Kras*-dependent lung tumorigenesis by involving in antagonism of Ras/MAPK signaling (Shaw et al., 2007).

By using *CCSP-rtTA;TetO-Cre;LSL-Kras(G12D)*mice Cho et al. (2011) established a dox-inducible, Kras(G12D)-driven lung AdCA to pursue the cellular origin and molecular processes involved in *Kras*-induced tumorigenesis. The EpCAM(+)MHCII(-) cells (bronchiolar origin) were more enriched with tumorigenic cells in generating secondary tumors than EpCAM(+)MHCII(+) cells (alveolar origin). In addition, secondary tumors derived from EpCAM(+)MHCII(-) cells showed diversity of tumor locations compared with those derived from EpCAM(+)MHCII(+) cells. Secondary tumors from EpCAM(+)MHCII(-) cells expressed differentiation marker, pro-SPC, consistent with the notion that cancer-initiating cells display not only the abilities for self-renewal, but also the features of differentiation to generate tumors of heterogeneous phenotypes. High level of ERK1/2 activation and colony-forming ability as well as lack of Sprouty-2 expression were also observed in EpCAM(+)MHCII(-) cells. Their data suggested that bronchiolar Clara cells are the origin of tumorigenic cells for Kras(G12D)-induced lung cancer.

## PI3K and Lung Cancer

Another important pro-survival pathway that is interlinked with RAS is PI3K/Akt signaling pathway. Phosphoinositide-3-kinase (PI3K) consists of a regulatory (p85) and a catalytic (p110) subunit. The overexpression of both subunits was reported in lung carcinomas (Samuels & Velculescu 2004;Wojtalla et al., 2011). Furthermore, selective *PIK3CA* amplification was found in lung squamous cell carcinomas (Angulo et al., 2008). To investigate the oncogenic potential of PIK3CA, transgenic mice were generated with a *tet*-inducible expression of an activated p110α mutant, H1047R, and it was crossed with CCSP-*rt*TA mice to generate *CCSP-rtTA;tetO7;PIK3CA(H1047R)* compound mice. Upon dox treatment of animals for 14 weeks, double transgenic mice developed AdCAs, which subsequently disappeared after dox withdrawal for 3 weeks (Engelman et al., 2008). To identify the effect of loss of PI3K signaling in *Kras*-induced lung tumorigenesis, PI3K activity was completely eliminated in *p85*knockouts (*Pik3r2$^{-/-}$;Pik3r1$^{-/-}$*), and a dramatic decrease in the number of lung tumors was observed in*LSL Kras$^{G12D}$;Pik3r2$^{-/-}$;Pik3r1$^{-/-}$* mice (Engelman et al., 2008). The clinical efficacy of NVP-BEZ235, a dual pan-PI3K and mammalian target of rapamycin (mTOR) inhibitor was also evaluated against p110α H1047R-induced mouse lung tumors. Application of this drug led to marked tumor regression. In contrast, NVP-BEZ235 barely had effect on mouse lung cancers driven by mutant *Kras*. However, a combination of NVP-BEZ235 and a MEK inhibitor ARRY-142886, had marked synergistic effect on tumor regression. These *in vivo* studies suggest that inhibitors of the PI3K-mTOR pathway when combined with MEK inhibitors, may effectively treat KRAS mutated lung cancers. Of note, Ras proteins directly interact with the p110α subunit of PI3K and introduction of specific mutations (T208D and K227A) in *PIK3CA* blocks this interaction (Gupta *et al*., 2007). To study the Ras-p110α interactions *in vivo* and its effects on tumorigenesis, these point mutations were introduced into the *Pik3ca* gene in the mice and these mice were crossed with $Kras^{LA2}$ alleles (Gupta *et al*., 2007). Interestingly, they were highly resistant to *Kras* induced lung tumor development, which suggest Ras-p110α interaction is required for Ras-driven tumorigenesis (Gupta *et al*., 2007). All these results emphasize the importance of PI3K signaling, not only in lung tumor induction, but also maintenance.

## Rac and Lung Cancer

Rac is a member of the Rho family of small GTPases, and it mediates the regulation of various important cellular processes including cell migration, proliferation and adhesion, all of which may contribute to tumorigenesis (Mack *et al*., 2011). The important role of Rac in Ras induced lung tumorigenesis was demonstrated in a mice model in which an oncogenic allele of *Kras* was activated by Cre-mediated recombination in the presence or absence of conditional deletion of *Rac1*. They showed that Rac1 function was required for tumorigenesis in lung carcinogenesis for mice with *Rac1* deletion had tumor regression and longer survival. These data showed a specific requirement for Rac1 function in cells expressing oncogenic *K-ras* (Kissil *et al*., 2007).

## Receptor-Type Protein Tyrosine Kinase and Lung Cancer – Roles of EGFR

### *EGFR and Lung Cancer*

Epidermal growth factor (EGF) receptor family is one type of RTKs, on which the tyrosine residues phosphorylation lead to activation of downstream TK signaling that contributes to cell proliferation, motility and invasion (Stella *et al*., 2012). The activation mutations on *EGFR* gene are found in about 10-20% of advanced NSCLC cases and its protein overexpression is found in more than 60% of all lung cancers (Lynch *et al*., 2004; Soria, *et al*., 2012). Lynch *et al*. reported that EGFR mutation correlated with clinical responsiveness to the tyrosine kinase inhibitor gefitinib (2004). Since these mutations lead to increased growth factor signaling with susceptibility to the inhibitor, screening for such mutations in lung cancers will identify patients who will have a response to gefitinib. To study a specific oncogenic potential of *EGFR* mutant, the variant III (vIII) deletion, Ji *et al.* (2006a) produced *Tet-op-EGFRvIII*; *CCSP-rtTA* mice, in which the EGFRvIII expression was induced in lung type II pneumocytes upon dox administration. Mice developed atypical adenomatous hyperplasia after 6-8 weeks of dox induction and progressed to lung adenocarcinomas after 16 weeks with high activation of AKT and ERK signaling pathways. De-induction of EGFRvIII resulted in significant tumor regression, supporting the requirement of continuous EGFRvIII expression in lung tumorigenesis. Furthermore, by using an EGFR/ERB2 inhibitor HKI-272, they found tumor volume in *EGFRvIII*; *CCSP-rtTA*; *Ink4a/Arf$^{/-}$* mice was dramatically decreased, suggesting a therapeutic strategy for lung cancers with *EGFRvIII* mutation by an irreversible EGFR inhibitor (Ji *et al.*, 2006a). Politi *et al.* (2006) also studied the role of EGFR mutations in the initiation and maintenance of lung cancer, and developed transgenic mice that express an exon 19 deletion mutant (EGFR($\Delta$L747-S752)) or the L858R mutant (EGFR(L858R)) in type II pneumocytes under the control of dox, and reported that expression of either EGFR mutant lead to the development of lung AdCa. Ji *et al.* (2006b) later created bitransgenic mice with inducible expression in type II pneumocytes of two common hEGFR mutants (hEGFR$^{DEL}$and hEGFR$^{L858R}$) seen in human lung cancer. Both bitransgenic lines developed lung AdCa with hEGFR mutant expression, confirming their oncogenic potential. Maintenance of transformed phenotypes of these lung cancers was dependent on sustained expression of the EGFR mutants. Treatment with small molecule inhibitors (erlotinib or HKI-272) as well as a humanized anti-hEGFR antibody (cetuximab) led to dramatic tumor regression (Ji *et al.*, 2006b). Thus persistent EGFR signaling is required for tumor maintenance in human lung AdCas expressing EGFR mutants. Li *et al.*(2007) generated another dox-inducible lung cancer mice model harboring both erlotinib sensitizing and resistance mutations L858R and T790M (*EGFR TL*). They found that specific expression of *EGFR TL* in lung compartments led to the development of typical bronchioloalveolar carcinoma after 4-5 weeks and peripheral adenocarcinoma after 7-9 weeks. Treatment of *EGFR TL*-driven tumors is most effective when using combined regimen of HKI-272 and rapamycin, suggesting that this combination therapy may benefit pateints harboring erlotinib resistance EGFR mutation (Li *et al.*, 2007).

## HER2 and Lung Cancer

The c-*ERBB2* gene is located on chromosome 17q11.2-12 and

encodes **H**uman **E**pidermal Growth Factor **R**eceptor **2** (HER2) (Hu *et al.*, 2011). This is a transmembrane glycoprotein receptor p185$^{HER2}$, which has been targeted by the humanized monoclonal antibody trastuzumab (Herceptin).*HER2* is amplified and overexpressed in approximately 25% of breast cancer patients and is associated with an aggressive clinical course and poor prognosis. HER2 protein overexpression without gene amplification happens in some cases, possibly due to promoter activation and/or protein stabilization. HER2 overexpression stimulates cell growth in *p53*-mutated cells while it inhibits cell proliferation in those with wild-type *p53*. The molecular mechanisms for these differential responses have recently been clarified: the *Dmp1* promoter was activated by HER2/neu through the PI3K-Akt-NF-κB pathway, which in turn stimulated *Arf* transcription and p53 activation to prevent tumorigenesis. Conversely HER2 simply stimulate cell proliferation in cells that lack *Dmp1*, *Arf*, or *p53* (Taneja *et. al.*, 2010).

HER2 receptor overexpression has been reported in 11% to 32% of NSCLC tumors, with gene amplification found in 2%-23% of cases (Hirsch *et al.*, 2009; Swanton *et al.*, 2006). High-level ERBB2 amplification occurs in a small fraction of lung cancers with a strong propensity to high-grade adenocarcinomas (Grob *et al.*, 2012). The frequency of *HER2* amplification in NSCLC and the widespread availability of HER2 fluorescence *in situ* hybridization analysis may justify a study of trastuzumab monotherapy in NSCLC cases. However, sensitivity to HER2-directed therapies is complex and involves expression not only of HER2, but also of other EGFR family members (HER1, HER2, and HER4), their ligands, and molecules that influence pathway activity (Swanton *et al.*, 2006). The role played by HER2 as a heterodimerization partner for other EGFR family members makes HER2 an attractive target regardless of receptor overexpression in lung cancer. However, targeted therapies in patients overexpressing HER2 have proven less successful in clinical trials for NSCLC. One reason to explain the failure is intratumoral heterogeneity of *ERBB2* amplification, which was found in 4 of 10 cases (Grob *et al.*, 2012). Of note, this heterogeneity is rare in breast cancer that responds relatively well to anti-HER2 therapy. Laboratory data indicate that forced expression of HER2 in a NSCLC line increases sensitivity to gefitinib. They speculated that this may result from the gefitinib-mediated inhibition of HER2/HER3 heterodimerization and HER3 phosphorylation. It might thus be expected that combinatorial approaches, such as EGFR inhibition (by gefitinib) together with HER2 dimerization blockade (by pertuzumab) may be even more effective. Preclinical data indicate this may be the case, with the combination of erlotinib and pertuzumab promoting more than additive antitumor activity in the NSCLC (Swanton *et al.*, 2006).

While HER2 is overexpressed in about 20% of lung cancers, mutations in HER2 also occur in about 2-3% of cases. HER2 mutations typically occur in adenocarcinomas and are more frequent in women and never-smokers (Pinder, 2011). Mutations in HER2 lead to constitutive activation of the HER2 receptor, similar to the situation with EGFR. In good contrast to what we experienced in breast cancer, early clinical trials of Herceptin combined with chemotherapy in lung cancer patients with HER2 overexpression did not show a benefit for patients. However, there are case reports of lung cancer with HER2 mutations who have responded well to treatment with Herceptin plus chemotherapy. For instance, BIBW2992 (a small molecule inhibitor of EGFR and HER2) has shown evidence of activity in lung cancer patients with HER2 mutations. Most of the patients described had cancers that had shown resistance to chemotherapy and/or EGFR inhibitors. More patients with SCLC should be screened for HER2 mutations since the number of patients described to date is too small to draw any definitive conclusions (Pinder, 2011).

## Cyclin D1 and Lung Cancer

The development of human lung carcinogenesis is very complex. Several oncogenes involved in this process have been identified, one of which is cyclin D1 (Meuwissen & Berns, 2005). Cyclin D1 is a crucial regulator in mammalian cell cycle, which drives cells to enter S phase by binding and activating CDK4/6. The cyclin D1/CDK4 complex phosphorylates the retinoblastoma protein (pRb), which releases E2F transcriptional factors from pRb constraint. The E2Fs can then activate genes that are required for the cell to enter S phase (Sherr, 1996, 2004). Cyclin D1 overexpression results in deregulation of phosphorylation of pRB, which can cause loss of growth control. In fact, Cyclin D1 gene and protein products are frequently overexpressed in a wide rang of cancers. In NSCLC, the *CCND1* locus at 11q13 is amplified in up to 32% of cases, and its protein is expressed at high level in average of 45% of all cases (Gautschi *et al.*, 2007).

The ability of cyclin D1 to cause malignant transformation has been demonstrated in breast cancer transgenic mice model, in which *MMTV-Cyclin D1* transgenic mice developed mammary AdCA (Wang *et al.*, 1994). Just like in breast cancer, *CCND1* is often found amplified and overexpressed in NSCLC patients. It has been shown that cyclin D1 overexpression is a marker for an increased risk of upper aerodigestive tract premalignant lesions for progressing to cancer (Kim *et al.*, 2011). A polymorphism, G/A870, has been identified in the *CCND1* gene and it results in an aberrantly spliced protein (Cyclin D1b) lacking the Thr-286 phosphorylation site necessary for nuclear export (Diehl *et al.*, 1997). It has been shown that the *MMTV-D1T286A* (analogous to Cyclin D1b in humans) mice developed mammary AdCAs at an increased rate relative to *MMTV-D1* mice. Even though cyclin D1b was detected in all NSCLC samples, and the G/A870 polymorphism in *CCND1* gene is predictive of the risk of lung malignancy (Gautschi et al., 2007), its impact on lung carcinogenesis has never been investigated. Thus creation of mouse models for aberrant cyclin D1 expression in lung epithelial tissue is needed to test whether it is a key factor in the development of lung carcinogenesis.

Cancer chemoprevention uses dietary or pharmaceutical agents to suppress or prevent carcinogenic progression to invasive cancer. In a recent study, it was shown that a combination of retinoid bexarotene and EGFR inhibitor erlotinib can suppress lung carcinogenesis in transgenic lung cancer cells as well as NSCLC patients in both early and advanced stages. Bexarotene can induce the proteasomal degradation of cyclin D1 and erlotinib can act as an inhibitor of EGFR which represses transcription of cyclin D1 (Kim et al., 2011). This finding implicates cyclin D1 as a chemopreventive target and the combination of bexarotene and erlotinib is an attractive candidate for lung cancer chemoprevention (Dragnev et al., 2011). Before using this regimen in clinical lung cancer chemoprevention, its activity should first be tested in clinically predictive cyclin D1 mouse lung cancer models.

## Pten and Lung Cancer

Since expression of phosphatase and tensin homologue deleted from chromosome 10 (PTEN; reviewed in Inoue et al., 2012) is often down regulated in NSCLC, several mice models have been generated in which *Pten* was inactivated in the bronchial epithelium (Yanagi et al., 2007; Iwanaga et al., 2008).*PTEN* is a tumor suppressor gene that acts by blocking the PI3K dependent activation of serine-threonine kinase Akt (Inoue et al., 2012). Since $Pten^{-/-}$ mice are embryonic lethal, one had to make use of floxed *Pten* alleles ($Pten^{flox/flox}$), combined with *CCSP-Cre* transgene, targeting *Pten* deletion into bronchial epithelial cells. However, these $Pten^{flox/flox}$;*CCSP-Cre* mice did not show any aberrant pulmonary development or phenotypic abnormalities even when mice were followed for more than 12 months (Iwanaga et al., 2008). This changed dramatically when the $Pten^{flox/flox}$;*CCSP-Cre* alleles were crossed with *LSLKras*$^{G12D}$. Lung tumor development was markedly accelerated compared in*Pten*$^{-/-}$;*Kras*$^{G12D}$ mice to that of single *LSLKras*$^{G12D}$ mice. *Pten*-deficient, *Kras* mutant tumors were often of the more advanced AdCA with higher vascularity (Iwanaga et al., 2008), suggesting that*Pten*-loss cooperates with *Kras* mutations in NSCLC. Contrary to these results were the findings of another study in which *Pten*-inactivation was targeted in bronchioalveolar epithelium with *SPC-rtTA;tetO7-Cre* (Yanagi et al., 2007). When dox was applied *in utero* at E10-16 during embryogenesis, most mice died postnatally from hypoxia. Their lungs showed an impaired alveolar epithelial cell differentiation with an overall lung epithelial cell hyperplasia. The few surviving mice developed spontaneous lung AdCAs. Post-natal dox application during P21-27 resulted in a mild bronchiolar and alveolar cell hyperplasia and increased cell size but no lethality. A majority of these animals developed AdCAs in comparison to WT controls. Prior addition of urethane induced an even higher amount of AdCAs. Interestingly, most $Pten^{-/-}$ AdCAs (33%), with or without urethane addition, showed spontaneous *Kras* mutations. The latter observation again indicates the importance of Kras activity in cooperating with *Pten*-loss during NSCLC development.

## LKB1 and Lung Cancer – A Novel Player

Mutations in liver kinase B1 (*LKB1*) are found in Peutz–Jeghers syndrome (PJS) patients and are characterized by intestinal polyps (hamartoma) and increased incidence of epithelial tumors, such as hamartomatous polyps in the gastrointestinal tract, as well as breast, colorectal, and thyroid cancers (Giardiello et al., 2000). It is a serine threonine kinase also known as *STK11* (Sanchez-Cespedes et al., 2002). LKB1 is a primary upstream kinase of adenine monophosphate-activated protein kinase (AMPK), a necessary element in cell metabolism that is required for maintaining energy homeostasis. It is now clear that LKB1 exerts its growth suppressing effects by activating a group of other ~14 kinases, creating a group of AMPK and AMPK-related kinases. Activation of AMPK by LKB1 suppresses cell growth and proliferation when energy and nutrient levels are low. The *LKB1* gene has been implicated in the regulation of multiple biological processes, signaling pathways (Wei et al., 2005), and tumorigenesis. It has been reported that LKB1 directly activates AMP-activated kinase and regulates apoptosis in response to energy stress (Shaw et al., 2004).

A large fraction of NSCLC cells have germ-line mutations and impaired expression of *LKB1*. LOH for*LKB1* has been reported in more than 50% in lung cancer (Makowski & Hayes, 2008) and thus *LKB1*inactivation is a common event for NSCLC (Sanchez-Cespedes et al., 2002, Sanchez-Cespedes, 2007). The highest numbers of mutations were found in AdCAs, especially in those with *KRAS* mutations (Matsumoto et al., 2007; Sanchez-Cespedes, 2007). *LKB1* inactivation cooperates with *KRAS*activation, suggesting a role for LKB1 as an active repressor of the KRAS downstream pathway (Ji et al., 2007). $Lkb1^{flox/flox}$;$LSLKras^{G12D}$ mice showed a broad spectrum of NSCLCs: the majority of lung tumors were AdCAs, but SqCLCs and large cell carcinoma (LCLC) also occurred. Conversely, no SqCLC or LCLC was detected in $p53^{flox/flox};LSLKras^{G12D}$ and $(Ink4a/Arf)^{flox/flox};LSLKras^{G12D}$ mice. Furthermore, 61% of AdCA in $Lkb1^{flox/flox};LSLKras^{G12D}$ mice developed metastases, but none found for SqCLC and LCLC. These results show that *LKB1*-loss permits squamous differentiation and facilitates metastases, but these two are independent events. AdCA from $Lkb1^{flox/flox};LSLKras^{G12D}$ mice had reduced pAMPK (phosphorylated, adenosyl monophosphate-activated protein kinase) and pACCA (phosphorylated, acetyl-CoA carboxylase α-subunit) levels and activated mTOR pathway. It is probable that *LKB1*-loss influences differentiation of NSCLC into subtypes by affecting discrete pathways (Shah *et al*., 2008). A large panel of human NSCLC showed *LKB1* mutations in AdCA (34%), SqCLC (19%), and LCC (16%) (Ji *et al*., 2007). Simultaneous mutations in *p53* and *LKB1* suggest non-overlapping roles in NSCLC. Moreover, reconstitution of LKB1 in human NSCLC cell lines showed anti-tumor effects independent of their *p53* or *INK4A/ARF* status (Ji *et al*., 2007). Finally, loss of LKB1 expression in alveolar adenomatous hyperplasia, precursor lesion for AdCA, suggests an early role of *LKB1*-inactivation during AdCA development (Ghaffar *et al*., 2003).

The same group conducted a mouse trial that mirrors a human clinical trial in patients with KRAS-mutant lung cancers (Chen *et al*., 2012). They demonstrated that simultaneous loss of either *p53* or*Lkb1*, strikingly weakened the response of *Kras*-mutant cancers to single therapy by docetaxel. Addition of selumetinib provided substantial benefit for mice with lung cancer caused by *Kras* and*Kras* and *p53* mutations, but not in mice with *Kras* and *Lkb1* mutations (Chen *et al*., 2012). Thus synchronous 'clinical' trials performed in mice, not only will be useful to anticipate the results of ongoing human clinical trials, but also to generate clinically-relevant hypotheses that will affect the analysis and design of human studies.

## miRNAs and Lung Cancer

Not only might genetic mutations in oncogenes and tumor suppressor genes affect their target gene expression during lung tumorigenesis, but also microRNAs (miRNAs) can also perform similar roles. microRNAs are evolutionarily conserved, endogenous, non-protein coding, 20–23 nucleotide, single-stranded RNAs that negatively regulate gene expression in a sequence-specific manner. In order to become active, small interfering RNA (siRNA) must undergo catalytic cleavage by the RNase DICER1. In human lung cancer, increased activities of DICER1 and variant regulations of miRNA clusters have been observed. For the latter, a frequent down regulation of the *let-7* miRNA family as well as an upregulation of *miR-17-92* have been reported (Hayashita *et al*., 2005). *miR-17-92* encodes a cluster of seven miRNAs transcribed as single primary transcript. To date, functional analyses of *Dicer1* and *let-7* have been performed in the background *Kras*-induced NSCLC models. A conditional deletion of *Dicer1* in the background of *LSLKras$^{G12D}$;Dicer1$^{flox/flox}$* mice let to a marked increase of tumor development (Kumar *et al.*, 2007). However, since the 3′ UTR region of *Kras* transcripts has been shown to be a direct target of *let-7* (Johnson *et al.*, 2005), it has become very tempting to increase *let-7* expression in *Kras$^{G12D}$* lung tumors. *let-7* inhibits the growth of multiple human lung cancer cell lines in culture, as well as the growth of lung cancer cell xenografts *in vivo*. Intranasal application of both adenoviral (Esquela-Kerscher *et al.*, 2008) and lentiviral (Kumar *et al.*, 2008) *let-7*miRNA caused a significant decrease of *Kras$^{G12D}$;p53$^{-/-}$* lung tumors. These findings provide direct evidence that *let-7* acts as a tumor suppressor gene in the lung and indicate that this miRNA might be useful as a novel therapeutic agent in lung cancer.

A large scale survey conducted by a different group to determine the miRNA signature of >500 lung, breast, stomach, prostate, colon, and pancreatic cancers and their normal adjacent tissue revealed that*miR-21* was the only miRNA up-regulated in all these tumors (Volinia *et al.*, 2006). Functional studies in cancer cell lines suggest that *miR-21* has oncogenic activity. Knockdown of *miR-21* in cultured glioblastoma cells activated caspases leading to apoptotic cell death, suggesting *miR-21* is an anti-apoptotic factor (Chan *et al.*, 2005). In MCF-7 cells, *miR-21* knock-down resulted in suppression of cell growth both *in vitro* and *in vivo* (Si *et al.*, 2007). Knock-down of *miR-21* in the breast cancer cells reduced invasion and metastasis (Zhu *et al.*, 2008). Targeted deletion of *miR-21* colon cancer cells resulted in tumorigenesis through compromising cell cycle progression and DNA damage-induced checkpoint function by targeting *Cdc25a* (Wang *et al.*, 2009). *miR-21* expression is increased and predicts poor survival in NSCLC. Hatley *et al.* used transgenic mice with loss-of-function and gain-of-function *miR-21* alleles combined with a model of NSCLC (*K-ras$^{LA2}$*) to determine the role of *miR-21*in lung cancer (Hatley *et al.*, 2010). They showed that overexpression of *miR-21* enhances lung tumorigenesis and that genetic deletion of *miR-21* protects against tumor formation. *miR-21* drives tumorigenesis through inhibition of negative regulators of the Ras/MEK/ERK pathway and inhibition of apoptosis (Hatley *et al.*, 2010). These studies indicate that knocking-down of *miR-21* expression in cancer cells results in phenotypes important for tumor biology.

Hennessey *et al.* (2012) conducted Phase I/II biomarker study to examine the feasibility of using serum miRNA as biomarkers for NSCLC. Examination of miRNA expression levels in serum from a multi-institutional cohort of 50 subjects (30 NSCLC patients and 20 healthy controls) identified differentially expressed miRNAs. They found that 140 candidate miRNA pairs distinguished NSCLC from healthy controls with a sensitivity and specificity of at least 80% each. Several miRNA pairs involving miRNAs-106a, miR-15b, miR-27b, miR-142-3p, miR-26b, miR-182, 126#, let7g, let-7i (described above) and miR-30e-5p exhibited a negative predictive value and a positive predictive value of 100%. Notably, a combination of two differentially expressed miRNAs *miR-15b* and *miR-27b*, was able to discriminate NSCLC from healthy volunteers with high sensitivity, specificity (Hennessey et al.,2012). Upon further testing on additional 130 subjects, this miRNA pair predicted NSCLC with a specificity of 84%, sensitivity of 100%. These data provide evidence that serum miRNAs have the potential to be sensitive, cost-effective biomarkers for the early detection of NSCLC.

## MOUSE MODELS FOR SQUAMOUS CELL LUNG CANCER (SQCLC)

So far genomic alterations in SqCLC have not been comprehensively characterized. The Cancer Genome Atlas group recently profiled 178 lung squamous cell carcinomas to provide a comprehensive view of genomic and epigenomic alterations (Hammerman et al., 2012). They showed that the SqCLC is characterized by hundreds of exonic mutations, genomic rearrangements, and gene copy number alterations. In addition to *TP53* mutations found in nearly all specimens, loss-of-function mutations were found in the *HLA-A* class I gene. In addition, *Nuclear factor (erythroid-derived 2)-like 2, Kelch-like ECH-associated protein 1, Squamous differentiation,* and *Phosphatidylinositol-3-OH kinase pathway* genes were frequently altered. *CDKN2A* and *RB1* genes were inactivated in as many as 72% of SqCLC cases. This comprehensive study identified a potential therapeutic target in most tumors, offering new avenues of investigation for the treatment of human SqCLC (Hammerman et al., 2012).

Although squamous cell carcinoma is a common type of lung cancer causing nearly 400,000 deaths per year worldwide, there is no established gene-engineered mouse model for squamous cell carcinoma of the lung. Human lung SqCLC is closely linked with smoking and shows a distinct order of pre-malignant changes in the bronchial epithelium from hyperplasia, metaplasia, dysplasia and carcinoma*in situ* to invasive and metastatic SqCLC (Brambilla et al., 2000). A better understanding of the cell of origin that give rise to SqCLC and identification of unique genetic alterations that are specific to lung squamous cell carcinoma as reported by the comprehensive study might help to create SqCLC mouse models. One important issue that should be taken into account is that normal human or mouse lungs do not contain squamous epithelium. Mice do not smoke, so only under pathological conditions does squamous metaplasia accompanied by high expression levels of keratins occur in the airway epithelium (Wistuba et al., 2002, 2003). Only a few mouse models reported the onset of SqCLC, mostly after carcinogen application. For instance, intratracheal intubation of methyl carbamate (Jetten et al., 1992) or extensive topical application of N-nitroso-compounds (Nettesheim et al., 1971; Rehm et al., 1991) caused SqCLC in mice. Wang et al. (2004) treated eight different inbred strains of mice with N-nitroso-tris-chloroethylurea by skin painting and found that this chemical induced SqCLCs in five strains (SWR, Swiss, A/J, BALB/c, and FVB), but not in the others (AKR, 129/svJ, and C57BL/6). Besides, specific loci for SqCLC susceptibility have been identified through linkage analyses in several mice strains (Wang et al., 2004), using 6,128 markers in publically available databases. Three markers (*D1Mit169, D3Mit178*, and *D18Mit91*) were found significantly associated with susceptibility to SqCLC. Interestingly, none of these sites overlapped with the major susceptibility loci associated with lung adenoma/adenocarcinomas in mice indicating that different sets of genes are responsible for SqCLC and AdCA. Their model can be used in determining genetic modifiers that contribute to susceptibility or resistance to SqCLC development.

The other group tried to induce SqCLC through constitutive expression of human K14 by creating*CC10-hK14* mice (Dakir et al., 2008). Although hK14 is highly expressed in bronchial epithelium, only precursor lesions varying from hyperplasia to squamous metaplasia were observed (Dakir et al., 2008). Clearly, the increased K14 expression and onset of squamous cell metaplasia alone was not sufficient to generate fully developed SqCLC. As far as transgenic/knockout mice models are concerned, only the $LSLKras^{G12D};Lkb1^{flox/flox}$ somatic mouse model has been able to generate advanced SqCLC. By using a somatically activatable mutant *Kras*-driven model of mouse lung cancer ($K$-$ras^{LA}$), Ji et al. (2007) compared the role of Lkb1 to other tumor suppressors in lung cancer. Although *Kras* mutation cooperated with loss of *p53* or *Ink4a/Arf* in this system, the strongest cooperation was seen with homozygous inactivation of *Lkb1*. *Lkb1*-deficient tumors demonstrated shorter latency, an expanded histological spectrum (adeno-, squamous, and large-cell carcinoma) and more frequent metastasis as compared to tumors lacking *p53* or *Ink4a/Arf*. Interestingly up to 60% of*Lkb1* deficient lung tumors had squamous or mixed squamous histology (Ji et al., 2007), which has not been reported in other mouse lung cancer models. Pulmonary tumorigenesis was also accelerated by hemizygous inactivation of *Lkb1*, confirming its haplo-insufficiency. Consistent with these findings, inactivation of *LKB1* was found in 34% and 19% of 144 human lung adenocarcinomas and squamous cell carcinomas, respectively. They also identified a variety of metastasis-promoting genes, such as *NEDD9*, *VEGFC* and *CD24*, as targets of LKB1 repression in lung cancer. These studies established LKB1 as a critical barrier to prevent lung carcinogenesis, controlling initiation, differentiation and metastasis (Ji *et al.*, 2007).

## CLINICAL IMPLICATIONS AND FUTURE DIRECTIONS FOR MOUSE LUNG CANCER MODELS

Xenograft models where manipulated human lung cancer cell lines are subcutaneously injected into nude mice have been extensively used for pre-clinical testing of novel drugs for lung cancer. The major issue for this approach is that lung cancer cell lines have already been adapted for long-term culture in a plastic dish with artificial medium and acquired stem-cell like phenotypes, and thus are not suitable for models of primary human lung cancer obtained by surgical resection. The more preferred method, however, have been orthotopical transplantation of human lung tumor cells in their lung cavity. To date, the results have shown that xenograft models do not accurately predict the clinical efficacy of anti-tumor drugs. Therefore, a question arises as to whether spontaneous and/or genetically-engineered mouse models for lung cancer would be more useful as tools for pre-clinical drug tests. It is obvious that there are differences in the lung anatomy and physiology between mice and humans, but some of the mouse models that we have described have a striking histological similarity, with an analogous genetic signature to that of human NSCLC. Importantly, genetically-engineered mouse model-derived tumors develop in an innate immune environment and, therefore, have all the tumor-stromal interactions, such as angiogenesis and degradation of the tissue matrix.

We have described two models for NSCLC in which either the continuous oncogenic activity of Kras (Fisher *et al*, 2001) or EGFR (Politi *et al*, 2006) are prerequisites of tumor maintenance since lung tumors underwent spontaneous regression with disappearance of the oncogene by dox withdrawal. This not only shows that tumor growth critically depends on the initiating active oncogenic pathways, but it also stresses the usefulness of these oncogenic pathways as therapeutic targets. Direct tumor intervention studies with tyrosine kinase inhibitors against EGFR mutations proved to be highly effective in several *hEGFR*-transgenic mouse models. TKIs such as gefitinib, erlotinib, and HKI-272 led to complete tumor regression (Politi *et al.*, 2006; Ji *et al.*, 2006a,b). In addition, treatment of lung cancer with humanized anti-hEGFR antibody (cetuximab) caused a significant tumor regression (Ji *et al.*, 2006a). Further studies will be needed to investigate the signaling cascades that determine the sensitivity and resistance to EGFR-related tyrosine kinase interventions. Other mouse models for NSCLC have also been used for targeted therapies. First, dox-induced overexpression of the PI3K p110α catalytic subunit PIK3CA, mutated in its kinase domain (H1047R) in *CCSP-rtTA;tetO7-PIK3CA(H1047R)* mice, induces adenocarcinomas (Engelman et al., 2008). Treatment of these lung tumors with NVP-BEZ235, a dual pan-PI3K and mammalian target of rapamycin (mTOR) inhibitor, caused a marked lung tumor regression. Interestingly, when this single agent NVP6-BEZ235 was tested on lung tumors in *CCSP-rtTA;tetO7-Kras*$^{G12D}$ mice, no regression was observed. However, when NVP-BEZ235 was combined with MEK inhibitor ARRY-142886, significant regression of *Kras*$^{G12D}$ tumors occurred (Engelman et al., 2008). Thus, two major RAS downstream effector pathways needed to be inactivated to get an irreversible regression in Ras mutated NSCLC.

Although *K-RAS* is mutated in ~30% of human NSCLC, direct targeting of RAS has been unsuccessful for lung cancer therapy. Many small molecules against Ras functions have been tested and farnesyl transferase inhibitors are the most marked examples of these failed attempts (Mahgoub et al., 1999;Omer et al., 2000). Recent results with lung cancer mouse models strongly suggest that KRAS4A, and not KRAS4B is driving the onset of NSCLC. An explanation for this failure can thus be attributed to the fact that only KRAS4B is farnesylated, but not its isoform KRAS4A. Although we still have to study if KRAS4A is important in the pathogenesis of human NSCLC, we can imagine the importance of *Kras* mouse models in testing functional inhibitiors for KRAS4A (To et al., 2008).

The use of optimized, genetically-modified mouse models for lung cancer for therapy research necessitates sophisticated non-invasive tools to follow tumor development and response to therapy *in vivo*. Measurement of tumor size as a function of time is the most obvious way of doing this and existing techniques such as computed-tomography imaging or magnetic resonance imaging for small animals are now in use (Engelman et al., 2008; Politi et al., 2006). However, these techniques are time-consuming and expensive, making them less suitable for large number of animals. Other techniques, such as fluorescence imaging and bioluminescence, can be used for measuring gene expression or tumor growth *in vivo* (Contag et al., 2000; Hadjantonakis et al., 2003). In case of latter studies, transgenic expression of luciferase allows accurate longitudinal monitoring and good quantification of tumor burden as has been shown in the *LSL Kras* lung tumor model (Jackson et al., 2001). These novel imaging techniques will greatly enhance the accuracy and reproducibility of mouse models. Transgenic lung cancer models created by Chen et al. (2002) can be applied to clinics by raising Ron-specific antibodies. O'Toole et al.

(2006) conducted an antibody phage display library to generate a human IgG1 antibody IMC-41A10 that binds with high affinity to RON and effectively blocks interaction with its ligand, macrophage-stimulating protein. They found IMC-41A10 to be a potent inhibitor of receptor and downstream signaling, cell migration, and tumorigenesis. It antagonized MSP-induced phosphorylation of RON, MAPK, and AKT in several cancer cell lines. In NCI-H292 lung cancer xenograft tumor models, IMC-41A10 inhibited tumor growth by 50% to 60% as a single agent. This antibody should be tested *in vivo* using the *SPC-RON* mice with developing lung AdCAs.

Recent strategies showed the importance of aberrant promoter methylation in lung cancer development, such a $p16^{INK4a}$, *Death-associated protein kinase 1,* and, *RAS association domain family 1A* (Shames*et al.*, 2006). Since chronic inflammations have been implicated in cancer pathogenesis (Shacter & Weitzman, 2002), altered methylation for lung surfactant proteins are good topics for future lung cancer studies; their signatures may serve as valuable markers in lung cancer detection. The lung surfactant protein (*SP*) genes, *SP-A* and *SP-D* have been identified with high throughput approach that showed an altered methylation pattern in lung cancer compared to normal lung tissue (Vaid & Floros, 2009). However, *SP-A*-deficient mice were able to survive with no apparent pathology in a sterile environment (Korfhagen et al., 1996), although their pulmonary immune responses were insufficient during immune challenge. *SP-D*-deficient mice, on the other hand, showed phenotypic abnormalities in alveolar macrophages and type II pneumocytes with increased lipid pools, indicating that *SP-D* has an important role in surfactant homeostasis (Botas et al., 1998). Paradoxically overexpression of *SP-A*and/or *SP-D* as a result of promoter hypomethylation has also been reported in lung cancer suggesting that it is critical to keep these protein levels within physiological ranges to prevent neoplastic transformation. Since the role of these lung surfactant proteins in lung carcinogenesis has never been studied *in vivo*, it will be worthwhile to cross lung surfactant-deficient mice with available transgenic/knockout strains to elucidate the roles of surfactant proteins in lung cancer initiation and development.

## ACKNOWLEDGEMENTS

K. Inoue has been supported by NIH/NCI 5R01CA106314, ACS RSG-07-207-01-MGO, and by WFUCCC Director's Challenge Award #20595. D. Maglic has been supported by DOD pre-doctoral fellowship BC100907. We thank K. Klein for editorial assistance.

## REFERENCES

1. Siegel R, Naishadham D, Jemal A. Cancer statistics, 2012. *CA Cancer J Clin* 2012;62:10-29.
2. Centers for Disease Control and Prevention (CDC). Racial/Ethnic disparities and geographic differences in lung cancer incidence --- 38 States and the District of Columbia, 1998-2006. *MMWR Morb Mortal Wkly Rep* 2010;59:1434-8.
3. Wisnivesky JP, McGinn T, Henschke C, Hebert P, Iannuzzi MC, Halm EA. Ethnic disparities in the treatment of stage I non-small cell lung cancer.*Am J Respir Crit Care Med* 2005;171:1158-63.
4. Swierzewski III, SJ. Lung Cancer Environmental Risk Factors. 1999. http://www.healthcommunities.com/lung-cancer/environmental.shtml
5. Menon P. Lung Cancer: Delayed Diagnosis Among Non-Smokers. 2012. http://trialx.com/curetalk/2012/06/lung-cancer-delayed-diagnosis-among-non-smokers/
6. Brambilla C, Fievet F, Jeanmart M, *et al*. Early detection of lung cancer: role of biomarkers. *Eur Respir J Suppl* 2003;39:36s-44s.
7. Yeung SC, Habra MA, Thosani SN. Lung cancer-induced paraneoplastic syndromes. *Curr Opin Pulm Med* 2011;17:260-8.
8. Moran CA. Pulmonary adenocarcinoma: The expanding spectrum of histologic variants. *Arch Pathol Lab Med* 2006;130:958-62.
9. Schiller JH. Current standards of care in small-cell and non-small-cell lung cancer. *Oncology* 2001;61, Suppl 1:3-13.
10. Travis WD. Pathology of lung cancer. *Clin Chest Med* 2002;23,65-81.
11. Zochbauer-Muller S, Gazdar AF, and Minna JD. Molecular pathogenesis of lung cancer. *Ann Rev Physiol* 2002;64,681-708.
12. Fong KM, Sekido Y, Gazdar AF, and Minna JD. Lung cancer. 9: Molecular biology of lung cancer: Clinical implications. *Thorax* 2003;58:892-900.
13. Meuwissen R and Berns A. Mouse models for human lung cancer. *Genes Dev* 2005;19:643-64.
14. Worden FP, Kalemkerian GP. Therapeutic advances in small cell lung cancer. *Expert Opin Investig Drugs* 2000;9:565-79.
15. Wakamatsu N, Devereux TR, Hong HH, *et al*. Overview of the molecular carcinogenesis of mouse lung tumor models of human lung cancer. *Toxicol Pathol* 2007;35:75-80.
16. Shimkin MB, Stoner GD. Lung tumors in mice: application to carcinogenesis bioassay. *Adv Cancer Res* 1975;21:1-58.

17. Pfeifer GP, Denissenko MF, Olivier M, *et al*. Tobacco smoke carcinogens, DNA damage and p53 mutations in smoking-associated cancers. *Oncogene* 2002;21:7435-51.
18. Sato M, Shames DS, Gazdar AF, *et al*. A translational view of the molecular pathogenesis of lung cancer. *J Thorac Oncol* 2007;2:327-43.
19. Malkinson AM. Primary lung tumors in mice as an aid for understanding, preventing, and treating human AdCA of the lung. *Lung Cancer* 2001;32:265-79.
20. DeMayo FJ, Finegold MJ, Hansen TN, *et al*. Expression of SV40 T antigen under control of rabbit uteroglobin promoter in transgenic mice. *Am J Physiol* 1991;261:L70-6.
21. Sandmoller A, Halter R, Gomez-La-Hoz E, *et al*. The uteroglobin promoter targets expression of the SV40 T antigen to a variety of secretory epithelial cells in transgenic mice. *Oncogene* 1994;9:2805-15.
22. Wikenheiser KA, Clark JC, Linnoila RI, *et al*. Simian virus 40 large T antigen directed by transcriptional elements of the human surfactant protein C gene produces pulmonary AdCAs in transgenic mice. *Cancer Res* 1992;52:5342-52.
23. Wikenheiser KA, Whitsett JA. Tumor progression and cellular differentiation of pulmonary AdCAs in SV40 large T antigen transgenic mice. *Am J Respir Cell Mol Biol* 1997;16:713-23.
24. Geick A, Redecker P, Ehrhardt A, *et al*. Uteroglobin promoter-targeted c-MYC expression in transgenic mice cause hyperplasia of Clara cells and malignant transformation of T-lymphoblasts and tubular epithelial cells. *Transgenic Res* 2001;10:501-11.
25. Ehrhardt A, Bartels T, Geick A, Klocke R, Paul D, Halter R. Development of pulmonary bronchiolo-alveolar AdCAs in transgenic mice overexpressing murine c-myc and epidermal growth factor in alveolar type II pneumocytes. *Br J Cancer* 2001;84:813-8.
26. Sunday ME, Haley KJ, Sikorski K, *et al*. Calcitonin driven v-Ha-ras induces multilineage pulmonary epithelial hyperplasias and neoplasms. *Oncogene* 1999;18:36-47.
27. Mallakin A, Sugiyama T, Taneja P, *et al*. Mutually exclusive inactivation of DMP1 and ARF/p53 in lung cancer. *Cancer Cell* 2007;12:381-94.
28. Morris GF, Hoyle GW, Athas GB, *et al*. Lung-specific expression in mice of a dominant negative mutant form of the p53 tumor suppressor protein. *J La State Med Soc* 1998;150:179-85.
29. Tchou-Wong KM, Jiang Y, Yee H, *et al*. Lung-specific expression of

dominant-negative mutant p53 in transgenic mice increases spontaneous and benzo(a)pyrene-induced lung cancer. *Am J Respir Cell Mol Biol* 2002;27:186-93.

30. Chen YQ, Zhou YQ, Fu LH, Wang D, Wang MH. Multiple pulmonary adenomas in the lung of transgenic mice overexpressing the RON receptor tyrosine kinase. Recepteur d›origine nantais. *Carcinogenesis* 2002;23:1811-9.

31. Sekido Y, Fong KM, Minna JD. Molecular genetics of lung cancer. *Annu Rev Med* 2003;54:73-87.

32. Jacks T, Fazeli A, Schmitt EM, et al. Effects of an Rb mutation in the mouse. *Nature* 1992;359:295-300.

33. Meuwissen R, Berns A. Mouse models for human lung cancer. *Genes Dev* 2005;19:643-64.

34. Olive KP, Tuveson DA, Ruhe ZC, et al. Mutant p53 gain of function in two mouse models of Li-Fraumeni syndrome. *Cell* 2004;119:847-60.

35. Lang GA, Iwakuma T, Suh YA, et al. Gain of function of a p53 hot spot mutation in a mouse model of Li-Fraumeni syndrome. *Cell* 2004;119:861-72.

36. Smith AJ, Xian J, Richardson M, et al. Cre-loxP chromosome engineering of a targeted deletion in the mouse corresponding to the 3p21.3 region of homozygous loss in human tumors. *Oncogene* 2002;21:4521-9.

37. Tommasi S, Dammann R, Zhang Z, et al. Tumor susceptibility of Rassf1a knockout mice. *Cancer Res* 2005;65:92–8.

38. Zanesi N, Mancini R, Sevignani C, et al. Lung cancer susceptibility in Fhit-deficient mice is increased by Vhl haploinsufficiency. *Cancer Res* 2005;65:6576-82.

39. Johnson L, Mercer K, Greenbaum D, et al. Somatic activation of the K-ras oncogene causes early onset lung cancer in mice. *Nature* 2001;410:1111-6.

40. Hirai H, Sherr CJ. Interaction of D-type cyclins with a novel myb-like transcription factor, DMP1. *Mol Cell Biol* 1996;16:6457-67.

41. Inoue K, Sherr CJ. Gene expression and cell cycle arrest mediated by transcription factor DMP1 is antagonized by D-type cyclins through a cyclin-dependent-kinase-independent mechanism. *Mol Cell Biol* 1998;18:1590-600.

42. Inoue K, Mallakin A, and Frazier DP. Dmp1 and tumor suppression. *Oncogene* 2007;26:4329-35. Review.

43. Sugiyama T, Frazier DP, Taneja P, et al. Signal transduction involving

the Dmp1 transcription factor and its alteration in human cancer. *Clinical Medicine Insights: Oncology* 2008a; 2:209-19.
44. Inoue K, Roussel MF, and Sherr CJ. Induction of ARF tumor suppressor gene expression and cell cycle arrest by transcription factor DMP1. *Proc Natl Acad Sci USA* 1999;96:3993-8.
45. Sreeramaneni R, Chaudhry A, McMahon M, Sherr CJ, and Inoue K. Ras-Raf-Arf signaling critically depends on the Dmp1 transcription factor. *Mol Cell Biol* 2005;25:220-32.
46. Inoue K, Wen R, Rehg JE, Adachi M, Cleveland JL, Roussel MF, and Sherr CJ. Functional loss of the ARF transcriptional activator DMP1 facilitates cell immortalization, ras transformation, and tumorigenesis. *Genes Dev* 2000;14:1797-809.
47. Inoue K, Zindy F, Randle DH, Rehg JE, and Sherr CJ. Dmp1 is haplo-insufficient for tumor suppression and modifies the frequencies of Arf and p53 mutations in Myc-induced lymphomas. *Genes Dev* 2001;15:2934-9.
48. Mallakin A, Sugiyama T, Taneja P, *et al*. Mutually exclusive inactivation of DMP1 and ARF/p53 in lung cancer. *Cancer Cell* 2007;12:381-94.
49. Sugiyama T, Frazier DP, Taneja P, *et al*. The role of Dmp1 and its future in lung cancer diagnostics. *Expert Rev Mol Diagn* 2008b;8:435-48.
50. Sweet-Cordero A, Mukherjee S, Subramanian A, *et al*. An oncogenic KRAS2 expression signature identified by cross-species gene-expression analysis. *Nat Genet* 2005;37:48-55.
51. Jackson EL, Willis N, Mercer K, *et al*. Analysis of lung tumor initiation and progression using conditional expression of oncogenic K-ras. *Genes Dev*2001;15:3243-8.
52. Jackson EL, Olive KP, Tuveson DA, Bronson R, Crowley D, Brown M, and Jacks T. The differential effects of mutant p53 alleles on advanced murine lung cancer. *Cancer Res* 2005;65:10280-8.
53. Ji H, Houghton AM, Mariani TJ, *et al*. K-ras activation generates an inflammatory response in lung tumors. *Oncogene* 2006;25:2105-12.
54. Jonkers J, Berns A. Conditional mouse models of sporadic cancer. *Nat Rev Cancer* 2002;2:251-65.
55. Lewandoski M. Conditional control of gene expression in the mouse. *Nat Rev Genet* 2001;2:743-55.
56. Gossen M, Bujard H. Tight control of gene expression in mammalian cells by tet-responsive promoters. *Proc Natl Acad Sci USA* 1992;89:5547-51.
57. Perl AK, Tichelaar JW, Whitsett JA. Conditional gene expression in the respiratory epithelium of the mouse. *Transgenic Res* 2002;11:21-9.

58. Floyd HS, Farnsworth CL, Kock ND, et al. Conditional expression of the mutant Ki-rasG12C allele results in formation of benign lung adenomas: development of a novel mouse lung tumor model. *Carcinogenesis* 2005;26:2196-206.

59. Tichelaar JW, Lu W, Whitsett JA. Conditional expression of fibroblast growth factor-7 in the developing and mature lung. *J Biol Chem*2000;275:11858–64.

60. Fisher GH, Wellen SL, Klimstra D, et al. Induction and apoptotic regression of lung adenocarcinomas by regulation of a K-Ras transgene in the presence and absence of tumor suppressor genes. *Genes Dev* 2001;15:3249-62.

61. Dutt A, Wong KK. Novel agents in the treatment of lung cancer: advances in EGFR-targeted agents: mouse models of lung cancer. *Clin Cancer Res*2006;12:4396s-402s.

62. Jackson EL, Willis N, Mercer K, et al. Analysis of lung tumor initiation and progression using conditional expression of oncogenic K-ras. *Genes Dev*2001;15:3243-8.

63. Meuwissen R, Linn SC, van der Vaulk M, et al. Mouse model for lung tumorigenesis through Cre/lox controlled sporadic activation of the K-Ras oncogene. *Oncogene* 2001;20:6551–58.

64. Guerra C, Mijimolle N, Dhawahir A, et al. Tumor induction by an endogenous K-ras oncogene is highly dependent on cellular context. *Cancer Cell*2003;4:111–120.

65. Ji H, Wang Z, Perera SA, et al. Mutations in BRAF and KRAS converge on activation of the mitogen-activated protein kinase pathway in lung cancer mouse models. *Cancer Res* 2007;67:4933-9.

66. Davies H, Bignell GR, Cox C, et al. Mutations of the BRAF gene in human cancer. *Nature* 2002;417:949-54.

67. Mercer KE, Pritchard CA. Raf proteins and cancer: B-Raf is identified as a mutational target. *Biochim Biophys Acta* 2003;1653:25-40. Review.

68. Dankort D, Filenova E, Collado M, Serrano M, Jones K, McMahon M. A new mouse model to explore the initiation, progression, and therapy of BRAFV600E-induced lung tumors. *Genes Dev* 2007;21:379-84.

69. Blasco RB, Francoz S, Santamaría D, et al. c-Raf, but not B-Raf, is essential for development of K-Ras oncogene-driven non-small cell lung carcinoma. *Cancer Cell* 2011;19:652-63.

70. Shaw AT, Meissner A, Dowdle JA, et al. Sprouty-2 regulates oncogenic K-ras in lung development and tumorigenesis. *Genes Dev* 2007;21:694-

707.

71. Cho HC, Lai CY, Shao LE, Yu J. Identification of tumorigenic cells in Kras(G12D)-induced lung AdCA. *Cancer Res* 2011;71:7250-8.
72. Samuels Y, Velculescu VE. Oncogenic mutations of PIK3CA in human cancers. *Cell Cycle* 2004;3:1221-4.
73. Wojtalla A, Arcaro A. Targeting phosphoinositide 3-kinase signalling in lung cancer. *Critical Reviews in Oncology/Hematology* 2011;80:278-290.
74. Angulo B, Suarez-Gauthier A, Lopez-Rios F, *et al*. Expression signatures in lung cancer reveal a profile for EGFR-mutant tumors and identify selective PIK3CA overexpression by gene amplification. *J Pathol* 2008;214:347-56.
75. Engelman JA, Chen L, Tan X, *et al*. Effective use of PI3K and MEK inhibitors to treat mutant Kras G12D and PIK3CA H1047R murine lung cancers.*Nat Med* 2008;14:1351-6.
76. Gupta S, Ramjaun AR, Haiko P, *et al*. Binding of ras to phosphoinositide 3-kinase p110alpha is required for ras-driven tumorigenesis in mice. *Cell*2007;129:957-68.
77. Mack NA, Whalley HJ, Castillo-Lluva S, Malliri A. The diverse roles of Rac signaling in tumorigenesis. *Cell Cycle* 2011;10:1571-81.
78. Kissil JL, Walmsley MJ, Hanlon L, *et al*. Requirement for Rac1 in a K-ras induced lung cancer in the mouse. *Cancer Res* 2007;67:8089-94.
79. Stella GM, Luisetti M, Inghilleri S, *et al*. Targeting EGFR in non-small-cell lung cancer: Lessons, experiences, strategies. *Resp Med* 2012;106,173-83.
80. Lynch TJ, Bell DW, Sordella R, *et al*. Activating mutations in the epidermal growth factor receptor underlying responsiveness of non-small-cell lung cancer to gefitinib. *N Engl J Med* 2004;350:2129-39.
81. Soria J-C, Mok TS, Cappuzzo F, *et al*. EGFR-mutated oncogene-addicted non-small cell lung cancer: Current trends and future prospects. *Cancer Treat Rev* 2012;38,416-30.
82. Ji H, Zhao X, Yuza Y, *et al*. Epidermal growth factor receptor variant III mutations in lung tumorigenesis and sensitivity to tyrosine kinase inhibitors. *Proc Natl Acad Sci USA*. 2006a;103:7817-22. Epub 2006 May 3.
83. Ji H, Li D, Chen L, *et al*. The impact of human EGFR kinase domain mutations on lung tumorigenesis and in vivo sensitivity to EGFR-targeted therapies. *Cancer Cell* 2006b;9:485-95.

84. Politi K, Zakowski MF, Fan PD, *et al*. Lung adenocarcinomas induced in mice by mutant EGF receptors found in human lung cancers respond to a tyrosine kinase inhibitor or to down-regulation of the receptors. *Genes Dev* 2006;20:1496-510.
85. Li D, Shimamura T, Ji H, Chen L *et al*. Bronchial and Peripheral Murine Lung Carcinomas Induced by T790M-L858R Mutant EGFR Respond to HKI-272 and Rapamycin Combination Therapy. *Cancer Cell* 2007;12:81-93.
86. Hu Y, Bandla S, Godfrey TE, Tan D, *et al*. HER2 amplification, overexpression and score criteria in esophageal adenocarcinoma. *Mod Pathol* 2011;24:899-907.
87. Taneja P, Maglic D, Kai F, *et al*. Critical role of Dmp1 in HER2/neu-p53 signaling and breast carcinogenesis. *Cancer Res* 70: 9084-94, 2010.
88. Hirsch FR, Varella-Garcia M, Cappuzzo F. Predictive value of EGFR and HER2 overexpression in advanced non-small-cell lung cancer. *Oncogene* 2009;28 Suppl 1:S32-7.
89. Swanton C, Futreal A, Eisen T. Her2-targeted therapies in non-small cell lung cancer. *Clin Cancer Res* 2006;12(14 Pt 2):4377s-83s.
90. Grob TJ, Kannengiesser I, Tsourlakis MC, *et al*. Heterogeneity of ERBB2 amplification in adenocarcinoma, squamous cell carcinoma and large cell undifferentiated carcinoma of the lung. *Mod Pathol* 2012 Aug 17. doi: 10.1038/modpathol.2012.125. [Epub ahead of print]
91. Pinder. Lesser Known Lung Cancer Mutations Part 1: HER2, a promising therapeutic target? http://cancergrace.org/lung/2011/03/19/her2-by-m/
92. Meuwissen R and Berns A. Mouse models for human lung cancer. *Genes Dev* 2005;19,643-64.
93. Sherr CJ. Cancer cell cycles. *Science* 1996;274:1672-7. Review.
94. Sherr CJ. Principles of tumor suppression. *Cell* 2004;116:235-46. Review.
95. Jiang W, Kahn SM, Zhou P, *et al*. Overexpression of cyclin D1 in rat fibroblasts causes abnormalities in growth control, cell cycle progression and gene expression. *Oncogene* 1993;8:3447-57.
96. Gautschi O, Ratschiller D, Gugger M, Betticher DC, Heighway J. Cyclin D1 in non-small cell lung cancer: A key driver of malignant transformation.*Lung Cancer* 2007;55:1-14.
97. Wang TC, Cardiff RD, Zukerberg L, *et al*. Mammary hyperplasia and carcinoma in MMTV-cyclin D1 transgenic mice. *Nature* 1994 ;369:669-71.
98. Kim ES, Lee JJ, Wistuba II. Cotargeting Cyclin D1 Starts a New

Chapter in Lung Cancer Prevention and Therapy. *Cancer Prevention Research* 2011;4:779-82.

99. Diehl JA, Zindy F, Sherr CJ. Inhibition of cyclin D1 phosphorylation on threonine-286 prevents its rapid degradation via the ubiquitin-proteasome pathway. *Genes Dev* 1997;11:957-72.

100. Dragnev KH, Ma T, Cyrus J, et al. Bexarotene Plus Erlotinib Suppress Lung Carcinogenesis Independent of KRAS Mutations in Two Clinical Trials and Transgenic Models. *Cancer Prev Res* 2011;4:818-28.

101. Yanagi S, Kishimoto H, Kawahara K, et al. Pten controls lung morphogenesis, bronchioalveolar stem cells, and onset of lung adenocarcinomas in mice. *J Clin Invest* 2007;117:2929–40.

102. Inoue K, Kulik G, Fry EA, Zhu S, and Maglic D. Recent progress in mouse models for tumor suppressor genes and its implications in human cancer (review). *Clinical Medicine Insights: Oncology,* submitted (2012).

103. Iwanaga K, Yang Y, Raso MG, et al. Pten inactivation accelerates oncogenic K-ras-initiated tumorigenesis in a mouse model of lung cancer. *Cancer Res* 2008;68:1119-27.

104. Yanagi S, Kishimoto H, Kawahara K, et al. Pten controls lung morphogenesis, bronchioalveolar stem cells, and onset of lung adenocarcinomas in mice. *J Clin Invest* 2007;117:2929–40.

105. Iwanaga K, Yang Y, Raso MG, et al. Pten inactivation accelerates oncogenic K-ras-initiated tumorigenesis in a mouse model of lung cancer. *Cancer Res* 2008;68:1119-27.

106. Giardiello FM, Brensinger JD, Tersmette AC, et al. Very high risk of cancer in familial Peutz-Jeghers syndrome. *Gastroenterology* 2000;119:1447–53.

107. Sanchez-Cespedes M, Parrella P, Esteller M, et al. Inactivation of LKB1/STK11 is a common event in AdCAs of the lung. *Cancer Res* 2002;62:3659–62.

108. Wei C, Amos CI, Stephens LC, et al. Mutation of Lkb1 and p53 genes exert a cooperative effect on tumorigenesis. *Cancer Res* 2005;65:11297-303.

109. Shaw RJ, Kosmatka M, Bardeesy N, et al. The tumor suppressor LKB1 kinase directly activates AMP-activated kinase and regulates apoptosis in response to energy stress. *Proc Natl Acad Sci USA* 2004;101:3329-35.

110. Makowski L, Hayes DN. Role of LKB1 in lung cancer development. *Br J Cancer* 2008;99:683-8.

111. Sanchez-Cespedes M. A role for LKB1 gene in human cancer beyond the

Peutz-Jeghers syndrome. *Oncogene* 2007;26:7825-32.

112. Matsumoto S, Iwakawa R, Takahashi K, *et al.* Prevalence and specificity of LKB1 genetic alterations in lung cancers. *Oncogene* 2007;26:5911-8.

113. Ji H, Ramsey MR, Hayes DN, *et al.* LKB1 modulates lung cancer differentiation and metastasis. *Nature* 2007;448:807-10.

114. Shah U, Sharpless NE, Hayes DN. LKB1 and lung cancer: more than the usual suspects. *Cancer Res* 2008;68:3562-65.

115. Ghaffar H, Sahin F, Sanchez-Cepedes M, *et al.* LKB1 protein expression in the evolution of glandular neoplasia of the lung. *Clin Cancer Res*2003;9:2998-3003.

116. Chen Z, Cheng K, Walton Z, *et al.* A murine lung cancer co-clinical trial identifies genetic modifiers of therapeutic response. *Nature* 2012;483:613-7.

117. Hayashita Y, Osada H, Tatematsu Y, *et al.* A polycistronic microRNA cluster, miR-17-92, is overexpressed in human lung cancers and enhances cell proliferation. *Cancer Res* 2005;65:9628-32.

118. Kumar MS, Lu J, Mercer KL, *et al.* Impaired microRNA processing enhances cellular transformation and tumorigenesis. *Nat Genet* 2007;39:673-7.

119. Johnson SM, Grosshans H, Shingara J, *et al.* RAS is regulated by the let-7 microRNA family. *Cell* 2005;120:635-47.

120. Esquela-Kerscher A, Trang P, Wiggins JF, *et al.* The let-7 microRNA reduces tumor growth in mouse models of lung cancer. *Cell Cycle* 2008;7:759-64.

121. Kumar MS, Erkeland SJ, Pester RE, *et al.* Suppression of non-small cell lung tumor development by the let-7 microRNA family. *Proc Natl Acad Sci USA* 2008;105:3903-8.

122. Volinia S, Calin GA, Liu CG, *et al.* A microRNA expression signature of human solid tumors defines cancer gene targets. *Proc Natl Acad Sci USA*2006;103:2257-61.

123. Chan JA, Krichevsky AM, Kosik KS. MicroRNA-21 is an antiapoptotic factor in human glioblastoma cells. *Cancer Res* 2005;65:6029-33.

124. Si ML, Zhu S, Wu H, Lu Z, Wu F, Mo YY. miR-21-mediated tumor growth. *Oncogene* 2007;26:2799-803.

125. Zhu S, Wu H, Wu F, Nie D, Sheng S, Mo YY. MicroRNA-21 targets tumor suppressor genes in invasion and metastasis. *Cell Res* 2008;18:350-9.

126. Wang P, Zou F, Zhang X, *et al.* microRNA-21 negatively regulates

Cdc25A and cell cycle progression in colon cancer cells. *Cancer Res* 2009;69:8157-65.
127. Hatley ME, Patrick DM, Garcia MR, *et al.* Modulation of K-Ras-dependent lung tumorigenesis by MicroRNA-21. *Cancer Cell* 2010;18:282-93.
128. Hennessey PT, Sanford T, Choudhary A, et al. Serum microRNA biomarkers for detection of non-small cell lung cancer. *PLoS One* 2012;7:e32307.
129. Hammerman PS, Lawrence MS, and Voet D *et al.* The Cancer Genome Atlas Research Network. Comprehensive genomic characterization of squamous cell lung cancers. *Nature* 2012 Sep 9. doi: 10.1038/nature11404. [Epub ahead of print].
130. Brambilla E, Lantuejoul S, Sturm N. Divergent differentiation in neuroendocrine lung tumors. *Semin Diagn Pathol* 2000;17:138–48.
131. Wistuba II, Mao L, Gazdar AF. Smoking molecular damage in bronchial epithelium. *Oncogene* 2002;21:7298–306.
132. Wistuba II, Gazdar AF. Characteristic genetic alterations in lung cancer. *Methods Mol Med* 2003;74:3-28.
133. Jetten AM, Nervi C, Vollberg TM. Control of squamous differentiation in tracheobronchial and epidermal epithelial cells: role of retinoids. *J Natl Cancer Inst Monogr* 1992;320:93-100.
134. Nettesheim P, Hammons AS. Induction of squamous cell carcinoma in the respiratory tract of mice. *J Natl Cancer Inst* 1971;47:697-701.
135. Rehm S, Lijinsky W, Singh G, *et al.* Mouse bronchiolar cell carcinogenesis. Histologic characterization and expression of Clara cell antigen in lesions induced by N-nitrosobis-(2-chloroethyl) ureas. *Am J Pathol* 1991;139:413-22.
136. Wang Y, Zhang Z, Yan Y, *et al.* A chemically induced model for squamous cell carcinoma of the lung in mice: histopathology and strain susceptibility.*Cancer Res* 2004;64:1647-54.
137. Dakir EL, Feigenbaum L, Linnoila RI. Constitutive expression of human keratin 14 gene in mouse lung induces premalignant lesions and squamous differentiation. *Carcinogenesis* 2008;29:2377-84.
138. Ji H, Ramsey MR, Hayes DN, *et al.* LKB1 modulates lung cancer differentiation and metastasis. *Nature* 2007;448:807-10.
139. Fisher GH, Wellen SL, Klimstra D, *et al.* Induction and apoptotic regression of lung adenocarcinomas by regulation of a K-Ras transgene in the presence and absence of tumor suppressor genes. *Genes Dev* 2001;15:3249-62.

140. Politi K, Zakowski MF, Fan PD, *et al*. Lung adenocarcinomas induced in mice by mutant EGF receptors found in human lung cancers respond to a tyrosine kinase inhibitor or to down-regulation of the receptors. *Genes Dev* 2006;20:1496-510.

141. Ji H, Li D, Chen L, *et al*. The impact of human EGFR kinase domain mutations on lung tumorigenesis and in vivo sensitivity to EGFR-targeted therapies. *Cancer Cell* 2006a;9:485-95.

142. Ji H, Zhao X, Yuza Y, *et al*. Epidermal growth factor receptor variant III mutations in lung tumorigenesis and sensitivity to tyrosine kinase inhibitors. *Proc Natl Acad Sci USA* 2006b;103:7817-22.

143. Engelman JA, Chen L, Tan X, *et al*. Effective use of PI3K and MEK inhibitors to treat mutant Kras G12D and PIK3CA H1047R murine lung cancers.*Nat Med* 2008;14:1351-6.

144. Mahgoub N, Taylor BR, Gratiot M, *et al*. In vitro and in vivo effects of a farnesyltransferase inhibitor on Nf1-deficient hematopoietic cells. *Blood*1999;94:2469-76.

145. Omer CA, Chen Z, Diehl RE, *et al*. Mouse mammary tumor virus-Ki-rasB transgenic mice develop mammary carcinomas that can be growth-inhibited by a farnesyl:protein transferase inhibitor. *Cancer Res* 2000;60:2680-8.

146. To MD, Wong CE, Karnezis AN, *et al*. Kras regulatory elements and exon 4A determine mutation specificity in lung cancer. *Nat Genet*2008;40:1240-4.

147. Contag CH, Jenkins D, Contag PR, *et al*. Use of reporter genes for optical measurements of neoplastic disease in vivo. *Neoplasia* 2000;2:41-52.

148. Hadjantonakis AK, Dickinson ME, Fraser SE, *et al*. Technicolour transgenics: imaging tools for functional genomics in the mouse. *Nature Rev Genet*2003;4:613-25.

149. Jackson EL, Willis N, Mercer K, *et al*. Analysis of lung tumor initiation and progression using conditional expression of oncogenic K-ras. *Genes Dev*2001;15:3243-8.

150. Chen YQ, Zhou YQ, Fu LH, Wang D, Wang MH. Multiple pulmonary adenomas in the lung of transgenic mice overexpressing the RON receptor tyrosine kinase. Recepteur d'origine nantais. *Carcinogenesis* 2002;23:1811-9.

151. O›Toole JM, Rabenau KE, Burns K, *et al*. Therapeutic implications of a human neutralizing antibody to the macrophage-stimulating protein receptor tyrosine kinase (RON), a c-MET family member. *Cancer*

*Res* 2006;66:9162-70.

152. Shames DS, Girard L, Gao B, *et al*. A genome-wide screen for promoter methylation in lung cancer identifies novel methylation markers for multiple malignancies. *PLoS Med* 2006; 3:e486.
153. Shacter E, Weitzman SA. Chronic inflammation and cancer. *Oncology (Williston Park)*. 2002;16:217-26, 229; discussion 230-2.
154. Vaid M, Floros J. Surfactant protein DNA methylation: a new entrant in the field of lung cancer diagnostics? *Oncol Rep* 2009;21:3-11.
155. Korfhagen TR, Bruno MD, Ross GF, *et al*. Altered surfactant function and structure in SP-A gene targeted mice. *Proc Natl Acad Sci USA* 1996;93:9594-9.
156. Botas C, Poulain F, Akiyama J, *et al*. Altered surfactant homeostasis and alveolar type II cell morphology in mice lacking surfactant protein D. *Proc Natl Acad Sci USA* 1998;95:11869–74.

# Chapter 6

## SPEECH ANALYSIS FOR DIAGNOSIS OF PARKINSON'S DISEASE USING GENETIC ALGORITHM AND SUPPORT VECTOR MACHINE

Mohammad Shahbakhi[1], Danial Taheri Far[1], Ehsan Tahami[2]

[1]Department of Biomedical Engineering, Dezful Branch, Islamic Azad University, Dezful, Iran

[2]Department of Biomedical Engineering, Mashhad Branch, Islamic Azad University, Mashhad, Iran

## ABSTRACT

Parkinson's disease (PD) is the most common disease of motor system degeneration that occurs when the dopamine-producing cells are damaged in substantia nigra. To detect PD, various signals have been investigated, including EEG, gait and speech. Since approximately 90 percent of the people with PD suffer from speech disorders, speech analysis is considered as the most common technique for this aim. This paper proposes a new algorithm for diagnosing of Parkinson's disease based on voice analysis. In the first step, genetic algorithm (GA) is undertaken for selecting optimized features from all extracted features. Afterwards a network based on support vector machine (SVM) is used for classification between healthy and people with Parkinson. The dataset of this research is composed of a range of biomedical voice signals from 31 people, 23 with Parkinson's disease and 8 healthy people. The subjects were asked to pronounce letter "A" for 3 seconds. 22 linear and non-linear features were extracted from the signals that 14 features were based on F0 (fundamental frequency or pitch), jitter, shimmer and noise to harmonics ratio, which are main factors in voice signal. Because changing in these factors is noticeable for the people with PD, optimized features were selected among them. Of the various numbers of optimized features, the data classification was investigated. Results show that the classification accuracy percent of 94.50 per 4 optimized features, the accuracy percent of 93.66 per 7 optimized features and the accuracy percent of 94.22 per 9 optimized features, could be achieved. It can be observed that the best classification accuracy may be achieved using Fhi (Hz), Fho (Hz), jitter (RAP) and shimmer (APQ5).

## INTRODUCTION

James Parkinson described Parkinson's disease (PD) in 1817 for the first time and the disease was named after him. Among neurological disorders, PD is the most common after Alzheimer and it is estimated that currently 4 to 6 million people suffer from it worldwide. Most people who are infected with PD are aged 50 or over but younger people can suffer it too [1]. Normally, there are brain cells (neurons) in the human brain that produce dopamine. These neurons concentrate in a particular area of the brain, called the substantia nigra. Dopamine is a chemical that relays messages between the substantia nigra and other parts of the brain to control movements of the human body. Dopamine helps humans to have smooth coordinated muscle movements [2]. Symptoms of PD typically begin appearing between the ages 50 and 60 and they develop slowly and often go unnoticed by the person who has them. Tremor is often the first symptom that people with PD or their family members notice. Initially, the tremor may appear in just one arm or leg or only on one side of the body. The tremor also may affect the chin, lips, and tongue. As the disease progresses, the tremor may spread to both sides of the body. Other symptoms may include depression and other motional changes: difficulty in swallowing, chewing, and speaking; urinary problems or constipation; skin problems; and sleep disruptions [3]. Currently, there is no cure for PD, although types of drugs called dopaminergic generally help reduce muscle rigidity, improve speed and coordination of movement and lessen tremor. Various signals, including EEG [4] speech [5] -[8] and gait, have been undertaken for diagnosis of PD. Voice signal recording is the earliest, easiest and most non-invasive technique for diagnosis of PD [9]. Since most of the people with PD suffer from speech disorders [10] [11], it could be considered as the most reasonable way for detection of PD [12] [13]. The PD dataset used in this article, has been studied by many professionals of voice analysis. It consists of 31 subjects: 23 suffer from PD and the rest are healthy. M. Ene [5] extracted 22 linear and nonlinear features out of the same data in this paper. Three types of probabilistic neural network (PNN), including incremental search (IS), Monte Carlo search (MCS) and hybrid search (HS), had been used for classification process. The concrete application had provided diagnosis accuracies ranging between 79% and 81%. The maximum classification accuracy of 81.28%, based on HS algorithm, was achieved. M. A. Little and his colleagues [6] extracted features similar to those used in [5]. They selected four optimized features based on the correlation equation and achieved a classification accuracy of 91.4% using the Support Vector Machine (SVM) method. M. F. Caglar and his colleagues [7] also selected four optimized features similar to those used in [5]. The process of selecting optimized features was based on Adaptive Neuro-Fuzzy Channel (ANFC) and the classification was investigated with Multi-Layer Perceptron (MLP), radial basis function (RBF) and ANFC networks. The classification accuracy for MLP was 89.69%, for RBF was 87.63% and for ANFC was 94.72%. D. Gil and M. Johnson [8] investigated ANN and SVM networks for diagnosis of PD with the same data in this paper. They could achieve classification accuracy of 90%. In this paper, a method based on combination of genetic algorithm (GA) and SVM, is investigated at which GA selects powerful features from all extracted features [14] and SVM network is used as the classifier. The GA is now widely recognized as an effective search paradigm in artificial intelligence, image processing, features extraction and many other areas. SVM is a computer algorithm that learns by example to assign labels to objects. SVMs have also been successfully applied to an increasingly wide variety of biological applications. A common biomedical application of support vector machines is the automatic classification of microarray gene expression profiles [15]. The remainder of the paper is organized as follows. In Section 2, genetic algorithm technique for selecting optimized features from the data, SVM network for classification, data acquisition and extracted features from the data are described, respectively. In Section 3, the effectiveness of our method with various numbers of optimized features is investigated. Afterwards, to evaluate the performance of classifier, three statistical parameters will be utilized. In the last section, we make a few concluding remarks.

## MATERIALS AND METHODS

### Overall Structure of the Proposed Method

In this section, we propose a new algorithm for detection of PD based on genetic algorithm and SVM network. In frist part, our strategy for selecting optimized features with genetic algorithm is described. In second part, SVM network and reasons that why it used for classification is explained.

### Genetic Algorithm

Genetic Algorithm (GA) is an adaptive heuristic search algorithm premised on the evolutionary ideas of natural selection and genetic. It is one of the most influential methods in the process of data classification, which is effectively used to select optimized features. In genetic algorithm, the solution is called chromosome or string. This method requires a population of chromosomes (strings) representing a combination of features from the solution set, and requires a cost function (called an evaluation or fitness function). This function calculates the fitness of each chromosome. The algorithm manipulates a finite set of chromosomes (the population), based loosely on the mechanism of evolution. In each generation, chromosomes are subjected to certain operators, such as crossover, inversion and mutation, which are analogous to processes, which occur in natural reproduction. Crossover of two chromosomes produces a pair of offspring chromosomes, which are synthesis of the traits of their parents. Mutation of a chromosome produces a nearly identical chromosome with only local alternations of some regions of the chromosome. The optimization process is performed in cycles called generations. During each generation, a set of new chromosomes is created using crossover, inversion, mutation and other operators. Since the population size is fixed, only the best chromosomes are allowed to survive to the next cycle of reproduction. The crossover rate usually assumes quite a high value (on the order of 80%), while the mutation rate is small (typically 1% - 15%) for efficient search. The cycle repeats until the population "converges", that is all the solutions are reasonably the same and further exploration seems fruitless, or until the answer is "good enough [16] -[18].

## Strategy for Selecting Optimized Features

The process of running this algorithm in order to select the optimized feature (pattern) is explained below [14].

1. Calculate each pattern's entropy by using Equation (1) and output (target) vector's entropy by using Equation (2)

$$H(X) = -\sum_{i=1}^{n} p(x_i) \log p(x_i) \tag{1}$$

$$H(Y) = -\sum_{i=1}^{n} p(y_i) \log p(y_i) \tag{2}$$

where x is the vector of features and y is the vector of targets, p(x) and p(y) are respectively density probability function of features and targets.

Measure mutual information between each pattern and every single output (target) via Equation (3)

$$I(X;Y) = H(X) - H(X|Y) = H(Y - H(Y|X))$$
$$= H(X) + H(Y) - H(X,Y) = H(X,Y) - H(X|Y) - H(Y|X) \tag{3}$$

In Equation (3), the patterns' entropy (H(X)), the target vector's entropy (H(Y)) and H(X,Y) are calculated by using Equation (4)

$$H(X,Y) = \sum_{i=1}^{N}\sum_{k=1}^{N} -\log(p(x_i,y_k))p(x_i,y_k) \qquad (4)$$

And

$$H(Y|X) = H(X,Y) - H(X) \qquad (5)$$

And ultimately H(Y|X) is measured via Equation (6)

$$H(Y|X) = \sum_{x \in X} p(x)H(Y|X=x) = \sum_{x \in X} p(x) \sum_{y \in Y} p(y|x) \log\left(\frac{1}{p(y|x)}\right)$$
$$= -\sum_{x \in X}\sum_{y \in Y} P(x,y)\log(y|x) = -\sum_{x \in X, y \in Y} P(x,y)\log(y|x) \qquad (6)$$

2. Initial population of genetic algorithm is produced randomly using 200 × n chromosomes, n is the number of features that need to be selected. Thus, each chromosome consists of n genes where the feature's number is placed randomly and it is possible for the feature number to be repeated randomly in a chromosome.

3. Measure the amount of relevance between patterns and targets for each chromosome using Equation (7)

$$V = \frac{1}{n}\sum_{i=1}^{n} I(X_i;Y_i) \qquad (7)$$

where I is the mutual information between features and targets. The amount of redundancy among patterns and targets is measured for each chromosome using Equation (8)

$$P = \frac{1}{n^2}\sum_{i=1}^{n}\sum_{j=1}^{n} I(X_i,Y_j) \qquad (8)$$

4. Assign the fitness value to each chromosome via Equation (9)
$$\phi = V - P \qquad (9)$$

Purpose of the suggested genetic algorithm is to maximize the fitness function of Equation (9).

5. Rearrange the chromosomes according to the given fitness function.
6. Select elite chromosomes as a parent.
7. Apply crossover and mutation and produce a new population.

The chromosomes which can maximize the fitness function will remain and the rest will be removed and then Steps 1 - 5 are repeated and this process continues as long as the changes in chromosomes' fitness is less than 0.02 or the algorithm reaches the predetermined number of iterations which is supposed to be 80 in this paper. Finally, the chromosome with the maximum fitness is chosen and the number of features in that chromosome is considered as selected features.

## Support Vector Machine

The most representative example of local neural network is the Support Vector Machine (SVM) of the Gaussian kernel function. It is a two layer neural network employing hidden layer of radial units and one output neuron. The procedure of creating this network and learning its parameters is organized in the way in which we deal only with kernel functions instead of direct processing of hidden unit signals [19]. Basic SVM is linear but it can be used for non-linear data by using kernel function to first indirectly map non-linear data into linear feature space. Basic SVM is also a two-class classifier however; with some modification, multiclass classifier can be obtained. If we consider a set of L linearly separable data and its class $\{(x_1, y_1), (x_2, y_2), L, (x_l, y_l)\}$ where $x_i$ Î $R^d$ and $y_i$ Î $\{\pm 1\}$ the maximum margin classifier is $f(x) = \text{sgn}(w \cdot x + b)$ where w and b are parameters that maximize the margin with respect to the two classes. A new form of the classifier, expressed with input and output vectors information is as follows:

$$f(x) = \text{sgn}\left(\sum_{i=1}^{L} \alpha_i y_i (x_i \cdot x) + b\right) \tag{10}$$

Here parameter α for each corresponding input vectors needs to be found in order to find the maximal margin classifier. The α values are mostly zero and those inputs with non-zero α's are called the support vectors and they contribute strongly towards the decision function. If the data set is not linear, the decision function used is:

$$f(x) = \text{sgn}\left(\sum_{i=1}^{L} \alpha_i y_i K(x_i \cdot x) + b\right) \tag{11}$$

where a kernel K is used in the mapping of the non-linear input space into a linear space. The maximal margin classifier is found in the linear space. There are a few possible kernels that can be chosen: Linear, Polynomial, Radial basis function, Hyperbolic tangent kernels [20]. Since the decision region is dependent on the data set, by using prior knowledge of the data and the characteristics of various kernels, we can achieve better performance. For example, if a data set is known to need closed decision regions, it is better to use an RBF kernel rather than a linear or a low order polynomial kernel [21], thus, RBF kernel has been preferred here.

$$k(x_i, x) = \exp\left(-\frac{\|x_i - x\|}{2\sigma^2}\right) \tag{12}$$

To define RBF kernel simpler, definition includes a parameter:

$$\lambda = -\frac{1}{2\sigma^2} \tag{13}$$

where $\|x_i - x\|$ is recognized as squared Euclidean distance between the two feature vectors and is a free parameter which is estimated empirically [22].

SVM training involves solving a convex quadratic programming (QP) problem with equality and inequality constraints obtained by the objective of margin maximization. The solution solves for nonzero parameters α's introduced in the formulation and extracts the support vectors corresponding to it. The values of α's obtained is constraint to be positive for perfectly separable case and between 0 and C in the case of non-linearly separable data. The value C is the penalty term and needs to be chosen prior to training the SVM. Other issues in SVM training include finding the best training model using appropriate kernel and the hyper parameters. In multiclass SVM, many two class SVMs are trained and in classification, voting schemes are used for selecting the correct class. SVM is preferred here because it can directly measure the extent to which people with Parkinson can be discriminated from healthy controls on the basis of measures of dysphonia, Addressing the problem of classifying subjects as healthy or PD. With such classification method, it is also possible to combine measures to create more effective discrimination in practice [6].

## DATASET

The dataset was created by Max Little of the University of Oxford, in collaboration with the National Centre for Voice and Speech, Denver, Colorado, who recorded the speech signals. The original study published the feature extraction methods for general voice disorders [23].

The data consists of 195 sustained vowel phonations from 31 male and female subjects, of which 23 were diagnosed with PD. The time since diagnoses ranged from 0 to 28 years, and the ages of the subjects ranged from 46 to 85 years (mean 65.8, standard deviation 9.8). Averages of six phonations were recorded from each subject, ranging from one to 36 seconds in length. The phonations were recorded in an IAC sound-treated booth using a head-mounted microphone (AKG C420) positioned at 8 cm from the lips. The voice signals were recorded directly to computer using CSL 4300B hardware (Kay Elemetrics), sampled at 44.1 kHz, with 16 bit resolution. Although amplitude normalization affects the calibration of the samples, the study is focused on measures insensitive to changes in absolute speech pressure level. Thus, to ensure robustness of the algorithms, all samples were digitally normalized in amplitude prior to calculation of the measures.Figure 1 illustrates speech signals of healthy and subject with PD, respectively [6] .

## Features Extraction

In this dataset, 22 linear and non-linear features were extracted from the data. Table 1 contains all the features and the brief descriptions [22] .

14 features are based on four factors: F0 (fundamental frequency or pitch), jitter, shimmer and noise to harmonics ratio, which are the most important factors of the voice signal.

It was concluded that the change in these factors is remarkable in people with Parkinson's disease compared to healthy people, therefore, optimized features are selected among them. Each feature is described below:

Fo (Hz): Average vocal fundamental frequency.

Fhi (Hz): Maximum vocal fundamental frequency.

Flo (HZ): Minimum vocal fundamental frequency.

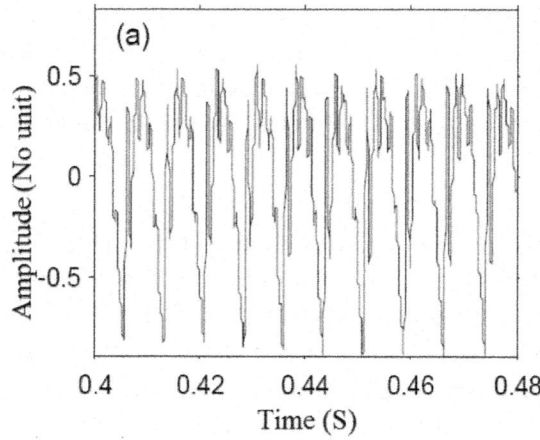

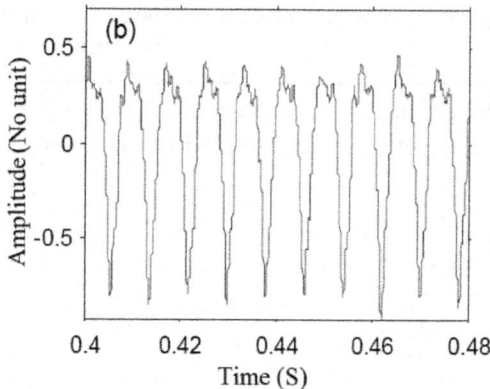

**Figure 1.** (a) healthy; (b) subject with Parkinson's disease.

**Table 1.** List of extracted features and their description

| Features | Description |
|---|---|
| MDVP: Fo (Hz) | Average vocal fundamental frequency |
| MDVP: Fhi (Hz) | Maximum vocal fundamental frequency |
| MDVP: Flo (Hz) | Minimum vocal fundamental frequency |
| Jitter (%) Jitter (Abs) MDVP: RAP MDVP: PPQ Jitter: DDP | Several measures of variation in fundamental frequency |
| Shimmer Shimmer (dB) Shimmer: APQ3 Shimmer: APQ5 MDVP: APQ Shimmer: DDA | Several measures of variation in amplitude |
| NHR HNR | Two measures of ratio of noise to tonal components in the voice |
| RPDE D 2 | Two nonlinear dynamical complexity measures |
| DFA | Signal fractal scaling exponent |

| spread 1 spread 2 PPE | Three nonlinear measures of fundamental frequency variation |
|---|---|

Jitter (%): This is the average absolute difference between consecutive periods of fundamental frequency, divided by the average period (expressed as a percentage)

$$\text{Jitter}(\%) = \frac{\frac{1}{N}\sum_{i=1}^{N-1}|T_i - T_{i-1}|}{\frac{1}{N}\sum_{i=1}^{N}T_i} \quad (14)$$

where $T_i$ is the period of fundamental frequencies of window number "i" and N is the total number of windows.

Jitter (ABS): Jitter absolute is the cycle-to-cycle variation of fundamental frequency, i.e. the average absolute difference between consecutive periods, expressed as:

$$\text{Jitter}(ABS) = \frac{1}{N-1}\sum_{i=1}^{N-1}|T_i - T_{i-1}| \quad (15)$$

where $T_i$ are the extracted F0 period lengths and N is the number of extracted F0 periods.

Jitter (RAP): it is defined as the Relative Average Perturbation, the average absolute difference between a period and the average of it and its two neighbours, divided by the average period.

Shimmer: This is the average absolute difference between the amplitudes of consecutive periods, divided by the average amplitude

$$\text{Shimmer} = \frac{\frac{1}{N-1}\sum_{i=1}^{N-1}|A_i - A_{i-1}|}{\frac{1}{N}\sum_{i=1}^{N}A_i} \quad (16)$$

Shimmer (APQ5): It is defined as the five-point Amplitude Perturbation Quotient, the average absolute difference between the amplitude of a period and the average of the amplitudes of it and its four closest neighbours, divided by the average amplitude.

HNR: harmonics to noise raito.

# RESULTS

Before the classification process between healthy people and people with Parkinson, various numbers of optimized features have been selected by genetic algorithm. In order to implement GA and SVM classification method, MATLAB software has been used. In classification process, after using genetic algorithm for determining powerful features, training and testing procedures are applied with them. Each column of the data has 195 different properties which divided into 2 parts, 75% for training and 25% for testing. The network was used 100 times for the classification per different numbers of features. The experimental results from Table 2 indicate that the maximum amount of classification accuracy, 94.50%, has been achieved with having Fhi (Hz), Fho (Hz), jitter (RAP) and shimmer (APQ5) features. It is also shown that a classification accuracy of 93.66% for Fhi (Hz), Fho (Hz), Flo (hz), jitter (RAP), shimmer (APQ5), Jitter (ABS), shimmer features and a classification accuracy of 94.22% for Fhi (Hz), Fho (Hz), Flo (hz), jitter (RAP), shimmer (APQ5), Jitter(ABS), shimmer, Jitter (%), HNR features. Table 2 contains data classification accuracy per different numbers of optimized features with SVM classifier.

To evaluate the performance of the proposed method, 3 statistical parameters: specificity, sensitivity and total classification accuracy, per various numbers of features, have been calculated.

$$\text{Specificity} = \frac{\text{NCPD}}{\text{NTPD}} \times 100 \tag{17}$$

$$\text{Sensitivity} = \frac{\text{NCH}}{\text{NTH}} \times 100 \tag{18}$$

$$\text{Total classification accuracy} = \frac{\text{NCCP}}{\text{NTP}} \times 100 \tag{19}$$

where NCPD, NTPD, NCH, NTH, NCCP and NTP are number of correct classified PD, number of total PD, number of correct classified healthy, number of total healthy, number of correct classified persons and number of total persons respectively. Table 3 shows the value of statistical parameters per various numbers of features. Figure 2 shows the plots of pair of first four prominent features which were extracted by genetic algorithm.

# DISCUSSION

The aim of this research was diagnosis of Parkinson's disease using voice analysis. In this paper, the dataset consisted of 31 people at which 23 subjects suffer from PD and the rest are healthy. Various linear and non-linear features were extracted from the data which among them, 14 features with emphasis on four main speech factors: fundamental frequency (pitch), jitter, shimmer and noise to harmonics ratio were extracted from the data. The recent researches showed that changing in these 14 factors is notable for people with Parkinson's disease compared to healthy people, thus, the process of extracting optimized features was done among them. Selecting powerful features is a primary step to improve classification accuracy. GA produces successive populations of alternate solutions that are represented by a chromosome, a solution to the problem, until acceptable results are obtained.

Table 2. Classification accuracy per different number of features

| Features | N = 4 | N = 7 | N = 9 |
|---|---|---|---|
| Accuracy | 94.50 ± 3.54 | 93.66 ± 3.61 | 94.22 ± 3.66 |

Table 3. Statistical parameters: specificity, sensitivity and total classification accuracy, per various numbers of features

| Feature Number | N = 4 | N = 7 | N = 9 |
|---|---|---|---|
| Specificity | 95.63 | 89.57 | 92.8 |
| Sensitivity | 88.46 | 80.04 | 70.12 |
| Total classification accuracy | 96.06 | 93.58 | 93.61 |

GA can deal with large search spaces efficiently, and hence has less chance to get local optimal solution than other algorithms. In first step, to extract most useful features for classification between normal people and people with PD, genetic algorithm had been undertaken. Afterwards, the classification process was done with various numbers of optimized features, which were extracted in the last step. Among different types of classifiers, support vector machine (SVM) was chosen. Indeed, by introducing the kernel, SVM gain flexibility in the choice of the form of the threshold separating healthy from people with Parkinson, which needs not be linear and even needs not have the same functional form for all data, since its function is non-parametric and operates locally. Furthermore, prior visual inspection of the layout and clustering of pairs of measures shows that it might be difficult to separate people with Parkinson from healthy people with linear or hyperplanes kernels thus, the kernel-SVM formulation, with Gaussian radial basis kernel functions was chosen [24]. As it was described, parameter should be determined experimentally, therefore, we systematically increased the value of from 0 to 1 and best classification accuracies were achieved when the value of was 0.452.

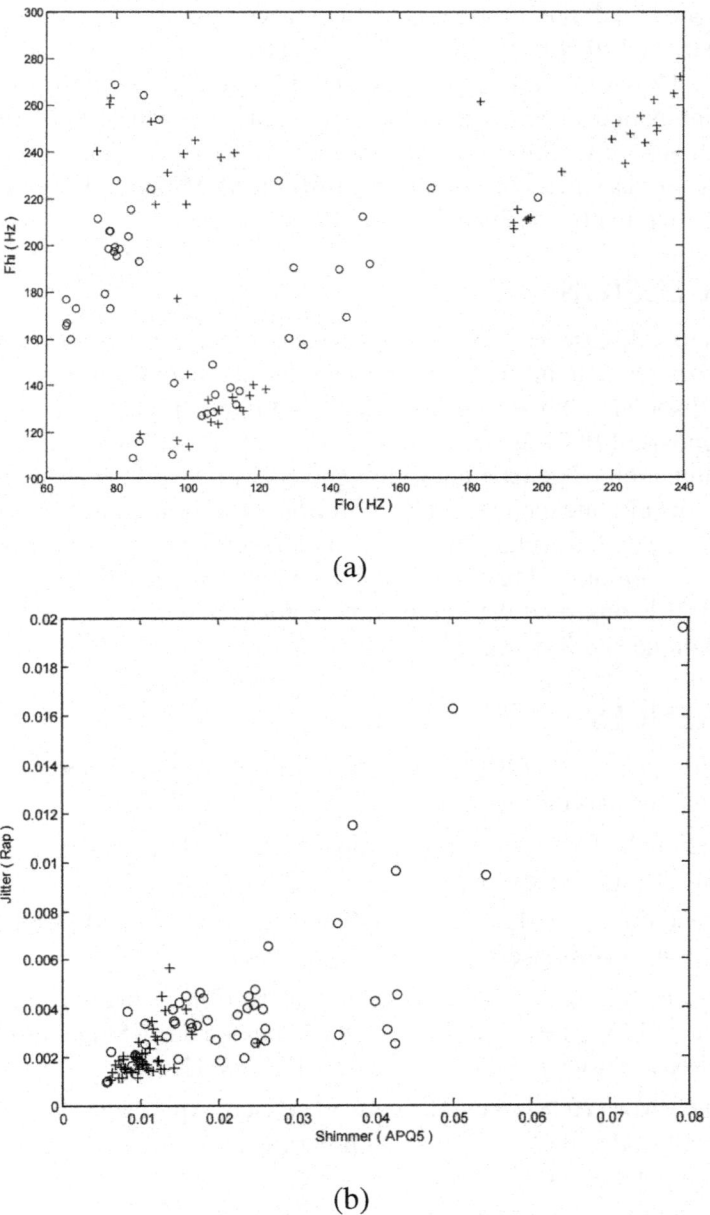

(a)

(b)

**Figure 2.** (a) demonstration of FHi (Hz)-Flo (Hz) database; (b) demonstration of Jitter

(RAP)- Shimmer (APQ5) database. The "O" marks are for healthy subjects, the "+" marks for Parkinson's subjects.

Per various numbers of features the classification process, had been investigated and the best classification accuracy was achieved with first four extracted features: Fhi (Hz), Fho (Hz), jitter (RAP) and shimmer (APQ5). However, it was possible to improve the classification accuracy if a combination of linear and non-linear features would be carried out. To this aim, we are working on a method, which uses both linear and non-linear features for classification. Moreover, a fusion SVM network will be used for the classification between healthy and PD subjects.

## CONCLUSION

Parkinson's disease is known as the second common neurological disorder after Alzheimer. It influxes several aspects of human's functions in which speech disorder is the most prominent. Several researches have been proposed for diagnosis of PD with voice analysis [5] -[8] . In this paper, a method based on combination of genetic algorithm and SVM network, for classification of healthy people and people with Parkinson of various numbers of features, was investigated. Results showed that the highest accuracy was achieved with extracting 4 optimized features: Fhi (Hz), Fho (Hz), jitter (RAP) and shimmer (APQ5). It is observed that there is no major difference between accuracy of our technique and Reference [7] .

## REFERENCES

1. Parkinson's UK (2013)http://www.parkinsons.org.uk/about_parkinsons/what_is_parkinsons.aspx
2. National Parkinson Foundation (2013) http://www.parkinson.org/parkinson-s-disease.aspx
3. Webmd (2013) http://www.webmd.com/parkinsonsdisease/tc/parkinsons-disease-symptoms
4. Pezard, L., Jech, R. and RuÊzÏicÏka, E. (2001) Investigation of Non-Linear Properties of Multichannel EEG in the Early Stages of Parkinson's Disease. Clinical Neurophysiology, 122, 38-45.
5. Ene, M. (2008) Neural Network-Based Approach to Discriminate Healthy People from Those with Parkinson's Disease. Mathematics and Computer Science Series, 35, 112-116.
6. Little, M.A., McSharry, P.E., Hunter, E.J., Spielman, J. and Ramig, L.O. (2008) Suitability of Dysphonia Measurements for Telemonitoring of

Parkinson's Disease. IEEE Transactions on Biomedical Engineering, 56, 1015-1022.http://dx.doi.org/10.1109/TBME.2008.2005954

7. Caglar, M.F., Cetisli, B. and Toprak, I.B. (2010) Automatic Recognition of Parkinson's Disease from Sustained Phonation Tests Using ANN and Adaptive Neuro-Fuzzy Classifier. Journal of Engineering Science Design, 1, 59-64.

8. Gil, D. and Johnson, M. (2009) Diagnosing Parkinson by Using Artificial Neural Networks and Support Vector Machines. Global Journal of Compute Science and Technology, 9, 63-71.

9. Duffy, R.J. (2005) Motor Speech Disorders: Substrates, Differential Diagnosis and Management. 2nd Edition, Elsevier Mosby, St. Louis.

10. Ho, A.K., Iansek, R., Marigliani, C., Bradshaw, J.L. and Gates, S. (1998) Speech Impairment in a Large Sample of Patients with Parkinson's Disease. Behaviour Neurology, 11, 131-137. http://dx.doi.org/10.1155/1999/327643

11. Logemann, J.A., Fisher, H.B., Boshses, B. and Blonsky, E.R. (1978) Frequency and Co-Occurrence of Vocal-Tract Dysfunctions in Speech of a Large Sample of Parkinson Patients. Journal of Speech Hearing Disorder, 43, 47-57.

12. Sapir, S., Spielman, J.L., Ramig, L.O., Story, B.H. and Fox, C. (2007) Effects of Intensive Voice Treatment (the Lee Silverman Voice Treatment [LSVT]) on Vowel Articulation in Dysarthric Individuals with Idiopathic Parkinson Disease: Acoustic and Perceptual Findings. Journal of Speech Lang Hearing Research, 50, 899-912.http://dx.doi.org/10.1044/1092-4388(2007/064)

13. Rahn, D.A., Chou, M., Jiang, J.J. and Zhang, Y. (2007) Phonatory Impairment in Parkinson's Disease: Evidence from Nonlinear Dynamic Analysis and Perturbation Analysis. Journal of Voice, 21, 64-71. http://dx.doi.org/10.1016/j.jvoice.2005.08.011

14. Ludwig, O., Nunes, U., Araujo, R., Schnitman, L. and Lepikson, H.A. (2009) Applications of Information Theory, Genetic Algorithms, and Neural Models to Predict Oil Flow. Communications in Nonlinear Science and Numerical Simulation, 14, 2870-2885.http://dx.doi.org/10.1016/j.cnsns.2008.12.011

15. Nobel, W. (2006) What Is a Support Vector Machine? Nature Biotechnology, 24, 1565-1568.

16. Holland, J. (1975) Adaptation in Natural and Artificial Systems. University of Michigan Press, Ann Arbor.

17. Goldberg, D. (1989) Genetic Algorithms in Search, Optimization and Machine Learning. Addison Wesley, New York.
18. Jong, K.A. and Spears, W.M. (1992) A Formal Analysis of the Role of Multi-Point Crossover in Genetic Algorithms. Annals of Mathematics and Artificial Intelligence, 5, 1-26. http://dx.doi.org/10.1007/BF01530777
19. Osowski, S., Siwek, K. and Markiewicz, T. (2004) MLP and SVM Networks—A Comparative Study. Proceedings of the 6th Nordic Signal Processing Symposium, Espoo, 37-40.
20. Ahmad, A.H., Viard-Gaudin, C., Khalid, M. and Poisson, E. (2004) Comparison of Support Vector Machine and Neural Network in Character Level Discriminant Training for Online Word Recognition, UNITEN Students Con on Research and Development, Malaisie.
21. Ganapathiraju, A., Hamaker, J. and Picone, J. (1998) Support Vector Machines for Speech Recognition. The Proceedings of the 5th International Conference on Spoken Language Processing, Incorporating the 7th Australian International Conference on Speech Science and Technology, Sydney.
22. Jean-Philippe, V., Tsuda, K. and Schölkopf, B. (2004) Kernel Methods in Computational Biology. MIT Press, Cambridge.
23. Center for Machine Learning and Intelligent Systems (2008) http://archive.ics.uci.edu/ml/datasets/Parkinsons

# Chapter 7

## PRAGMATIC ISSUES IN BIOMARKER EVALUATION FOR TARGETED THERAPIES IN CANCER

Armand de Gramont[1], Sarah Watson[2], Lee M. Ellis[3], Jordi Rodón[5], Josep Tabernero[5], Aimery de Gramont[6] and Stanley R. Hamilton[4]

[1]New Drug Evaluation Laboratory, Centre of Experimental Therapeutics, Department of Oncology, Centre Hospitalier Universitaire Vaudois (CHUV), 1011 Lausanne, Switzerland.
[2]INSERM U830, Genetics and Biology of Paediatric Tumours Group, Institut Curie, France.
[3]Departments of Surgical Oncology, and Molecular and Cellular Oncology, University of Texas MD Anderson Cancer Center, USA.
[4]Division of Pathology and Laboratory Medicine, University of Texas MD Anderson Cancer Center, USA.
[5]Medical Oncology, Vall d'Hebron University Hospital, Vall d'Hebron Institute of Oncology (VHIO) and Universitat Autonoma de Barcelona (UAB), Spain.
[6]Medical Oncology Department, Institut Hospitalier Franco-Britannique, France.

## ABSTRACT

Predictive biomarkers are becoming increasingly important tools in drug development and clinical research. The importance of using both guidelines for specimen acquisition and analytical methods for biomarker measurements that are standardized has become recognized widely as an important issue, which must be addressed in order to provide high-quality, validated assays. Herein, we review the major challenges in biomarker validation processes, including pre-analytical (sample-related), analytical, and post-analytical (data-related) aspects of assay development. Recommendations for improving biomarker assay development and method validation are proposed to facilitate the use of predictive biomarkers in clinical trials and the practice of oncology.

## INTRODUCTION

The development of cancer therapies is increasingly dependent on our understanding of tumour biology, and biomarkers—especially predictive biomarkers—are crucial tools in the field of personalized medicine and health economics, in particular, as they enable definition of the populations of patients who are most likely to benefit from targeted therapies. More-effective patient selection than is possible at present is mandatory to improve the success rate of new therapies, which are sometimes prohibitively expensive, and thereby increase their cost–utility; thus, delineating reliable predictive biomarkers is essential if we are to achieve this objective.

One commonly used definition of a biomarker is a measurable indicator that is used to distinguish precisely, reproducibly and objectively either a normal biological state from a pathological state, or the response to a specific therapeutic intervention.[1] In fact, biomarkers are used for numerous purposes: to predict survival (prognostic biomarkers); to assess drug safety and evaluate target engagement and the immediate consequence on biological processes (pharmacodynamics biomarkers), to identify patients who are more likely to benefit from a treatment (predictive biomarkers; more generally termed companion biomarkers when associated with a specific therapeutic agent); to predict outcome given the response to therapy (surrogate biomarkers); and to monitor disease progression or therapeutic efficacy (monitoring biomarkers). Identification and widespread use of biomarkers will help ensure that patients receive the best possible therapeutic strategies, thereby avoiding unnecessary treatments and associated toxicities, and eventually reducing total health costs.

Most cancer therapies, especially those developed in unselected patient populations, offer only limited clinical benefits. As an example, in phase III trials that enrolled patients with metastatic colorectal cancer (mCRC), bevacizumab was associated with a median overall survival advantage of 1.4–4.7 months when added to first-line chemotherapy and 2.1 months with second-line chemotherapy, for a median overall survival duration of more than 20 months; therefore, the cost–utility benefit of adding bevacizumab to chemotherapy is potentially marginal, in some cases.[2,3,4] Thus, biomarkers that clearly define a subgroup of patients with mCRC who are most likely to benefit from the addition of bevacizumab to chemotherapy would enable the use of this agent to be focused more effectively, which would be equally important for patients and health payers. Considering that biomarkers are nowadays integrated into most drug development programmes, from target identification and validation to clinical practice, robust measurements and assay validation for analyses of biological samples have become essential. Without a robust methodological foundation and pertinent biological interpretation, the number of reliable biomarkers that emerge will probably be limited, and their potential utility in the evaluation of novel treatments and customization of clinical strategies will be underexploited. In fact, despite the large volume of research that has been devoted to identifying cancer biomarkers and the vast quantity of candidate biomarkers studied, only a small number of cancer biomarkers per year have been approved for use by the FDA in the USA and the European Medicine Agency (EMA).[5]

Development and validation of biomarkers is as difficult as the development and approval of a new drug; indeed, approximately 30–50% of biomarkers are coupled to drug development programmes and only 3–5% reach the clinic.[6, 7, 8, 9] When co-developing a drug and a biomarker, it is relevant to use a ‹fit-for-purpose› approach to biomarker method validation (Box 1), in which methods can be refined throughout the development phases of the experimental agent.[10, 11, 12] This key conceptual methodology enables the developers to focus on the specific requirements of biomarker method validation in a timely manner, depending on the purpose of the biomarker (predictive versus pharmacodynamic, for instance), the type of bioanalytical method and purpose of the clinical trial, and the information that needs to be collected as part of the drug-development process. In light of the high drop-out rate in biomarker development, a fit-for-purpose strategy for method validation might also be economically relevant to progress biomarker assays and achieve regulatory approval. The importance of robust methodology is heightened by the fact that new drugs often display modest benefits and that many potential biomarkers—such as gene-copy number and gene and/or protein expression—are continuous variables, the application of which relies heavily on interpretation of data, with the risk of subjectivity, to establish thresholds.[13]Hence, robust and validated biomarker cut-points that can accurately quantify drug benefits, stratify patient populations, and predict patient responses to treatment are required. A key goal is, therefore, to classify cancers not only according to their molecular profiles (such as mutational status), but also, more importantly, based on their response to therapies (that is, according to individual biomarkers or composite clinical, radiological, and/or biological biomarkers that clearly define the beneficial therapeutic windows of a treatment). A combination of cancer molecular and pharmacological profiles is likely to be the most-successful strategy for guiding therapeutic interventions.

## Box 1: Fit-For-Purpose Biomarker Method Validation: An Overview

Adopting a fit-for-purpose approach to method validation for biomarker assays relies on acknowledgement of the fact that evaluation of the technical performance of an assay should reflect the intended purpose of the biomarker and the nature of the bioanalytical methods used to generate data. In the development process of biomarkers, the intended purpose for method validation will be intertwined with the development phases of a potential drug. As pharmacodynamic, monitoring, prognostic, predictive or surrogate biomarkers have different intended use, it follows that the stringency of the assay will depend on the intended use of the biomarker, and will increase with each developmental phase, from discovery to validation for the intended purpose. Stringency in biomarker assay validation also needs to integrate the nature of the bioanalytical methods—whether quantitative or qualitative. As proposed by Lee and colleagues,[12, 185] bioanalytical methods can be divided in four categories: definitive quantitative assays, such as the use of mass spectrometry assays to quantify circulating insulin concentrations; relative quantitative assays, such as enzyme-linked immunosorbent assays (ELISAs); quasi-quantitative assays, such as real-time quantitative reverse-transcription PCR (qRT-PCR)-based assays; and qualitative assays, which include most immunohistochemistry assays.

According to the position paper by Lee and colleagues,[12] the fit-for-purpose biomarker assay validation can be separated in four continuous iterative activities:

- The pre-validation process that defines the intended purpose of the biomarker, considering pre-analytical variables and bioanalytical method feasibility
- The exploratory validation process that assesses the basic assay performance
- The advanced validation process that characterizes the formal performance of the assay with regard to its intended use
- The in-study validation process that ensures that the assay method performs robustly across studies according to predefined specifications and facilitates the establishment of definitive acceptance criteria

Recognizing the above challenges and the absence of robust standards for evaluation and adoption of biomarkers, an ongoing trend—involving academia, professional organizations, and industry—has been to improve standardization of procedures for biomarker development in oncology.[14, 15, 16] Although these joint efforts have yielded some technological improvements in terms of specimen acquisition and processing, assay automation, production of qualified reagents, and standardization of laboratory procedures, much work remains to be done to achieve universal and robust methodologies. In particular, although the quality and consistency of technological assays have improved, less progress has been made in ensuring the quality of biospecimens and harmonization of tissue collection, processing and storage procedures, attributable largely to the long-standing success of formalin-fixed paraffin-embedded tissue analysis as the standard in diagnostic pathology. Although continued technological advancement would be beneficial, further effort should be made to standardize methodologies as well as quality control and quality assurance procedures, and to rigorously apply such standards in clinical practice.

Herein, we discuss different technical and logistical challenges that must be addressed in the process of standardization of biomarker measurements. Recognizing these challenges, we also outline key considerations for validation of pre-analytical, analytical, and post-analytical processes in biomarker assays. Recommendations for optimizing biomarker evaluation are provided.

## CHALLENGES FOR STANDARDIZATION

Continuing progress in the widespread implementation of valid biomarkers into clinical trials and clinical practice as elements of the development of effective targeted therapies presents biological, clinical, and logistical challenges. The challenges in each of these areas are discussed in the following sections.

### Biological Challenges

A major challenge in biomarker development is the inherent biological complexity underlying tumour response to treatments (treatment sensitivity, and primary and/or acquired treatment resistance). A complex network of multiple interacting molecular pathways, with adaptive feedback and cross-talk loops, clearly hinder the ability of a single biomarker to capture responses of the system as a whole. Thus, to improve upon the limited predictive power of individual biomarker candidates, a panel of multiple markers will generally be required to generate more-sensitive and more-specific composite biomarkers for characterizing system functions, and predicting treatment responses and outcomes. Further biological obstacles are the multiple diverse functions of potential drug targets, as well as the various mechanisms of action and biological effects of individual treatments (cytostatic or cytotoxic), each of which necessitate the development of evidence-based and disease-tailored biomarkers. In addition, intratumoural heterogeneity—characterized by both genetic diversity of tumour cells and the heterotypic matrix comprised of tumours cells, nontumour cells of different types, and the extracellular matrix—represents a universal feature of solid tumours that must be factored into analyses in the search for robust predictive biomarkers. The challenge posed to biomarker standardization by intratumoural heterogeneity is emphasized by differences in their expression between primary and metastatic tumours. Several studies have performed comparative analyses of gene expression and mutation status of key biomarker oncogenes—*HER2*, *KRAS*, and *BRAF*—between primary and metastatic sites (Tables 1 and 2).[17, 18, 19, 20, 21, 22, 23, 24, 25, 26, 27, 28, 29, 30, 31, 32, 33, 34, 35, 36, 37, 38, 39, 40, 41, 42, 43, 44, 45, 46, 47, 48, 49, 50, 51, 52, 53, 54, 55, 56, 57, 58, 59, 60, 61, 62, 63, 64, 65, 66, 67, 68, 69, 70, 71, 72, 73, 74, 75, 76] Overall, the clinical concordance between expression of these genes in primary tumours and disseminated tumour cells ranged between 53% and 100%; however, most studies demonstrate concordance rates of 85–100% (55 out 75 concordance evaluations; Tables 1 and 2). We identified a series of 35 evaluations that compared HER2 expression by immunohistochemistry, *HER2* amplification using fluorescence *in situ* hybridization (FISH), or both, between primary tumours and metastases or recurrence lesions; among these evaluations, the overall concordance estimation was around 87% (Table 1). Given that HER2 positivity occurs in roughly 25% of primary breast tumours (23.2% based on the data in Table 1), however, concordance in the population of patients with HER2-positive primary tumours might be reduced due to the predominance of HER2-negative tumours with HER2-negative disseminated disease. For example, considering the pooled data of Curigliano et al.[23] and Dieci et al.,[24] we calculated that HER2 was expressed in 25.8% of the primary tumours, and 31.6% of the combined primary and disseminated tumour samples. The overall concordance in the whole population across these two studies was 86.9%, whereas the concordance of HER2-positive metastases in patients with HER2-positive primary tumours was 72.0%, and this concordance was further decreased to 58.7% when both HER2-positive primary and disseminated tumours were considered. Furthermore, two studies in the series that studied only patients with HER2-positive primary tumours displayed a calculated overall concordance of 78.6% (based on a 76.4% concordance in one study of 182 patients,[38] and 84.9% concordance in the second study of 66 patients[47]). Overexpression of HER2 assessed by immunohistochemistry had a 100% concordance rate in only one study.[41] The difference in concordance between FISH-detected *HER2* amplification and HER2 overexpression by immunohistochemistry is further elaborated on in the 'Analytical standardization' section of this manuscript. Similarly, in a series of 40 studies comparing the mutation status of *KRAS* or *BRAF* between primary tumours and metastases, the overall concordance reached 93% (Table 2). Sequencing was the most-frequently applied method of biomarker analysis, and even though the overall degree of the agreement was high (66–100%), only four out of the 40 studies we reviewed reported 100% concordance (Table 2). Details on the techniques used for detection of *KRAS* or *BRAF* mutations are beyond the scope of the Review and have been published elsewhere;[77, 78, 79, 80] however, the methodologies used in determining any fraction of a biomarker alteration among studies—specific technologies or cut-off points, for instance—are critical for a pertinent biomarker evaluation and, therefore, are discussed further herein.

**Table 1**: HER2 status* **concordance in matched primary** tumours and metastases from patients with breast cancer

| Study (year of publication) | Method of biomarker assessment | Number of patients | | | Timing of metastasis | Concordance rate‡(%) |
|---|---|---|---|---|---|---|
| | | $n_{total}$ | $n_{primary}$(%) | $n_{disseminated}$(%) | | |
| Aktas et al. (2011)[17] | IHC | 86 | 7 (8) | 27§ (32) | MC | 79‖ |
| Amir et al. (2012)[18] | FISH | 83 | 10 (12) | 14 (17) | MC | 90 |
| Aoyama et al. (2010)[19] | FISH | 60 | 18 (30) | 15¶ (25) | SC | 92 |
| Botteri et al. (2012)[20] | IHC and FISH | 60 | 17 (28) | 25 (42) | MC | 87 |
| Cardoso et al. (2001)[21] | IHC | 334 | 36 (11) | 40¶ (12) | SC | 98 |
| Chang et al. (2011)[22] | IHC and FISH | 56 | 15 (27) | 18 (32) | SC and MC | 88 |
| Curigliano et al. (2011)[23] | IHC and FISH | 172 | 54 (31) | 44 (26) | MC | 86 |
| Dieci et al. (2013)[24] | IHC and FISH | 119 | 21 (18) | 27 (23) | SC and MC | 88 |
| Duchnowska et al. (2012)[25] | IHC and FISH | 119 | 58 (49) | 61 (51) | SC and MC | 86 |
| Fabi et al. (2011)[26] | IHC and FISH | 137 | 25 (18) | 36 (26) | MC | 90 |
| Fehm et al. (2009)[27] | RT-PCR | 58 | 9 (16) | 22§ (38) | MC | 53 |
| Fuchs et al. (2006)[28] | IHC and FISH | 48 | 8 (17) | 6¶ (13) | MC | 79 |
| Gancberg et al. (2002)[29] | IHC | 93 | 13 (14) | 19¶ (20) | MC | 94‖ |
| Gancberg et al. (2002)[29] | FISH | 68 | 16 (24) | 17¶ (25) | MC | 93 |
| Gong et al. (2005)[30] | FISH | 60 | 20 (33) | 18¶ (30) | SC and MC | 97‖ |
| Guarneri et al. (2008)[31] | IHC and FISH | 75 | 14 (19) | 22 (29) | MC | 84 |
| Jensen et al. (2012)[32] | IHC and FISH | 114 | 10 (9) | 16 (14) | SC and MC | 91 |

| Lear-Kaul et al. (2003)[33] | IHC and FISH | 12 | 4 (33) | 5[#] **(42)** | MC | 92 |
|---|---|---|---|---|---|---|
| Lindström et al. (2012)[34] | IHC and FISH | 104 | 29 (28) | 26[#] **(25)** | MC | 86 |
| Lower et al. (2009)[35] | IHC | 382 | 140 (37) | 87[¶] **(23)** | MC | 66 |
| Macfarlane et al. (2012)[36] | IHC and FISH | 154 | 29 (19) | 25[¶] **(16)** | MC | 95 |
| Montagna et al. (2012)[37] | IHC | 174 | 51 (29) | 52[#] **(30)** | MC | 96 |
| Niikura et al. (2012)[38] | IHC and FISH | 182[**] | 182 (100) | 139 (76) | SC and MC | 76 |
| Regitnig et al. (2004)[39] | IHC | 31 | 3 (10) | 6[#] **(19)** | MC | 77 |
| Regitnig et al. (2004)[39] | FISH | 18 | 2 (11) | 6[#] **(33)** | MC | 78 |
| Santinelli et al. (2008)[40] | IHC and FISH | 54 | 16 (30) | 13[‡‡] **(24)** | SC and MC | 94[∥] |
| Santinelli et al. (2008)[40] | IHC and FISH | 65 | 12 (18) | 17[#] **(26)** | SC and MC | 78 |
| Shimizu et al. (2000)[41] | IHC | 21 | 8 (38) | 8[¶] **(38)** | MC | 100[∥] |
| Simon et al. (2001)[42] | IHC and FISH | 125 | 31 (25) | 24[¶] **(19)** | SC | 95[∥] |
| Simmons et al. (2009)[43] | FISH | 25 | 4 (16) | 6 (24) | MC | 92 |
| Thompson et al. (2010)[44] | IHC and FISH | 137 | 14 (10) | 16 (12) | MC | 97 |
| Vincent-Salomon et al. (2002)[45] | IHC | 44 | 11 (25) | 9[¶] **(20)** | SC | 95[∥] |
| Wilking et al. (2011)[46] | IHC and FISH | 151 | 43 (28) | 41 (27) | SC and MC | 90 |
| Xiao et al. (2011)[47] | IHC and FISH | 66[**] | 66 (100) | 56[¶] **(85)** | MC | 85 |
| Zidan et al. (2005)[48] | IHC | 58 | 14 (24) | 20[¶] **(34)** | MC | 86 |
| All | NA | NA | NA | NA | NA | 87[§§] |

*Overexpression or amplification.

‡As calculated by the authors of this manuscript or within the original publication, unless otherwise noted.

§Circulating tumour cells.

∥(Over)-estimated using the calculation $1-[(n_{primary}-n_{disseminated})/n_{total}]$.

¶Lymph-node or distant metastases.

#Recurrence or distant metastases.

**HER2-positive population only.

‡‡Lymph-node involvement only.

§§Calculated using the equation [Σ(ntotal × Concordance rate)]/Σntotal.

Abbreviations: FISH, fluorescence in situ hybridization; IHC, immunohistochemistry; MC, metachronous; NA, not applicable; ndisseminated, number of patients with HER2-positive disseminated cells; nprimary, number of patients with a HER2-positive primary tumour; ntotal, number of patients in the study; RT-PCR, reverse-transcription PCR; SC, synchronous.

**Table 2:** *KRAS* and *BRAF* mutation status concordance in matched primary tumours and metastases in CRC

| Study (year of publication) | Method of biomarker (mutation status) assessment | Number of patients | | | Timing of metastasis | Concordance rate* (%) |
|---|---|---|---|---|---|---|
| | | $n_{total}$ | $n_{primary}$(%) | $n_{disseminated}$(%) | | |
| **KRAS** | | | | | | |
| Albanese et al. (2004)[49] | SSCP | 30 | 14 (47) | 13‡ **(43)** | SC and MC | 70 |
| Al-Mulla et al. (1998)[50] | ASO/Seq | 47 | NR | NR‡ | NR | 83 |
| Artale et al. (2008)[51] | NR | 48 | 11 (23) | 12‡ **(25)** | SC and MC | 94 |
| Baldus et al. (2010)[52] | Seq/PyroSeq | 55 | 29 (53) | 16§ **(29)** | NR | 69 |
| Baldus et al. (2010)[52] | Seq/PyroSeq | 20 | 9 (45) | 9‖ **(45)** | NR | 90 |
| Cejas et al. (2009)[53] | Seq | 110 | 37 (34) | 40‡ **(36)** | SC and MC | 94 |
| Cejas et al. (2012)[54] | Seq | 117 | 47 (40) | NR | NR | 91 |
| Etienne-Grimaldi et al. (2008)[55] | PCR-RFLP | 48 | 16 (33) | 16‡ **(33)** | NR | 100 |
| Finkelstein et al. (1993)[56] | Seq | 23 | 12 (52) | 12§ **(52)** | NR | 100 |
| Garm Spindler et al. (2009)[57] | Seq/qPCR | 31 | 11 (35) | 9‡ **(29)** | NR | 94 |
| Italiano et al. (2010)[58] | Seq | 62 | 24 (39) | 25‡ **(40)** | SC and MC | 95 |
| Knijn et al. (2011)[59] | Seq | 305 | 108 (35) | 104‡ **(34)** | SC and MC | 96 |
| Losi et al. (1992)[60] | AS-PCR | 35 | 25 (71) | 25‖ **(71)** | MC | 100 |
| Loupakis et al. (2009)[61] | Seq | 43 | NR | NR‡ | NR | 95 |
| Mariani et al. (2010)[62] | Seq/ARMS | 38 | 20 (53) | 19‡ **(50)** | SC and MC | 97 |
| Melucci et al. (2010)[63] | Seq | 62 | NR | NR | NR | 94 |

| Study | Method | N | Col (%) | Met (%) | Site | % |
|---|---|---|---|---|---|---|
| Molinari et al. (2009)[64] | Seq | 37 | 16 (43) | 15‡ (41) | SC and MC | 92 |
| Mostert et al. (2013)[65] | AS-PCR/Seq | 43 | 9 (21) | 10∥ (23) | NR | 79 |
| Mostert et al. (2013)[65] | AS-PCR/Seq | 42 | 9 (21) | 5 (12)¶ | NR | 71 |
| Oliveira et al. (2007)[66] | SSCP/Seq | 28 | 18 (64) | 23§ (82) | NR | 68 |
| Oltedal et al. (2011)[67] | PNA-PCR | 91# | 0 (0) | 7§ (8) | NR | 92** |
| Oudejans et al. (1991)[68] | ASO | 31 | 15 (48) | 17‡ (55) | NR | 87 |
| Park et al. (2011)[69] | NR | 69 | 19 (28) | NR | NR | 76 |
| Perrone et al. (2009)[70] | Seq | 29‡‡ | 4/22 (18) | 4/17‡ (24) | SC and MC | 80 |
| Santini et al. (2008)[71] | Seq | 99 | 38 (38) | 36‡ (36) | SC and MC | 96 |
| Schimanski et al. (1999)[72] | PCR-RFLP/Seq | 32 | 14 (44) | 7‡ (22) | NR | 78** |
| Thebo et al. (2000)[73] | AS-PCR | 20§§ | 20 (100) | 16§ (80) | SC | 80 |
| Watanabe et al. (2011)[74] | AS-PCR/Seq | 43 | 15 (35) | 18‡ (42) | SC and MC | 88 |
| Zauber et al. (2003)[75] | SSCP | 42 | 22 (52) | 22‡ (52) | SC | 100 |
| All | NA | NA | NA | NA | NA | 90∥∥ |
| **BRAF** | | | | | | |
| Artale et al. (2008)[51] | NR | 48 | 2 (4) | 1‡ (2) | SC and MC | 98 |
| Baldus et al. (2010)[52] | Seq/PyroSeq | 55 | 5 (9) | 3§ (5) | NR | 96 |
| Baldus et al. (2010)[52] | Seq/PyroSeq | 20 | 1 (5) | 1∥ (5) | NR | 100 |
| Cejas et al. (2012)[54] | NR | 70# | 1 (1.4) | 1‡ (1.4) | NR | 100 |
| Italiano et al. (2010)[58] | Seq | 57 | 1 (2) | 3‡ (5) | SC and MC | 98 |
| Mostert et al. (2013)[65] | AS-PCR/Seq | 43 | 3 (7) | 4∥ (9) | NR | 93 |
| Mostert et al. (2013)[65] | AS-PCR/Seq | 40 | 2 (5) | 1 (3)¶ | NR | 98 |
| Oliveira et al. (2007)[66] | SSCP/Seq | 28¶¶ | 7 (25) | 10§ (36) | NR | 89 |
| Park et al. (2011)[69] | NR | 71 | 5 (7) | NR | NR | 90 |
| Perrone et al. (2009)[70] | Seq | 29 | 2 (7) | 1‡ (3) | SC and MC | 91 |
| Santini et al. (2010)[76] | NR | 208# | 13 (6) | 9‡ (4) | SC and MC | 97 |
| All | NA | NA | NA | NA | NA | 96∥∥ |

*As calculated by the authors of this manuscript or within the original publication, unless otherwise noted.

‡Lymph-node or distant metastases.

§Lymph node.

∥Recurrence or distant metastases.

¶Circulating tumour cells studied.

#Wild-type KRAS population only.

**(Over)-estimated using the calculation $1-[(\text{nprimary}-\text{ndisseminated})/\text{ntotal}]$.

‡‡Although 29 patients were included in this study, KRAS-mutation status was know for only 22 of the primary tumours and 17 of the disseminated tumours examined in this study.

§§Mutant KRAS population only.

||||Calculated using the equation $[\Sigma(\text{ntotal} \times \text{Concordance rate})]/\mathbf{\Sigma ntotal}$.

¶¶Microsatellite-stable population only.

Abbreviations: ARMS, amplication-refractory mutation system; ASO, allele-specific oligonucleotide hybridization; AS-PCR, allele-specific based polymerase chain reaction; MC, metachronous; NA, not applicable; ndisseminated, number of patients with KRAS/BRAF-mutated disseminated cells;nprimary, number of patients with a KRAS/BRAF-mutated primary tumour; NR, not reported; ntotal, number of patients in the study; PCR-RFLP, restriction fragment length polymorphism PCR; PNA-PCR, peptide nucleic acid clamp PCR; PyroSeq, pyrosequencing; qPCR, quantitative PCR; SC, synchronous; Seq, Sanger-based sequencing using various amplification methods; SSCP, single-strand conformation polymorphism.

Among recognized positive or negative predictive biomarkers, genetic alterations such as mutations, amplifications, or translocations seemed to be more concordant between primary tumours and associated metastases than protein or gene-expression levels or signatures.[61] This observation probably reflects the introduction of increased analytical variation in gene-expression methodologies and the complexity of protein biochemistry, including post-translational modifications and catabolism. However, whereas high concordance occurs for many recognized genetic biomarkers (including recurrent *TP53* mutation), such is not the case for many genetic modifications that are nonrecurrent and probably represent passenger alterations.[81] Given these findings, the relevance of primary resection specimens to evaluate biomarkers when planning treatment in the metastatic setting has been questioned. In fact, during the course of the disease, a number of factors could potentially influence biomarker concordance (Table 3), and might, therefore, affect biomarker evaluation and challenge therapeutic decisions. It has long been recognized that biomarker status can be discordant due to inherent intratumoural and intertumoural heterogeneity, clonal evolution during tumour progression due to genomic instability, or treatment effects that result in elimination of sensitive tumour cells and/or adaptation of tumour cells in response to therapeutic agents. As a result, the discrepancy between the first tumour evaluation (typically based on resected primary tumour or core needle-biopsy specimens) and assessments of subsequent samples, either from the same site or distant metastatic sites, might be evident and could have been introduced by intervening treatment.[82, 83, 84] Pre-analytical and analytical factors, such as the sensitivity of the laboratory method used for biomarker evaluation, might also be involved in such discrepancies, and can have important clinical implications. For example, detection of rare *KRAS/NRAS*-mutant clones that will ultimately become predominant can predict eventual resistance of cancerous lesions to EGFR-targeted monoclonal antibody therapies.

Table 3: Factors that might affect biomarker concordance during the course of disease

| Causal factors | Examples of the effect of the causal factor on biomarkers concordance | References |
|---|---|---|
| **Clinical and biological** | | |
| Biomarker type: genetic vs protein | Difference in detection frequency of the *EML4–ALK* **gene rearrangements in NSCLC specimen by IHC, FISH and RT-PCR** Difference in detection frequency of *EGFR* **gene mutation in NSCLC specimen by IHC, direct sequencing and** qPCR | Teixidó et al. (2014)[130] Angulo et al. (2012)[131] |
| Biomarker type: radiological vs biological | Difference in evaluation of disease response between AFP level monitoring and RECIST criteria in hepatocellular carcinoma | Personeni et al. (2012)[132] |
| Biological rhythms | Modification in expression of ERGs during the menstrual cycle in ER⁺ **breast cancer** | Haynes et al. (2013)[133] |
| Prior neoadjuvant therapy | Difference between post-treatment and pre-treatment Ki-67 in breast cancer | Von Minckwitz et al. (2013)[134] |
| Prior adjuvant therapy | Change in ER, PR and/or HER2 status between primary and relapsed tumours in breast cancer | Lindström et al. (2012)[34] |
| Prior interval therapy | Increased incidence of PTEN loss and *PI3K* **mutation after anti-HER2 therapy in breast cancer** | Chandarlapaty et al. (2012)[135] |
| **Logistical and technical** | | |
| Tissue origin: distant vs lymph-node metastases | Differences in HER2-evaluation method (IHC and FISH) among primary tumours, lymph-node metastases, and distant metastases in breast cancer | Regitnig et al. (2004)[39] |
| Sampling origin: surgical specimen vs CNB | Discordance in grade, and ER, PR and HER2 status in breast cancer when comparing surgical specimen and CNB | Lorgis et al. (2011)[136] Arnedos et al. (2009)[137] |
| Sampling origin: CNB vs FNA | Variation relating to the use of IHC vs ICC for analysis of ER, PR and Ki-67 status in breast cancer | Stalhammar et al. (2014)[138] |
| Tissue and antigen preservation: specimen fixation and conservation | Pre-analytical variables for IHC or FISH analysis of FFPE specimens | Engel et al. (2011)[139] Khoury et al. (2009)[140] |

| Tissue and antigen preservation: FFPE vs frozen or fresh | Introduction of mutation artefacts when starting with an old or low abundance DNA sample (demonstrated during assessment of *EGFR* **mutations in colon cancer**) | Marchetti *et al.* (2006)[141] |
|---|---|---|
| Analytical method: specificity and sensitivity | More than 100% circulating-tumour-cell recovery in spike-and-recovery control experiments | Punnoose *et al.* (2010)[142] |
| Scoring method | Change in Ki-67 evaluation in breast cancer Improve classification of patients likely to benefit from sorafenib using Choi criteria instead of RECIST criteria | Voros *et al.* (2013)[143] Ronot *et al.* (2014)[144] |
| Laboratory experience: central or reference vs local | *KRAS, BRAF, NRAS, PI3KCA* **mutation assessment in** *KRAS* **wild-type colorectal cancer population** HER2 testing (FISH or IHC) in the N9831 breast cancer adjuvant trial | André *et al.* (2013)[126] Perez *et al.* (2006)[117] |

Abbreviations: CNB, core needle biopsy; ER, oestrogen receptor; ERG, oestrogen-regulated genes; FFPE, formalin-fixed paraffin-embedded; FNA, fine-needle aspiration; ICC, immunocytochemistry; IHC, immunohistochemistry; NSCLC, non-small-cell lung cancer; PR, progesterone receptor; qPCR, quantitative PCR; RECIST, Response Evaluation Criteria in Solid Tumours; RT-PCR, reverse-transcription PCR; vs, versus.

Clearly, both laboratory techniques as well as biomarker heterogeneity (with regard to expression and types of mutation), must be considered when incorporating biomarkers, especially predictive biomarkers, into future clinical trials or routine patient care. Detection of selected mutations in circulating cell-free DNA (cfDNA) by BEAMing (beads, emulsion, amplification, magnetics) or droplet digital PCR represents an interesting potential strategy for future therapeutic decision-making.[85] At present, treatment decisions in the metastatic setting are based on analysis of earlier primary tumour samples, or data from emerging techniques such as evaluation of cfDNA or circulating tumour cells (CTCs) for select mutations. Additional studies should be performed to establish whether any concordance between analyses of primary tumours, cfDNA and metastases are clinically relevant for predicting treatment outcome. However, ethical considerations (potential complications and inconvenience to the patient, for example) and costs must be recognized, and will necessarily limit these types of investigations.

## Clinical Challenges

A validated predictive biomarker can identify patients who are likely to have a favourable clinical outcome—that is, the population with a high response rate or improved survival—after treatment with a specific therapy, hence differentiating responders from nonresponders. The low objective response rate for many emerging therapeutic agents and lack of survival benefit with some targeted therapies represent challenges to the validation of biomarkers that could inform treatment decisions. As a result, the predictive biomarkers currently available are validated for only a small percentage of patients with solid tumours (Table 4). A substantial hurdle for biomarker discovery is that agents produced by different pharmaceutical and biotechnology companies are often used in combination regimens, rather than as stand-alone treatments, in order to enhance therapeutic efficacy. Combination therapy obscures the association between any one agent used in the treatment regimen and the biomarkers under consideration. Combination therapies also raise questions about data sharing, collaborations, intellectual property of the integral use of biomarkers (including biomarker analysis methodologies) developed by different companies, and the approaches to validating such biomarkers in clinical trials.

**Table 4:** FDA-approved targeted agents with demonstrated activity and an effective predictive biomarker of efficacy in solid cancers*

| Year of approval | Drug | Clinical biomarker(s) | Target(s) | FDA-approved indication(s) | Patient population positive for biomarker | RR to treatment |
|---|---|---|---|---|---|---|
| 1998 | Trastuzumab | HER2 overexpression | HER2 | HER2-positive mBC: single agent in second-line therapy, and in combination with paclitaxel in first-line treatment | 18–20% (HER2-positive population) | 15–50%[145,146] |
| 2003 | Imatinib | KIT (CD117) | KIT, ABL and PDGFR | In unresectable and/or KIT-positive mGIST | CD117-positive: 95% *KIT*-mutation-positive: 80% | 45–83%[147,148] |
| 2004 | Cetuximab | EGFR-protein expression‡ | EGFR | With irinotecan or as single agent (2007) for EGFR-positive mCRC refractory to irinotecan | 60–80% | 11–55%[149,150] |
| 2006 | Trastuzumab | HER2 overexpression | HER2 | With adjuvant treatment for node-positive, HER2-positive BC | 18–20% (HER2-positive population) | 38% DFS increase[145,151] |

| 2006 | Panitu-mumab | Wild-type§ KRAS(specifically at codons 12 or 13 in exon 2) | EGFR | EGFR-expressing mCRC with disease progression on chemotherapy regimens | 40–60% | 17–58%[92,152] |
|---|---|---|---|---|---|---|
| 2007 | Lapa-tinib | HER2 overexpression | HER2; EGFR | In combination with capecitabine in pretreated HER2-positive mBC | 18–20% (HER2-positive population) | 24–41%[153,154] |
| 2008 | Imatinib | COL1A1–PDGFB-fusion | KIT, ABL and PDGFR | For COL1A1–PDGFB gene-fusion-negative metastatic DFSP (or DFSP with unknown mutation status), and as adjuvant therapy in KIT-positive GIST | >95% | 36–100%[155,156] |
| 2009 | Gefitinib | EGFR-activating mutations | EGFR | NSCLC with EGFR mutations that respond to or had prior response to gefitinib (limited approval by FDA) | 10–15% of white patients and 30–35% of East Asian patients | 37–78%[157,158] |
| 2010 | Lapa-tinib | HER2 overexpression | HER2; EGFR | With letrozole in postmenopausal women with hormone-receptor-positive and HER2-positive mBC | 18–20% (HER2-positive population) | 8–48%[159,160] |
| 2010 | Trastu-zumab | HER2 overexpression | HER2 | With cisplatin and fluoropyrimidine in the first-line treatment of HER2-positive metastatic GC and GEC | 7–34% | 47%[161] |
| 2011 | Crizo-tinib | EML4–ALKtranslocation | ALK; MET | ALK-positive locally advanced or metastatic NSCLC | 1–7% | 50–65%[162,163] |

| 2011 | Vemurafenib | BRAF V600E mutation | BRAF | Metastatic melanoma with BRAF V600E mutation | 80–90% of *BRAF*-mutated population | 48–57%[164,165] |
|---|---|---|---|---|---|---|
| 2012 | Cetuximab | Wild-type§ *KRAS* | EGFR | In combination with FOLFIRI for the first-line treatment of *KRAS*-wild-type patients with EGFR-positive mCRC | 40–60% | 47–61%[166,167] |
| 2012 | Pertuzumab | *HER2* amplification | HER2 | In combination with trastuzumab and docetaxel as first-line therapy for HER2-positive mBC | 18–20% (HER2-positive population) | 24–63%[168,169] |
| 2013 | Ado-trastuzumab emtansine | HER2 overexpression | HER2 | HER2-positive mBC with prior exposure to trastuzumab and/or a taxane | 18–20% (HER2-positive population) | 26–64%[170,171] |
| 2013 | Afatinib | *EGFR* exon 19 deletions or exon 21 mutation (L858R) | EGFR, HER2 and HER4 | First-line treatment of metastatic NSCLC with *EGFR* exon 19 deletions or exon 21 mutations | 45% with *EGFR* exon 19 deletion and 41% with *EGFR* exon 21 mutation | 56–67%[172,173] |
| 2013 | Ceritinib | *ALK* rearrangement | ALK | ALK-positive NSCLC that progressed during or after treatment with crizotinib | 2–5% | 56%[174,175] |
| 2013 | Erlotinib | *EGFR* exon 19 deletion or exon 21 mutation (L858R) | EGFR | First-line treatment of metastatic NSCLC with *EGFR* exon 19 deletions or exon 21 mutations | 45% with *EGFR* exon 19 deletion and 41% with *EGFR* exon 21 mutation | 54–83%[176,177] |
| 2013 | Pertuzumab | *HER2* amplification | HER2 | As neoadjuvant treatment with trastuzumab and docetaxel for HER2-positive advanced, inflammatory or early-stage BC | 18–20% (HER2-positive population) | 24–62%[178,179] |

| 2013 | Trametinib | BRAF V600E/K mutations | MEK | Unresectable/ metastatic $BRAF^{V600E/K}$-mutated melanoma | $BRAF^{V600E}$-mutated: 80–90%; $BRAF^{V600K}$-mutated: 20% | 22–25%[180,181] |
|------|------------|------------------------|-----|-----------------------------------------------------------|------------------------------------------------------------|-----------------|
| 2014 | Dabrafenib | BRAF V600E/K mutations | BRAF | With trametinib for metastatic melanoma with BRAF V600E/K mutations | $BRAF^{V600E}$-mutated: 80–90%; $BRAF^{V600K}$-mutated: 20% | 31–76%[180,182,183] |

*Data taken from the FDA website95 on 15th June 2014 and completed using EPAR from the EMA product information.184

‡EGFR expression was not confirmed as a predictive biomarker in mCRC.

§EMA restricted panitumumab and cetuximab therapy to KRAS and NRAS wild-type mCRC in 2013.

Abbreviations: BC, breast cancer; DFS, disease-free survival; DFSP, dermatofibrosarcoma protuberans; EMA, European Medicine Agency; EPAR, European public assessments reports; FOLFIRI, 5-fluorouracil, folinic acid and irinotecan; GC, gastric cancer; GEC, gastroesophageal cancer; mBC, metastatic breast cancer; mCRC, metastatic colorectal cancer; mGIST, metastatic gastrointestinal stromal tumours; NSCLC, non-small-cell lung cancer; RR, response rate.

Given that biomarker development is moving oncology toward personalized medicine, the future progress in drug and biomarker research lies in the choice of ideal populations that might benefit from a particular treatment. However, population stratification in clinical trials narrows the landscape of drug development and, as such, the potential market share for the drug. In fact, the clinical integration of some cancer medications on the market benefited from retrospective biomarker analysis to overcome the difficulties encountered during clinical development, such as limited responses in unselected patients owing to inherent drug resistance. For example, such studies in mCRC identified genetic aberrations that predict outcome of treatment with the anti-EGFR antibodies cetuximab and panitumumab.[86,87,88,89,90,91,92] The reduction in the potential market share must be compensated by acceleration of validation and reduction in the cost of drug development. However, the use of selected clinical trial populations raises a challenge for the validation of the biomarker assay itself, as comparison of the outcomes of a potentially biomarker-guided treatment between the biomarker-positive and biomarker-negative populations is ultimately required to assess assay performance. Therefore, although the development of biomarker-based diagnostics is recognized as an important paradigm, technical and economic considerations relating to the standardization of biomarker evaluation and validation must be taken into account.

## Logistical Challenges

Several ongoing logistical hurdles are linked to integration of biomarkers into clinical trials and the practice of oncology. These include the need for well-managed, centralized specimen biobanks for high-quality biomarker studies and standardization of sample collection, processing, and storage among the facilities, as these factors are critical determinants of the reliability of biomarker analysis. It is clear that statistical analysis of biomarker data is also an important logistical component of the validation process; statistical evaluation is challenging in terms of achieving uniformity in data management, bioinformatics, and biostatistics methodologies.[93] Optimizing outcomes assessment requires multidisciplinary effort and fit-for-purpose statistical methods that rely on a synergy between statistics and biological understanding. Given the inherent methodological challenges of conducting prospective studies to confirm the validity of predictive biomarkers, well-designed retrospective studies, using existing well-characterized samples, can be of great value: such studies can be used to accumulate evidence of biomarker effectiveness more rapidly—albeit lower-level evidence than is provided by prospective studies—and, therefore, support the transfer of candidate biomarkers into clinical practice. However, to yield convincing evidence, so-called retrospective–prospective study designs must be pre-planned (including cut-points and statistical methods) and conducted with reference to standardized guidelines. Moreover, translation of biomarkers from the research laboratory into the real-world setting without loss of analytical performance and standardization is often time-consuming and difficult, as the sensitivity and specificity of biomarkers developed in the research laboratory have to be feasible in the clinical laboratory with regulatory compliance, and meaningful for decision-making in order to guide patient care.[94] Unfortunately, many promising biomarkers fail to meet these requirements and are never used outside of limited applications, such as proof-of-concept testing. Thus, it is crucial that the biomarker validation process is performed in settings mirroring closely the clinical environment.

## CONSIDERATIONS FOR METHOD VALIDATION

The global issue for biomarker development is the robustness of the laboratory methodology in all analytical aspects, including assay precision, accuracy, sensitivity, specificity, reproducibility, linearity, reliability, and generalizability. Unfortunately, highly standardized assays for biomarker identification and analysis are rare. In fact, most of the FDA-recognized pharmacogenomic biomarkers[95] are not validated *in vitro* diagnostics (IVDs), but are rather laboratory-developed tests (LDTs). LDTs represent ‹in-house› tests that might be subject to considerable interlaboratory variability despite accreditations such as ISO 15189 in the European Union, or Clinical Laboratory Improvement Amendments (CLIA) and Investigational Device Exemption (IDE) in the USA, which are discussed in more detail in a following section. Of note, substantial differences exist in the requirements for accreditation between Europe and the USA.[96]

Biomarker sensitivity and specificity can be interpreted in terms of analytical or clinical performance. Analytical performance must be optimized for three different aspects of the biomarker validation process: pre-analytical, analytical, and post-analytical (Box 2).

## Box 2: Considerations for Procedure Standardization

### *Pre-Analytical Standardization*

- Patient factors: anaesthetic agents; hydration; stress responses; drugs; concomitant diseases or co-morbidities; tissue ischaemia; sample-processing delays (phosphorylation); and other unknown factors
- Tissue factors: collection (device/process, tissue versus serum based specimen, sample volume, contamination); fixation (type, time, penetration); processing (methods, times for each step, temperature); storage; and stability and integrity

### *Analytical Standardization*

- Tissue factors: analyte differences (DNA, RNA, protein); antigen retrieval (for immunohistochemistry); antibody variability; detection reagents (chromagens); inconsistencies relating to kits and automation; control selection; and quality control
- Scoring systems for staining: intensity; extent; topography; nonlinearity of methodologies; and computerized image analysis ('precise measurement of the imprecise')

### *Post-Analytical Standardization*

- Effects of volume of testing by laboratories: high-volume testing laboratories, such as central laboratories, usually have more expertise and proficiency than low-volume local laboratories

- Data interpretation: dichotomous variables; continuous variables (cut-points relevant to clinical decisions); and reproducibility
- Collaborative role of professional pathology organizations: at the international level, to define standards; at the local level, to facilitate implementation of these standards

## Pre-Analytical Standardization

Pre-analytical processing is generally considered the greatest challenge in the biomarker standardization process. Indeed, several pre-analytical variables influence the effective assessment of biospecimens, the reliability of the analyses, and the final results of the biomarker evaluation that ultimately influence the patient's care and outcome. These variables include patient factors, such as physiological variables and pathological states, as well as 'specimen and sample factors' that relate to the clinical procedures that are used to obtain the biospecimens (the collection and handling processes), including patient identification; sample labelling or mislabelling; volume of usable material; collection, transport and storage conditions; and processing delays (Box 2).

### *Guidelines for Standardization of Samples*

To improve standardization of specimens, the US National Cancer Institute (NCI) has published best-practice guidelines for biospecimen resources,[16] as has the International Society for Biological and Environmental Repositories.[97] These documents provide a comprehensive approach to the procedures for tissue collection, processing, banking, retrieval, analysis, and dissemination, as well as issues of ethics, informed consent, privacy, and intellectual property. These reports are oriented predominantly at research use. In the clinical trial setting, reliance on standard pathological material and collection techniques is usually greater than in other research settings, but if the NCI maxims are adopted and applied with rigour, a more-successful biomarkers programme is likely to emerge. In addition, the College of American Pathologists (CAP) has initiated an Accreditation for Biorepositories Program with a clinical perspective.[98] It should be noted, however, that the logistics and cost for achieving better standardization are likely to be burdensome for many institutions, and in some cases might prove to be prohibitively expensive. For instance, 24-hour pathology laboratories could ensure that analyses are performed routinely at the same time interval after samples are obtained, but meeting the cost of establishing such facilities is unrealistic for most centers, given the current constraints on health-care expenditures and research funding.

## Important Sources of Pre-Analytical Variation

The first step in the pre-analytical standardization of any biomarker assessment is the selection of a meaningful sample that is easy to obtain and optimal for analysis, because sample origin can influence the validation process.[99] In addition, sample and reagent integrity (from sampling through processing), processing conditions, and the elapsed time from sample collection to both processing and analysis can have major impacts on biomarker data. For instance, sample stability is influenced by freezing–thawing, storage duration and temperature, consistency of temperature, and specimen-container types and stabilizers.[100] Time is routinely an influential factor throughout the biospecimen collection and processing period, especially for proteins and peptides that are highly labile and subject to various alterations—in phosphorylation status, for example.[101] By contrast, nucleic acids, in particular DNA, are more stable and, therefore, less sensitive to variation in sample processing times.[102, 103] Tissue fixation parameters might also markedly affect the results of biomarker analyses by changing the molecular profile of the analytes:[104] formalin fixation has been shown to substantially reduce DNA and RNA solubility and induce a high frequency of sequence alterations.[105, 106, 107] Thus, new methodologies have been developed in attempts to avoid the cumbersome sample freezing process and provide appropriately stabilized fixed tissue with unchanged and well-preserved analytes (DNA, mRNA, and proteins).[108, 109, 110, 111]

In addition, patient factors such as the level of hydration, tissue-ischaemia time, stress responses, and concomitant drug and anaesthetic agent effects, as well as heterogeneity of samples that might be composed of normal, tumoural, and/or necrotic tissues, can affect expression of potential biomarkers and their analysis, particularly when samples are obtained during surgery.[112, 113] Another important factor is the analyte volumes available for testing: tissue specimens obtain through small biopsies and fine-needle aspiration can limit the analysis. Currently available methods for amplification of material might have utility in overcoming this limitation, but could introduce analysis artefacts.[90]

## Addressing Pre-Analytical Variables

Many pre-analytical factors, including those pertaining to the patient as well as others such as the time of day at which an operation is scheduled, cannot realistically be controlled. Therefore, it is important that disease-related and patient-related characteristics (demographics, clinical condition before medical intervention, ischaemic time, and treatment-related variables), and the pre-analytical procedures used are annotated as completely as possible to enable their possible influences on assay results to be considered on a patient-by-patient basis during statistical analysis. Once the sample is obtained from the patient, greater potential for standardization exists. Tissue preparation protocols relating to the timing of fixation, the specific type of fixative and its penetration into tissue, as well as sample processing protocols that outline the timing of each step, procedural temperatures, and subsequent microtomy sectioning and slide mounting of fixed specimens have been addressed.[99] However, despite the rigorous application of protocols, lack of cross-institutional uniformity of procedures remains an issue. For instance, in a comparison of protocols for the pathological examination of prostate cancer needle-biopsy specimens from the 11 institutions enrolled in the NCI Specialized Programs of Research Excellence (SPORE), none of the centres used precisely the same protocol.[114] Lack of uniformity of standards between technology platforms for molecular and pathological analysis has also been an issue recognized by the French National Cancer Institute (INCa) centres.[115] The general lack of standardization is due, in part, to a number of technical limitations, such as differences in performance and instrumentation. As an example, all the operational steps required in tissue sample preparation are performed by tissue processing machines, but in adherence with standard operating procedures that are customized locally to account for specific factors, such as the timing of pathological evaluation. Consequently, the procedural steps for which the standard protocols must be re-optimized by laboratories to make them applicable locally can be numerous, leading to substantial variation in the data obtained, which ultimately are not uniform or shareable.

## Analytical Standardization

The applicability of a qualified biomarker relies on the development of a robust, validated assay with high sensitivity, specificity, precision, and accuracy. Many of the techniques that are currently used in the development of biomarkers for patient stratification, such as immunohistochemistry, FISH or silver *in situ* hybridization (SISH), real-time quantitative reverse-transcription PCR (qRT-PCR), microarrays, epigenetic assays, sequencing, and mutation analyses, continue to lack high-level performance and robust evaluation processes. Development of robust and accurate analytical standards is mostly constrained by tissue availability and the complexity of the biological samples containing DNA, RNA and proteins. For instance, conventional immunohistochemistry, the most widely used platform for biomarker assessment in diagnostic surgical pathology, has been faced with several practical limitations when applied to biomarker examination, such as the selection of the ideal antigen, antibody, detection reagents, kits, and positive and negative controls, and difficulties in quantification with reference standards.

## Ensuring Reproducibility and Concordance

An assessment of accuracy and reproducibility of the diagnostic evaluation of HER2 by immunohistochemistry between two Breast Cancer International Research Group (BCIRG) central laboratories and local laboratories showed an overall concordance of 77.5%;[116] however, a concordance rate as low as 51.7% (281 of 543) was observed for HER2-positive (2+ or 3+) immunostaining patterns.[116] Concordance figures were slightly better in the North Central Cancer Treatment Group (NCCTG) N9831 phase III adjuvant trial, with an overall concordance of 82%.[117] Of note, these data were all derived from CLIA-accredited laboratories. An FDA-cleared kit has been shown to yield similar discordance in HER2 positivity between different laboratories.[118] Such findings indicate that attempts to standardize biomarker methods are essential to ensure uniformity and quality of data collected. Given that laboratory-based evaluation approaches vary worldwide, that pre-analytical standardization is difficult to achieve among centres, and that standardized reagents and analytes are unavailable for most assays, more stringency is clearly needed. For example, an integrated network of high-volume clinical laboratories with proven expertise and proficiency should perform biomarker validation, and establish the baseline for the reference standards proposed by the CAP and ASCO for HER2 testing.[119] Although several aspects of biomarker method standardization have been addressed, and standardized kits and automation have resulted in some marked improvements, immunohistochemical qualitative evaluation and many pivotal analytical procedures, such as fixation and antigen retrieval, remain problematic.[13, 15, 120] To evaluate the robustness of such evaluations and to enable the clinical application of biomarkers, reference centres could be established to coordinate the activities across centralized laboratories, as has been done in the UK and in Canada.

The lack of sufficient intraplatform and interplatform studies on the concordance of qualitative and quantitative data has also been of major concern for standardization procedures, recognizing that the results for the same biomarker, under similar conditions, could vary substantially among laboratories and across platforms. Given the frequency of biomarker discordance between primary tumours and corresponding metastases assessed by different methods (Tables 1 and 2), questions arise as to which techniques and what concordance levels should be required to ensure consistency. A review of HER2 immunohistochemical test performance among laboratories in patients with invasive breast carcinoma established that an overall ≥90% consensus between all the laboratories, which was achieved for 69% of the samples analysed, was a reasonable indicator of assay performance, even if considerable discordances were observed between the results of tests performed by multiple laboratories using the same standardized equipment and reagents.[118] As illustrated by this study, thresholds for sufficient concordance rates are often arbitrary. It is, therefore, mandatory to collect sufficient data and address this issue further in future attempts to define universally expected concordance rates for biomarkers. The inherent limitations regarding the performance characteristics of laboratory methods (sensitivity, specificity, reproducibility, accuracy, and linearity) make resolution of this problem difficult.

## Scoring Systems

Scoring is another potential source of variability for which improvements in standardization are required. Despite the widespread immunohistochemical assessment of HER2, oestrogen receptor (ER), and progesterone receptor (PR) expression in routine diagnostic practice, and the availability of antibodies recognizing mutated KRAS and BRAF that enable assessment of *KRAS* and *BRAF* mutation status, no universally accepted scoring standardization for these markers has been realized. Recommended scoring procedures described in immunohistochemistry kits are not always followed closely, leading to decreased reproducibility and sensitivity of the methods; that some laboratory-developed scoring systems might perform better than those recommended by the vendor is also possible. These assays are often evaluated on the basis of archived tissue samples in which storage characteristics can influence protein-expression levels.[121] Moreover, the presence of large numbers of non-neoplastic cell types in needle-biopsy specimens can limit the analysis of tumour cells and result in incorrect biomarker evaluation. Finally, immunohistochemical methods are notoriously nonlinear, and scoring systems are generally vulnerable to heterogeneity among intensity, extent, and topography of staining.[12] Unlike mutational analyses, immunohistochemical studies are not dichotomous, which complicates their role in clinical decision-making. Thus, both the proportion and type of cells positive for the targeted antigen as well as both the intensity and pattern of the immunoreactivity should be measured and standardized against reference values.

Computerized image analysis is potentially of value in the scoring of biomarkers. Image analysis can be criticized on philosophical grounds, considering that it provides accurate and precise measurements of data from an inherently imprecise assay method; however, reasonable levels of concordance can be achieved with respect to basic interpretation.[13] A challenge with immunohistochemical scoring of HER2 status was demonstrated in a study in which *HER2* status was evaluated across five laboratories in Europe.[122] Although the laboratories were fully concordant with regard to the interpretation of HER2 status (positive or negative), considerable divergence in scoring (according to the 0 to 3+ scale) was observed, particularly in cases with ambiguous immunochemistry and borderline FISH results.[122]

## The Influence of Technological Improvement

Technological improvements generally have a great impact on clinical practice; despite being highly desirable, these improvements can, however, result in confusion regarding clinical decisions. As an example, it has been shown that microfluidic droplet-based PCR technology for the identification of gene mutations has greatly improved the sensitivity of detection for mutations and/or alterations affecting *KRAS*, *BRAF*, and *HER2*.[123, 124, 125] The technique enabled the determination and precise quantification of a mutant *KRAS* gene in the presence of a 200,000-fold excess of unmutated *KRAS* DNA (sensitivity of detection of approximately 0.0005%), whereas conventional methods such pyrosequencing or the amplification refractory mutation system (ARMS) gave a sensitivity of detection of approximately 1–10%.[123, 124, 125] One study revealed discrepancies in tumour mutation-status assessment by standard methods at local laboratories compared with a central evaluation process.[126] In this study, tissue samples from 60 patients with mCRC that were assessed locally were defined as wild-type *KRAS* codon 12 and codon 13; however, central evaluation showed that 10% of the tumour samples in fact harboured *KRAS* mutated at codon 12, and around 20% displayed rare *KRAS* mutations, or *BRAF* and *NRAS* mutations.[126] Clinicians are thus faced with the question of whether or not the differences in sensitivity, considering analytical and biological variables, represent clinically meaningful information that should influence medical decisions. For instance, should the detection of a low-frequency *KRAS* mutation be a contraindication to anti-EGFR antibody therapy in patients with metastatic colon cancer? To make such decisions easier, widely established cut-off points should be implemented in the interpretation of data on continuous variables, to distinguish meaningful measures that can be transformed into dichotomized decisions: positive versus negative; mutant versus wild type; or eligibility for a treatment or trial versus ineligibility. This type of clinical validation poses a major challenge in clinical research because of the low frequency of patients with such equivocal assay results and, therefore, the large starting population needed to generate a sufficient sample size, as well as the length of follow up needed to complete and analyse trials in order to draw conclusions.

A variety of high-throughput technologies, such as transcriptomic, proteomic, and metabolomic modalities, enable large-scale analysis of complex biological systems to identify candidate biomarkers and characterize relevant pathways. To date, most of these types of analyses have generated valuable research results, pinpointing potential biomarkers for further development.

However, the complex workflow of these approaches and the inability to verify some candidate markers in subsequent studies provide evidence that such analyses are, in general, insufficiently robust to be translated into the clinical arena to guide therapeutic choices at present, and intermediate values from such biomarker assay techniques are difficult to interpret.[127] Nevertheless, comprehensive broad-scale assessments that rely on a series of measurements, ideally of different parameters within a multivariate framework, have the potential to provide more-extensive and/or more-robust predictive data, and thus these analytical approaches hold promise in advancement of the current state of the art in clinical practice.

## Post-Analytical Standardization

The post-analytical phase of biomarker evaluations involves reporting of the assay results, including normalization procedures and interpretation. Although thought to be less common than pre-analytical and analytical methodological issues, post-analytical errors, especially those that produce inconsistent values, might affect biomarker performance. Therefore, adequate measures must be taken to ensure a post-analytical phase that is as error-free as possible. Dichotomous variables are relatively straightforward to incorporate into calculations of data sensitivity and specificity. However, most variables in the setting of cell biology are continuous, which raises the problem of consensus with respect to clinically relevant cut-off points for diagnostic testing. This issue was exemplified by the finding that almost 10% of the women with breast cancer included in the National Surgical Adjuvant Breast and Bowel Project (NSABP) B-31 trial had neither *HER2* amplification nor overexpression of HER2 based on centrally reviewed testing for this biomarker, but nevertheless benefited from adjuvant therapy with trastuzumab.[117, 128] This result raised question about the current definition of HER2 positivity as an indication for trastuzumab treatment and provided the rationale for the NSABP B-47 trial that is currently investigating whether women with 1+ (‹HER2-negative›) and non-*HER2*-amplified 2+ (HER2-low) breast cancers benefit from addition of trastuzumab to adjuvant chemotherapy.[129] Data based on continuous variables also tend to be less reproducible than information on dichotomous parameters. Thus, successful standardization of post-analytical biomarker methods in this setting requires close collaboration between professional medical associations, investigators, clinicians, and statisticians. In fact, so-called 'dichotomous' variables have been made dichotomous because a cut-off point has been established either actively, as was the case for gene-copy numbers in FISH analysis of *HER2*, or passively enforced by technical limitations such as sensitivity of qRT-PCR and sequencing in the detection of mutations.

## RECOMMENDATIONS FOR BIOMARKER STUDIES

One should not be dismayed at the long list of varied challenges to biomarker method development, and instead thoughtfully acknowledge and address these issues to drive continuous improvements, as rigorous assay validation is expensive in terms of time, materials, financial costs, and biological specimens. Depending on the complexity and the intended purpose of the biomarker, its development will take several years and costs might rise to over US$100 million, for companion biomarkers in particular, owing to the requirement for large retrospective studies and prospective validation trials. As biomarker measurement and standardization can be assessed at several levels, different types of considerations should be addressed in order to maximize successful biomarker evaluation. Examples of such preliminary theoretical, clinical, technical, and logistical considerations are provided in Box 3. Ultimately, data should support the cost–utility of biomarker methodologies to ensure cost-effective clinical decision-making.

### Box 3: Recommendations for Optimizing Biomarker Evaluation

*Preliminary Considerations*

- Biomarker studies should be based on sound biology and a thorough understanding of the biological relevance of the biomarker and underlying biology
- Consider composite biomarkers panels to improve sensitivity, specificity, and predictive power

*Clinical Considerations*

- Ensure the specimen source (archived tissues, fresh biopsy tissues, metastatic lesions, etc.) most relevant to trial goal, design and ethical standards is used
- Record patient factors and clinical procedural variables that are relevant for each biospecimen
- Perform high-quality correlative studies in clinical trials (use preplanned cut-points and statistical methods with reference to standardized guidelines), obtaining consent for biospecimen-banking to support a wide range of scientific investigation
- Adapt biomarker assays according to the clinical stage of drug development as well as the information that needs to be gathered for both biomarker-assay and drug development: consider using a fit-for-

purpose strategy to avoid premature lock-down of biomarker-assay development
- Centralize specimens in a well-managed biobank and biomarker evaluation in core-credentialed laboratories, with a reference centre that will coordinate the activities among the evaluating laboratories
- Ensure rigorous pursuit of defined standards through optimized studies rather than limiting research to clinical data from trials
- Perform biomarker validation in settings mirroring closely the clinical environment

## *Technical Considerations*

- Implement best practices for biospecimen resources based on the available guidelines[15]
- Use a specimen source for which easy collection, appropriate volumes and optimal analysis are feasible
- Ensure quality and integrity of biospecimens throughout all processes by using relevant newly developed methodologies
- Concentrate on the specific quality-control and quality-assurance practices for appropriate procurement, formalin fixation, and paraffin embedding
- Consider DNA markers, as these are the most resistant to degradation and alteration, and are more likely to yield a dichotomous end point
- Standardize sample handling—harmonization of collection factors, such as sample labelling, volumes, transport, stabilization and storage—and processing methodology, including delays and data collection/annotation
- Ensure fit-for-purpose approaches
- Develop procedures with rigorous quality assurance, reproducibility, and control procedures built-in
- Collect data on the effects of methodological variables on assay performance to construct calibrators and control materials for routine real-world consensus performance
- Pre-define a threshold (cut-off point) for designating the status of the potential biomarker
- Consider using computerized image analysis

## Logistical Considerations

- Foster collaboration between professional medical associations, investigators, clinicians and statisticians for their diverse and valuable inputs in assay standardization and validation
- Use fit-for-purpose statistical methods that rely on a synergy between statistics and biological understanding

# CONCLUSIONS

As biomarkers have increasingly important roles in drug development and clinical trials, quality assurance and method validation have become crucial, and highlight the necessity of establishing standardized methodological guidelines. The ultimate goal for a biomarker is the establishment of clinical utility that guides patient care, but attempts to reach this goal must be preceded by analytical and clinical validation of the 'locked-down' biomarker assay. Substantial progress has been made in biomarker research, from discovery to development, standardization, and clinical application. However, major challenges regarding integrated and harmonized processes, spanning pre-analytical, analytical and post-analytical phases of development, remain. In the era of targeted therapies, the need for standardized approaches for biomarker validation has become widely recognized as an important issue to overcome. Several joint collaborative initiatives across different sectors in the USA and worldwide have emerged to address the lack of standardized guidelines in biomarker validation, specifically regarding biological specimens and assay methodologies. Although these efforts have contributed to the promotion of standardized procedures, sustained and continued commitment to ensure worldwide standards and harmonization are required.

It is important to recognize the fact that, even if all of the above recommendations are addressed, several additional factors will continue to pose major challenges: the complexity of the biological systems under investigation; the marginal effects provided by many drugs; the continuous nature of the data from assays for many potential biomarkers; and the inevitable variability among patients. Therefore, that basic assessment methodologies are robustly qualified, and are applied with rigorous adherence to high methodological standards and close attention to guidelines at each successive step of the validation process, including pre-study and in-study method validation, is essential if we are to obtain reliably validated biomarkers for routine use. Investment of effort and resources in the development of these biomarkers will expand their roles as valid end points for assessing patient outcome.

## CONTRIBUTIONS

Armand de Gramont, S.W., Aimery de Gramont and S.R.H. researched the data from the article; Armand de Gramont, L.M.E., J.R., J.T. Aimery de Gramont and S.R.H. contributed substantially to writing the article; and all authors contributed to discussion of content and review/editing of the manuscript before submission.

## ACKNOWLEDGEMENTS

The work of the authors has been supported by the Aide et Recherche en Cancérologie Digestive (ARCAD) foundation, a not-for-profit organization, and editorial assistance was provided by M. Benetkiewicz.

## REFERENCES

1. Biomarkers Definitions Working Group. Biomarkers and surrogate endpoints: preferred definitions and conceptual framework. *Clin. Pharmacol. Ther.* **69**, 89–95 (2001).
2. Hurwitz, H. *et al.* Bevacizumab plus irinotecan, fluorouracil, and leucovorin for metastatic colorectal cancer. *N. Engl. J. Med.* **350**, 2335–2342 (2004).
3. Giantonio, B. J. *et al.* Bevacizumab in combination with oxaliplatin, fluorouracil, and leucovorin (FOLFOX4) for previously treated metastatic colorectal cancer: results from the Eastern Cooperative Oncology Group Study E3200. *J. Clin. Oncol.* **25**, 1539–1544 (2007).
4. Saltz, L. B. *et al.* Bevacizumab in combination with oxaliplatin-based chemotherapy as first-line therapy in metastatic colorectal cancer: a randomized phase III study. *J. Clin. Oncol.* **26**, 2013–2019 (2008).
5. Taube, S. E. *et al.* A perspective on challenges and issues in biomarker development and drug and biomarker codevelopment. *J. Natl Cancer Inst.* **101**, 1453–1463 (2009).
6. DiMasi, J. A., Feldman, L., Seckler, A. & Wilson, A. Trends in risks associated with new drug development: success rates for investigational drugs. *Clin. Pharmacol. Ther.* **87**, 272–277 (2010).
7. Arrowsmith, J. Trial watch: phase II failures: 2008–2010. *Nat. Rev. Drug Discov.* **10**, 328–329 (2011).
8. Arrowsmith, J. Trial watch: phase III and submission failures: 2007–2010. *Nat. Rev. Drug Discov.* **10**, 87 (2011).
9. Huriez, A. *Personalized medicine, introduction, business impact on the healthcare sector and regulatory aspects* [online], (2013).

10. Cummings, J., Raynaud, F., Jones, L., Sugar, R. & Dive, C. Fit-for-purpose biomarker method validation for application in clinical trials of anticancer drugs. *Br. J. Cancer* **103**,1313–1317 (2010).
11. Garcia, V. M., Cassier, P. A. & de Bono, J. Parallel anticancer drug development and molecular stratification to qualify predictive biomarkers: dealing with obstacles hindering progress. *Cancer Discov.* **1**, 207–212 (2011).
12. Lee, J. W. *et al*. Fit-for-purpose method development and validation for successful biomarker measurement. *Pharm. Res.* **23**, 312–328 (2006).
13. Walker, R. A. Quantification of immunohistochemistry—issues concerning methods, utility and semiquantitative assessment, I. *Histopathology* **49**, 406–410 (2006).
14. Boenisch, T. Can a more selective application of antigen retrieval facilitate standardization in immunohistochemistry? *Appl. Immunohistochem. Mol. Morphol.* **12**, 172–176 (2004).
15. Goldstein, N. S., Hewitt, S. M., Taylor, C. R., Yaziji, H. & Hicks, D. G. Recommendations for improved standardization of immunohistochemistry. *Appl. Immunohistochem. Mol. Morphol.***15**, 124–133 (2007).
16. Office of Biorepositories and Biospecimen Research. *NCI Best Practices for Biospecimen Resources* [online], (2014).
17. Aktas, B. *et al*. Comparison of estrogen and progesterone receptor status of circulating tumor cells and the primary tumor in metastatic breast cancer patients. *Gynecol. Oncol.* **122**,356–360 (2011).
18. Amir, E. *et al*. Prospective study evaluating the impact of tissue confirmation of metastatic disease in patients with breast cancer. *J. Clin. Oncol.* **30**, 587–592 (2012).
19. Aoyama, K., Kamio, T., Nishikawa, T. & Kameoka, S. A comparison of HER2/neu gene amplification and its protein overexpression between primary breast cancer and metastatic lymph nodes. *Jpn J. Clin. Oncol.* **40**, 613–619 (2010).
20. Botteri, E. *et al*. Biopsy of liver metastasis for women with breast cancer: impact on survival.*Breast* **21**, 284–288 (2012).
21. Cardoso, F. *et al*. Evaluation of HER2, p53, bcl-2, topoisomerase II-α, heat shock proteins 27 and 70 in primary breast cancer and metastatic ipsilateral axillary lymph nodes. *Ann. Oncol.* **12**, 615–620 (2001).
22. Chang, H. J. *et al*. Discordant human epidermal growth factor receptor 2 and hormone receptor status in primary and metastatic breast cancer and

response to trastuzumab. *Jpn J. Clin. Oncol.* **41**, 593–599 (2011).
23. Curigliano, G. *et al.* Should liver metastases of breast cancer be biopsied to improve treatment choice? *Ann. Oncol.* **22**, 2227–2233 (2011).
24. Dieci, M. V. *et al.* Discordance in receptor status between primary and recurrent breast cancer has a prognostic impact: a single-institution analysis. *Ann. Oncol.* **24**, 101–108(2013).
25. Duchnowska, R. *et al.* Conversion of epidermal growth factor receptor 2 and hormone receptor expression in breast cancer metastases to the brain. *Breast Cancer Res.* **14**, R119(2012).
26. Fabi, A. *et al.* HER2 protein and gene variation between primary and metastatic breast cancer: significance and impact on patient care. *Clin. Cancer Res.* **17**, 2055–2064 (2011).
27. Fehm, T. *et al.* Detection and characterization of circulating tumor cells in blood of primary breast cancer patients by RT-PCR and comparison to status of bone marrow disseminated cells. *Breast Cancer Res.* **11**, R59 (2009).
28. Fuchs, I. B. *et al.* Epidermal growth factor receptor changes during breast cancer metastasis. *Anticancer Res.* **26**, 4397–4401 (2006).
29. Gancberg, D. *et al.* Comparison of HER-2 status between primary breast cancer and corresponding distant metastatic sites. *Ann. Oncol.* **13**, 1036–1043 (2002).
30. Gong, Y., Booser, D. J. & Sneige, N. Comparison of HER-2 status determined by fluorescence *in situ* hybridization in primary and metastatic breast carcinoma. *Cancer* **103**,1763–1769 (2005).
31. Guarneri, V. *et al.* Comparison of HER-2 and hormone receptor expression in primary breast cancers and asynchronous paired metastases: impact on patient management.*Oncologist* **13**, 838–844 (2008).
32. Jensen, J. D., Knoop, A., Ewertz, M. & Laenkholm, A. V. ER, HER2, and TOP2A expression in primary tumor, synchronous axillary nodes, and asynchronous metastases in breast cancer. *Breast Cancer Res. Treat.* **132**, 511–521 (2012).
33. Lear-Kaul, K. C., Yoon, H. R., Kleinschmidt-DeMasters, B. K., McGavran, L. & Singh, M.*HER-2/neu* status in breast cancer metastases to the central nervous system. *Arch. Pathol. Lab. Med.* **127**, 1451–1457 (2003).
34. Lindström, L. S. *et al.* Clinically used breast cancer markers such as estrogen receptor, progesterone receptor, and human epidermal growth factor receptor 2 are unstable throughout tumor progression. *J. Clin.*

Oncol. **30**, 2601–2608 (2012).
35. Lower, E. E., Glass, E., Blau, R. & Harman, S. HER-2/*neu* expression in primary and metastatic breast cancer. *Breast Cancer Res. Treat.* **113**, 301–306 (2009).
36. Macfarlane, R. *et al*. Molecular alterations between the primary breast cancer and the subsequent locoregional/metastatic tumor. *Oncologist* **17**, 172–178 (2012).
37. Montagna, E. *et al*. Breast cancer subtypes and outcome after local and regional relapse.*Ann. Oncol.* **23**, 324–331 (2012).
38. Niikura, N. *et al*. Loss of human epidermal growth factor receptor 2 (HER2) expression in metastatic sites of HER2-overexpressing primary breast tumors. *J. Clin. Oncol.* **30**, 593–599(2012).
39. Regitnig, P., Schippinger, W., Lindbauer, M., Samonigg, H. & Lax, S. F. Change of HER-2/neu status in a subset of distant metastases from breast carcinomas. *J. Pathol.* **203**,918–926 (2004).
40. Santinelli, A., Pisa, E., Stramazzotti, D. & Fabris, G. HER-2 status discrepancy between primary breast cancer and metastatic sites. Impact on target therapy. *Int. J. Cancer* **122**,999–1004 (2008).
41. Shimizu, C. *et al*. c-erbB-2 protein overexpression and p53 immunoreaction in primary and recurrent breast cancer tissues. *J. Surg. Oncol.* **73**, 17–20 (2000).
42. Simon, R. *et al*. Patterns of HER-2/neu amplification and overexpression in primary and metastatic breast cancer. *J. Natl Cancer Inst.* **93**, 1141–1146 (2001).
43. Simmons, C. *et al*. Does confirmatory tumor biopsy alter the management of breast cancer patients with distant metastases? *Ann. Oncol.* **20**, 1499–1504 (2009).
44. Thompson, A. M. *et al*. Prospective comparison of switches in biomarker status between primary and recurrent breast cancer: the Breast Recurrence In Tissues Study (BRITS).*Breast Cancer Res.* **12**, R92 (2010).
45. Vincent-Salomon, A. *et al*. HER2 status in patients with breast carcinoma is not modified selectively by preoperative chemotherapy and is stable during the metastatic process.*Cancer* **94**, 2169–2173 (2002).
46. Wilking, U. *et al*. HER2 status in a population-derived breast cancer cohort: discordances during tumor progression. *Breast Cancer Res. Treat.* **125**, 553–561 (2011).
47. Xiao, C., Gong, Y., Han, E. Y., Gonzalez-Angulo, A. M. & Sneige, N. Stability of HER2-positive status in breast carcinoma: a comparison

between primary and paired metastatic tumors with regard to the possible impact of intervening trastuzumab treatment. *Ann. Oncol.* **22**, 1547–1553 (2011).

48. Zidan, J. *et al.* Comparison of HER-2 overexpression in primary breast cancer and metastatic sites and its effect on biological targeting therapy of metastatic disease. *Br. J. Cancer* **93**, 552–556 (2005).

49. Albanese, I. *et al.* Heterogeneity within and between primary colorectal carcinomas and matched metastases as revealed by analysis of Ki-ras and p53 mutations. *Biochem. Biophys. Res. Commun.* **325**, 784–791 (2004).

50. Al-Mulla, F. *et al.* Heterogeneity of mutant versus wild-type Ki-ras in primary and metastatic colorectal carcinomas, and association of codon-12 valine with early mortality. *J. Pathol.* **185**, 130–138 (1998).

51. Artale, S. *et al.* Mutations of *KRAS* and *BRAF* in primary and matched metastatic sites of colorectal cancer. *J. Clin. Oncol.* **26**, 4217–4219 (2008).

52. Baldus, S. E. *et al.* Prevalence and heterogeneity of *KRAS*, *BRAF*, and *PIK3CA* mutations in primary colorectal adenocarcinomas and their corresponding metastases. *Clin. Cancer Res.* **16**, 790–799 (2010).

53. Cejas, P. *et al. KRAS* mutations in primary colorectal cancer tumors and related metastases: a potential role in prediction of lung metastasis. *PLoS ONE* **4**, e8199 (2009).

54. Cejas, P. *et al.* Analysis of the concordance in the EGFR pathway status between primary tumors and related metastases of colorectal cancer patients: implications for cancer therapy. *Curr. Cancer Drug Targets* **12**, 124–131 (2012).

55. Etienne-Grimaldi, M. C. *et al.* K-Ras mutations and treatment outcome in colorectal cancer patients receiving exclusive fluoropyrimidine therapy. *Clin. Cancer Res.* **14**, 4830–4835 (2008).

56. Finkelstein, S. D., Sayegh, R., Christensen, S. & Swalsky, P. A. Genotypic classification of colorectal adenocarcinoma. Biologic behavior correlates with K-ras-2 mutation type. *Cancer* **71**, 3827–3838 (1993).

57. Garm Spindler, K. L. *et al.* The importance of *KRAS* mutations and *EGF*61A>G polymorphism to the effect of cetuximab and irinotecan in metastatic colorectal cancer. *Ann. Oncol.* **20**, 879–884 (2009).

58. Italiano, A. *et al. KRAS* and *BRAF* mutational status in primary colorectal tumors and related metastatic sites: biological and clinical implications. *Ann. Surg. Oncol.* **17**, 1429–1434 (2010).

59. Knijn, N. *et al. KRAS* mutation analysis: a comparison between primary

tumours and matched liver metastases in 305 colorectal cancer patients. *Br. J. Cancer* **104**, 1020–1026(2011).
60. Losi, L., Benhattar, J. & Costa, J. Stability of K-ras mutations throughout the natural history of human colorectal cancer. *Eur. J. Cancer* **28A**, 1115–1120 (1992).
61. Show context
62. Loupakis, F. *et al.* PTEN expression and *KRAS* mutations on primary tumors and metastases in the prediction of benefit from cetuximab plus irinotecan for patients with metastatic colorectal cancer. *J. Clin. Oncol.* **27**, 2622–2629 (2009).
63. Mariani, P. *et al.* Concordant analysis of *KRAS* status in primary colon carcinoma and matched metastasis. *Anticancer Res.* **30**, 4229–4235 (2010).
64. Melucci, E. *et al.* Relationship between K-Ras mutational status and EGFR expression evaluated using Allred score in primary and metastatic colorectal cancer [abstract]. *J. Clin. Oncol.* **28** (Suppl.), a3568 (2010).
65. Molinari, F. *et al.* Differing deregulation of EGFR and downstream proteins in primary colorectal cancer and related metastatic sites may be clinically relevant. *Br. J. Cancer* **100**,1087–1094 (2009).
66. Mostert, B. *et al. KRAS* and *BRAF* mutation status in circulating colorectal tumor cells and their correlation with primary and metastatic tumor tissue. *Int. J. Cancer* **133**, 130–141(2013).
67. Oliveira, C. *et al. KRAS* and *BRAF* oncogenic mutations in MSS colorectal carcinoma progression. *Oncogene* **26**, 158–163 (2007).
68. Oltedal, S. *et al.* Heterogeneous distribution of K-ras mutations in primary colon carcinomas: implications for EGFR-directed therapy. *Int. J. Colorectal Dis.* **26**, 1271–1277(2011).
69. Oudejans, J. J., Slebos, R. J., Zoetmulder, F. A., Mooi, W. J. & Rodenhuis, S. Differential activation of ras genes by point mutation in human colon cancer with metastases to either lung or liver. *Int. J. Cancer* **49**, 875–879 (1991).
70. Park, J. H. *et al.* Analysis of *KRAS, BRAF, PTEN, IGF1R, EGFR* intron 1 CA status in both primary tumors and paired metastases in determining benefit from cetuximab therapy in colon cancer. *Cancer Chemother. Pharmacol.* **68**, 1045–1055 (2011).
71. Perrone, F. *et al.* PI3KCA/PTEN deregulation contributes to impaired responses to cetuximab in metastatic colorectal cancer patients. *Ann. Oncol.* **20**, 84–90 (2009).

72. Santini, D. et al. High concordance of *KRAS* status between primary colorectal tumors and related metastatic sites: implications for clinical practice. *Oncologist* **13**, 1270–1275 (2008).
73. Schimanski, C. C., Linnemann, U. & Berger, M. R. Sensitive detection of K-ras mutations augments diagnosis of colorectal cancer metastases in the liver. *Cancer Res.* **59**, 5169–5175 (1999).
74. Thebo, J. S., Senagore, A. J., Reinhold, D. S. & Stapleton, S. R. Molecular staging of colorectal cancer: K-ras mutation analysis of lymph nodes upstages Dukes B patients. *Dis. Colon Rectum* **43**, 155–159; discussion 159–162 (2000).
75. Watanabe, T. et al. Heterogeneity of *KRAS* status may explain the subset of discordant*KRAS* status between primary and metastatic colorectal cancer. *Dis. Colon Rectum* **54**, 1170–1178 (2011).
76. Zauber, P., Sabbath-Solitare, M., Marotta, S. P. & Bishop, D. T. Molecular changes in the Ki-ras and APC genes in primary colorectal carcinoma and synchronous metastases compared with the findings in accompanying adenomas. *Mol. Pathol.* **56**, 137–140 (2003).
77. Santini, D. et al. High concordance of *BRAF* status between primary colorectal tumours and related metastatic sites: implications for clinical practice. *Ann. Oncol.* **21**, 1565 (2010).
78. Jancik, S. et al. A comparison of direct sequencing, pyrosequencing, high resolution melting analysis, TheraScreen DxS, and the K-ras StripAssay for detecting *KRAS* mutations in non small cell lung carcinomas. *J. Exp. Clin. Cancer Res.* **31**, 79 (2012).
79. Davidson, C. J. et al. Improving the limit of detection for Sanger sequencing: a comparison of methodologies for *KRAS* variant detection. *Biotechniques* **53**, 182–188 (2012).
80. Ihle, M. A. et al. Comparison of high resolution melting analysis, pyrosequencing, next generation sequencing and immunohistochemistry to conventional Sanger sequencing for the detection of p.V600E and non-p.V600E *BRAF* mutations. *BMC Cancer* **14**, 13 (2014).
81. Suchy, B., Zietz, C. & Rabes, H. M. K-ras point mutations in human colorectal carcinomas: relation to aneuploidy and metastasis. *Int. J. Cancer* **52**, 30–33 (1992).
82. Vignot, S. et al. Next-generation sequencing reveals high concordance of recurrent somatic alterations between primary tumor and metastases from patients with non-small-cell lung cancer. *J. Clin. Oncol.* **31**, 2167–2172 (2013).

83. Gerlinger, M. et al. Intratumor heterogeneity and branched evolution revealed by multiregion sequencing. *N. Engl. J. Med.* **366**, 883–892 (2012).
84. Burrell, R. A., McGranahan, N., Bartek, J. & Swanton, C. The causes and consequences of genetic heterogeneity in cancer evolution. *Nature* **501**, 338–345 (2013).
85. Junttila, M. R. & de Sauvage, F. J. Influence of tumour micro-environment heterogeneity on therapeutic response. *Nature* **501**, 346–354 (2013).
86. Schwarzenbach, H., Hoon, D. S. & Pantel, K. Cell-free nucleic acids as biomarkers in cancer patients. *Nat. Rev. Cancer* **11**, 426–437 (2011).
87. Camp, E. R. et al. Molecular mechanisms of resistance to therapies targeting the epidermal growth factor receptor. *Clin. Cancer Res.* **11**, 397–405 (2005).
88. Laurent-Puig, P. et al. Analysis of *PTEN*, *BRAF*, and *EGFR* status in determining benefit from cetuximab therapy in wild-type *KRAS* metastatic colon cancer. *J. Clin. Oncol.* **27**,5924–5930 (2009).
89. De Roock, W. et al. Effects of *KRAS*, *BRAF*, *NRAS*, and *PIK3CA* mutations on the efficacy of cetuximab plus chemotherapy in chemotherapy-refractory metastatic colorectal cancer: a retrospective consortium analysis. *Lancet Oncol.* **11**, 753–762 (2010).
90. Douillard, J. Y. et al. Panitumumab-FOLFOX4 treatment and RAS mutations in colorectal cancer. *N. Engl. J. Med.* **369**, 1023–1034 (2013).
91. Bredel, M. et al. Amplification of whole tumor genomes and gene-by-gene mapping of genomic aberrations from limited sources of fresh-frozen and paraffin-embedded DNA. *J. Mol. Diagn.* **7**, 171–182 (2005).
92. Karapetis, C. S. et al. K-ras mutations and benefit from cetuximab in advanced colorectal cancer. *N. Engl. J. Med.* **359**, 1757–1765 (2008).
93. Amado, R. G. et al. Wild-type *KRAS* is required for panitumumab efficacy in patients with metastatic colorectal cancer. *J. Clin. Oncol.* **26**, 1626–1634 (2008).
94. Buyse, M. et al. Integrating biomarkers in clinical trials. *Expert Rev. Mol. Diagn.* **11**,171–182 (2011).
95. Buyse, M., Sargent, D. J., Grothey, A., Matheson, A. & de Gramont, A. Biomarkers and surrogate end points—the challenge of statistical validation. *Nat. Rev. Clin. Oncol.* **7**,309–317 (2010).
96. Food and Drug Administration. *Table of Pharmacogenomic Biomarkers in Drug Labels* [online], (2014).
97. Libeer, J. C. & Ehrmeyer, S. ISO 15189: a worldwide standard for

medical laboratories.*Point of Care* **3**, 5–7 (2004).

98. International Society for Biological and Environmental Repositories. *ISBER Best Practices for Repositories* [online], (2014).

99. College of American Pathologists. *Accreditation and Laboratory Improvement* [online], (2014).

100. National Cancer Institute. *NCI Biospecimen Research Database* [online], (2014).

101. Mitchell, B. L., Yasui, Y., Li, C. I., Fitzpatrick, A. L. & Lampe, P. D. Impact of freeze-thaw cycles and storage time on plasma samples used in mass spectrometry based biomarker discovery projects. *Cancer Inform.* **1**, 98–104 (2005).

102. Siddiqui, S. & Rimm, D. L. Pre-analytic variables and phospho-specific antibodies: the Achilles heel of immunohistochemistry. *Breast Cancer Res.* **12**, 113 (2010).

103. Johnsen, I. K. et al. Evaluation of a standardized protocol for processing adrenal tumor samples: preparation for a European adrenal tumor bank. *Horm. Metab. Res.* **42**, 93–101(2010).

104. Chung, J. Y. et al. Factors in tissue handling and processing that impact RNA obtained from formalin-fixed, paraffin-embedded tissue. *J. Histochem. Cytochem.* **56**, 1033–1042 (2008).

105. Medeiros, F., Rigl, C. T., Anderson, G. G., Becker, S. H. & Halling, K. C. Tissue handling for genome-wide expression analysis: a review of the issues, evidence, and opportunities.*Arch. Pathol. Lab. Med.* **131**, 1805–1816 (2007).

106. Douglas, M. P. & Rogers, S. O. DNA damage caused by common cytological fixatives.*Mutat. Res.* **401**, 77–88 (1998).

107. Williams, C. et al. A high frequency of sequence alterations is due to formalin fixation of archival specimens. *Am. J. Pathol.* **155**, 1467–1471 (1999).

108. Wong, S. Q. et al. Sequence artefacts in a prospective series of formalin-fixed tumours tested for mutations in hotspot regions by massively parallel sequencing. *BMC Med. Genomics* **7**, 23 (2014).

109. Guo, H. et al. An efficient procedure for protein extraction from formalin-fixed, paraffin-embedded tissues for reverse phase protein arrays. *Proteome Sci.* **10**, 56 (2012).

110. Sprung, R. W. Jr et al. Equivalence of protein inventories obtained from formalin-fixed paraffin-embedded and frozen tissue in multidimensional liquid chromatography-tandem mass spectrometry shotgun proteomic

analysis. *Mol. Cell. Proteomics* **8**, 1988–1998 (2009).

111. Fedorowicz, G., Guerrero, S., Wu, T. D. & Modrusan, Z. Microarray analysis of RNA extracted from formalin-fixed, paraffin-embedded and matched fresh-frozen ovarian adenocarcinomas. *BMC Med. Genomics* **2**, 23 (2009).

112. Kalmar, A. *et al.* Gene expression analysis of normal and colorectal cancer tissue samples from fresh frozen and matched formalin-fixed, paraffin-embedded (FFPE) specimens after manual and automated RNA isolation. *Methods* **59**, S16–S19 (2013).

113. Xie, R. *et al.* Factors influencing the degradation of archival formalin-fixed paraffin-embedded tissue sections. *J. Histochem. Cytochem.* **59**, 356–365 (2011).

114. Liu, N. W. *et al.* Impact of ischemia and procurement conditions on gene expression in renal cell carcinoma. *Clin. Cancer Res.* **19**, 42–49 (2013).

115. De Marzo, A. M., Fine, S. & Trock, B. J. *Impact of Pre-Analytic Variation on Tissue Analysis: Issues & Practical Applications*. Presented at the 2008 Biospecimen Research Network Symposium (2008).

116. Blons, H. & Laurent-Puig, P. Technical considerations for *KRAS* testing in colorectal cancer. The biologist's point of view [French]. *Bull. Cancer* **96** (Suppl.), S47–S56 (2009).

117. Press, M. F. *et al.* Diagnostic evaluation of HER-2 as a molecular target: an assessment of accuracy and reproducibility of laboratory testing in large, prospective, randomized clinical trials. *Clin. Cancer Res.* **11**, 6598–6607 (2005).

118. Perez, E. A. *et al.* HER2 testing by local, central, and reference laboratories in specimens from the North Central Cancer Treatment Group N9831 intergroup adjuvant trial. *J. Clin. Oncol.* **24**, 3032–3038 (2006).

119. Fitzgibbons, P. L., Murphy, D. A., Dorfman, D. M., Roche, P. C. & Tubbs, R. R. Interlaboratory comparison of immunohistochemical testing for HER2: results of the 2004 and 2005 College of American Pathologists HER2 Immunohistochemistry Tissue Microarray Survey. *Arch. Pathol. Lab. Med.* **130**, 1440–1445 (2006).

120. Wolff, A. C. *et al.* American Society of Clinical Oncology/College of American Pathologists guideline recommendations for human epidermal growth factor receptor 2 testing in breast cancer. *J. Clin. Oncol.* **25**, 118–145 (2007).

121. Taylor, C. R. & Levenson, R. M. Quantification of immunohistochemistry—issues concerning methods, utility and semiquantitative assessment II.

*Histopathology* **49**, 411–424 (2006).

122. Pauletti, G. et al. Assessment of methods for tissue-based detection of the HER-2/neu alteration in human breast cancer: a direct comparison of fluorescence *in situ* hybridization and immunohistochemistry. *J. Clin. Oncol.* **18**, 3651–3664 (2000).
123. Dowsett, M. et al. Standardization of HER2 testing: results of an international proficiency-testing ring study. *Mod. Pathol.* **20**, 584–591 (2007).
124. Lievre, A. et al. *KRAS* mutations as an independent prognostic factor in patients with advanced colorectal cancer treated with cetuximab. *J. Clin. Oncol.* **26**, 374–379 (2008).
125. Hindson, B. J. et al. High-throughput droplet digital PCR system for absolute quantitation of DNA copy number. *Anal. Chem.* **83**, 8604–8610 (2011).
126. Pekin, D. et al. Quantitative and sensitive detection of rare mutations using droplet-based microfluidics. *Lab Chip* **11**, 2156–2166 (2011).
127. Andre, T. et al. Panitumumab combined with irinotecan for patients with *KRAS* wild-type metastatic colorectal cancer refractory to standard chemotherapy: a GERCOR efficacy, tolerance, and translational molecular study. *Ann. Oncol.* **24**, 412–419 (2013).
128. van't Veer, L. J. & Bernards, R. Enabling personalized cancer medicine through analysis of gene-expression patterns. *Nature* **452**, 564–570 (2008).
129. Paik, S., Kim, C. & Wolmark, N. HER2 status and benefit from adjuvant trastuzumab in breast cancer. *N. Engl. J. Med.* **358**, 1409–1411 (2008).
130. Fehrenbacher, L. et al. NSABP B-47: a randomized phase III trial of adjuvant therapy comparing chemotherapy alone to chemotherapy plus trastuzumab in women with node-positive or high-risk node-negative HER2-low invasive breast cancer [abstract]. *J. Clin. Oncol.* **31** (Suppl.), TPS1139 (2013).
131. Teixidó, C., Karachaliou, N., Peg, V., Gimenez-Capitan, A. & Rosell, R. Concordance of IHC, FISH and RT-PCR for *EML4–ALK* rearrangements. *Transl. Lung Cancer Res.* **3**, 70–74 (2014).
132. Angulo, B. et al. A comparison of EGFR mutation testing methods in lung carcinoma: direct sequencing, real-time PCR and immunohistochemistry. *PLoS ONE* **7**, e43842 (2012).
133. Personeni, N. et al. Usefulness of alpha-fetoprotein response in patients treated with sorafenib for advanced hepatocellular carcinoma. *J. Hepatol.*

57, 101–107 (2012).
134. Haynes, B. P. *et al.* Expression of key oestrogen-regulated genes differs substantially across the menstrual cycle in oestrogen receptor-positive primary breast cancer. *Breast Cancer Res. Treat.* **138**, 157–165 (2013).
135. von Minckwitz, G. *et al.* Ki67 measured after neoadjuvant chemotherapy for primary breast cancer. *Clin. Cancer Res.* **19**, 4521–4531 (2013).
136. Chandarlapaty, S. *et al.* Frequent mutational activation of the PI3K–AKT pathway in trastuzumab-resistant breast cancer. *Clin. Cancer Res.* **18**, 6784–6791 (2012).
137. Lorgis, V. *et al.* Discordance in early breast cancer for tumour grade, estrogen receptor, progesteron receptors and human epidermal receptor-2 status between core needle biopsy and surgical excisional primary tumour. *Breast* **20**, 284–287 (2011).
138. Arnedos, M. *et al.* Discordance between core needle biopsy (CNB) and excisional biopsy (EB) for estrogen receptor (ER), progesterone receptor (PgR) and HER2 status in early breast cancer (EBC). *Ann. Oncol.* **20**, 1948–1952 (2009).
139. Stalhammar, G., Rosin, G., Fredriksson, I., Bergh, J. & Hartman, J. Low concordance of biomarkers in histopathological and cytological material from breast cancer. *Histopathology* **64**, 971–980 (2014).
140. Engel, K. B. & Moore, H. M. Effects of preanalytical variables on the detection of proteins by immunohistochemistry in formalin-fixed, paraffin-embedded tissue. *Arch. Pathol. Lab. Med.* **135**, 537–543 (2011).
141. Khoury, T. *et al.* Delay to formalin fixation effect on breast biomarkers. *Mod. Pathol.* **22**, 1457–1467 (2009).
142. Marchetti, A., Felicioni, L. & Buttitta, F. Assessing EGFR mutations. *N. Engl. J. Med.* **354**, 526–528 (2006).
143. Punnoose, E. A. *et al.* Molecular biomarkers analyses using circulating tumor cells. *PLoS ONE* **5**, e12517 (2010).
144. Voros, A., Csorgo, E., Nyari, T. & Cserni, G. An intra- and interobserver reproducibility analysis of the Ki-67 proliferation marker assessment on core biopsies of breast cancer patients and its potential clinical implications. *Pathobiology* **80**, 111–118 (2013).
145. Ronot, M. *et al.* Alternative Response Criteria (Choi, European association for the study of the liver, and modified Response Evaluation Criteria in Solid Tumors [RECIST]) Versus RECIST 1.1 in patients with advanced hepatocellular carcinoma treated with sorafenib. *Oncologist* **19**, 394–402 (2014).

146. Slamon, D. J. *et al.* Use of chemotherapy plus a monoclonal antibody against HER2 for metastatic breast cancer that overexpresses HER2. *N. Engl. J. Med.* **344**, 783–792 (2001).

147. Cobleigh, M. A. *et al.* Multinational study of the efficacy and safety of humanized anti-HER2 monoclonal antibody in women who have HER2-overexpressing metastatic breast cancer that has progressed after chemotherapy for metastatic disease. *J. Clin. Oncol.* **17**, 2639–2648 (1999).

148. Heinrich, M. C. *et al.* Kinase mutations and imatinib response in patients with metastatic gastrointestinal stromal tumor. *J. Clin. Oncol.* **21**, 4342–4349 (2003).

149. Blanke, C. D. *et al.* Long-term results from a randomized phase II trial of standard- versus higher-dose imatinib mesylate for patients with unresectable or metastatic gastrointestinal stromal tumors expressing KIT. *J. Clin. Oncol.* **26**, 620–625 (2008).

150. Kang, M. J. *et al.* Biweekly cetuximab plus irinotecan as second-line chemotherapy for patients with irinotecan-refractory and *KRAS* wild-type metastatic colorectal cancer according to epidermal growth factor receptor expression status. *Invest. New Drugs* **30**, 1607–1613 (2012).

151. Cunningham, D. *et al.* Cetuximab monotherapy and cetuximab plus irinotecan in irinotecan-refractory metastatic colorectal cancer. *N. Engl. J. Med.* **351**, 337–345 (2004).

152. Romond, E. H. *et al.* Trastuzumab plus adjuvant chemotherapy for operable HER2-positive breast cancer. *N. Engl. J. Med.* **353**, 1673–1684 (2005).

153. Schwartzberg, L. S. *et al.* Analysis of *KRAS/NRAS* mutations in PEAK: a randomized phase II study of FOLFOX6 plus panitumumab (pmab) or bevacizumab (bev) as first-line treatment (tx) for wild-type (WT) *KRAS* (exon 2) metastatic colorectal cancer (mCRC) [abstract]. *J. Clin. Oncol.* **31** (Suppl.), a3631 (2013).

154. Cameron, D. *et al.* A phase III randomized comparison of lapatinib plus capecitabine versus capecitabine alone in women with advanced breast cancer that has progressed on trastuzumab: updated efficacy and biomarker analyses. *Breast Cancer Res. Treat.* **112**, 533–543 (2008).

155. Martin, M. *et al.* A phase two randomised trial of neratinib monotherapy versus lapatinib plus capecitabine combination therapy in patients with HER2+ advanced breast cancer. *Eur. J. Cancer* **49**, 3763–3772 (2013).

156. McArthur, G. A. *et al.* Molecular and clinical analysis of locally advanced dermatofibrosarcoma protuberans treated with imatinib: Imatinib Target

Exploration Consortium Study B2225. *J. Clin. Oncol.* **23**, 866–873 (2005).

157. Kerob, D. *et al.* Imatinib mesylate as a preoperative therapy in dermatofibrosarcoma: results of a multicenter phase II study on 25 patients. *Clin. Cancer Res.* **16**, 3288–3295 (2010).

158. Hirsch, F. R. *et al.* Molecular predictors of outcome with gefitinib in a phase III placebo-controlled study in advanced non-small-cell lung cancer. *J. Clin. Oncol.* **24**, 5034–5042(2006).

159. Sutani, A. *et al.* Gefitinib for non-small-cell lung cancer patients with epidermal growth factor receptor gene mutations screened by peptide nucleic acid-locked nucleic acid PCR clamp.*Br. J. Cancer* **95**, 1483–1489 (2006).

160. Villanueva, C. *et al.* Phase II study assessing lapatinib added to letrozole in patients with progressive disease under aromatase inhibitor in metastatic breast cancer-Study BES 06.*Target. Oncol.* **8**, 137–143 (2013).

161. Johnston, S. *et al.* Lapatinib combined with letrozole versus letrozole and placebo as first-line therapy for postmenopausal hormone receptor-positive metastatic breast cancer. *J. Clin. Oncol.* **27**, 5538–5546 (2009).

162. Bang, Y. J. *et al.* Trastuzumab in combination with chemotherapy versus chemotherapy alone for treatment of HER2-positive advanced gastric or gastro-oesophageal junction cancer (ToGA): a phase 3, open-label, randomised controlled trial. *Lancet* **376**, 687–697(2010).

163. Shaw, A. T. *et al.* Crizotinib versus chemotherapy in advanced ALK-positive lung cancer. *N. Engl. J. Med.* **368**, 2385–2394 (2013).

164. Malik, S. M. *et al.* U. S. Food and Drug Administration approval: crizotinib for treatment of advanced or metastatic non-small cell lung cancer that is anaplastic lymphoma kinase positive. *Clin. Cancer Res.* **20**, 2029–2034 (2014).

165. Chapman, P. B. *et al.* Improved survival with vemurafenib in melanoma with BRAF V600E mutation. *N. Engl. J. Med.* **364**, 2507–2516 (2011).

166. McArthur, G. A. *et al.* Safety and efficacy of vemurafenib in $BRAF^{V600E}$ and $BRAF^{V600K}$ mutation-positive melanoma (BRIM-3): extended follow-up of a phase 3, randomised, open-label study. *Lancet Oncol.* **15**, 323–332 (2014).

167. Bokemeyer, C. *et al.* Addition of cetuximab to chemotherapy as first-line treatment for *KRAS* wild-type metastatic colorectal cancer: pooled analysis of the CRYSTAL and OPUS randomised clinical trials. *Eur. J. Cancer* **48**, 1466–1475 (2012).

168. Tveit, K. M. *et al.* Phase III trial of cetuximab with continuous or intermittent fluorouracil, leucovorin, and oxaliplatin (Nordic FLOX) versus FLOX alone in first-line treatment of metastatic colorectal cancer: the NORDIC-VII study. *J. Clin. Oncol.* **30**, 1755–1762 (2012).

169. Baselga, J. *et al.* Phase II trial of pertuzumab and trastuzumab in patients with human epidermal growth factor receptor 2-positive metastatic breast cancer that progressed during prior trastuzumab therapy. *J. Clin. Oncol.* **28**, 1138–1144 (2010).

170. Swain, S. M. *et al.* Pertuzumab, trastuzumab, and docetaxel for HER2-positive metastatic breast cancer (CLEOPATRA study): overall survival results from a randomised, double-blind, placebo-controlled, phase 3 study. *Lancet Oncol.* **14**, 461–471 (2013).

171. Hurvitz, S. A. *et al.* Phase II randomized study of trastuzumab emtansine versus trastuzumab plus docetaxel in patients with human epidermal growth factor receptor 2-positive metastatic breast cancer. *J. Clin. Oncol.* **31**, 1157–1163 (2013).

172. Burris, H. A. 3rd *et al.* Phase II study of the antibody drug conjugate trastuzumab-DM1 for the treatment of human epidermal growth factor receptor 2 (HER2)-positive breast cancer after prior HER2-directed therapy. *J. Clin. Oncol.* **29**, 398–405(2011).

173. Sequist, L. V. *et al.* Phase III study of afatinib or cisplatin plus pemetrexed in patients with metastatic lung adenocarcinoma with EGFR mutations. *J. Clin. Oncol.* **31**, 3327–3334(2013).

174. Wu, Y. L. *et al.* Afatinib versus cisplatin plus gemcitabine for first-line treatment of Asian patients with advanced non-small-cell lung cancer harbouring EGFR mutations (LUX-Lung 6): an open-label, randomised phase 3 trial. *Lancet Oncol.* **15**, 213–222 (2014).

175. Shaw, A. T. *et al.* Ceritinib in *ALK*-rearranged non-small-cell lung cancer. *N. Engl. J. Med.* **370**, 1189–1197 (2014).

176. Vansteenkiste, J. F. Ceritinib for treatment of *ALK*-rearranged advanced non-small-cell lung cancer. *Future Oncol.* **10**, 1925–1939 (2014).

177. Zhou, C. *et al.* Erlotinib versus chemotherapy as first-line treatment for patients with advanced EGFR mutation-positive non-small-cell lung cancer (OPTIMAL, CTONG-0802): a multicentre, open-label, randomised, phase 3 study. *Lancet Oncol.* **12**, 735–742 (2011).

178. Kim, S. T. *et al.* Randomized phase II study of gefitinib versus erlotinib in patients with advanced non-small cell lung cancer who failed previous chemotherapy. *Lung Cancer* **75**, 82–88 (2012).

179. Gianni, L. et al. Efficacy and safety of neoadjuvant pertuzumab and trastuzumab in women with locally advanced, inflammatory, or early HER2-positive breast cancer (NeoSphere): a randomised multicentre, open-label, phase 2 trial. *Lancet Oncol.* **13**, 25–32 (2012).
180. Schneeweiss, A. et al. Pertuzumab plus trastuzumab in combination with standard neoadjuvant anthracycline-containing and anthracycline-free chemotherapy regimens in patients with HER2-positive early breast cancer: a randomized phase II cardiac safety study (TRYPHAENA). *Ann. Oncol.* **24**, 2278–2284 (2013).
181. Flaherty, K. T. et al. Combined BRAF and MEK inhibition in melanoma with BRAF V600 mutations. *N. Engl. J. Med.* **367**, 1694–1703 (2012).
182. Kim, K. B. et al. Phase II study of the MEK1/MEK2 inhibitor trametinib in patients with metastatic *BRAF*-mutant cutaneous melanoma previously treated with or without a BRAF inhibitor. *J. Clin. Oncol.* **31**, 482–489 (2013).
183. Long, G. V. et al. Dabrafenib in patients with Val600Glu or Val600Lys BRAF-mutant melanoma metastatic to the brain (BREAK-MB): a multicentre, open-label, phase 2 trial.*Lancet Oncol.* **13**, 1087–1095 (2012).
184. Sosman, J. A. et al. BRAF inhibitor (BRAFi) dabrafenib in combination with the MEK1/2 inhibitor (MEKi) trametinib in BRAFi-naive and BRAFi-resistant patients (pts) with BRAF mutation-positive metastatic melanoma (MM) [abstract]. *J. Clin. Oncol.* **31** (Suppl.), a9005(2013).
185. European Medicines Agency (EMA). *European public assessment reports* [online], (2014).
186. Lee, J. W. et al. Method validation and measurement of biomarkers in nonclinical and clinical samples in drug development: a conference report. *Pharm. Res.* **22**, 499–511(2005).

# Chapter 8

## ENRICHMENT OF G2/M CELL CYCLE PHASE IN HUMAN PLURIPOTENT STEM CELLS ENHANCES HDR-MEDIATED GENE REPAIR WITH CUSTOMIZABLE ENDONUCLEASES

DianeYang[1], Marissa A Scavuzzo[2], Jolanta Chmielowiec[3,4], Robert Sharp[3,4], Aleksandar Bajic[6] and Malgorzata Borowiak[1,5]

[1] Molecular and Cellular Biology Department, Baylor College of Medicine, One Baylor Plaza, Houston, TX 77030, USA

[2] Program in Developmental Biology, Baylor College of Medicine, One Baylor Plaza, Houston, TX 77030, USA

[3] Stem Cells and Regenerative Medicine Center, Baylor College of Medicine, One Baylor Plaza, Houston, TX 77030, USA

[4] Center for Cell and Gene Therapy, Baylor College of Medicine, Texas Children's Hospital and Houston Methodist Hospital, Houston, TX 77030, USA

[5]McNair Medical Institute, Houston, TX 77030, USA

[6]Jan and Dan Duncan Neurological Research Institute, Texas Children's Hospital, 1250 Moursund Street, Houston, TX 77030, USA

## ABSTRACT

Efficient gene editing is essential to fully utilize human pluripotent stem cells (hPSCs) in regenerative medicine. Custom endonuclease-based gene targeting involves two mechanisms of DNA repair: homology directed repair (HDR) and non-homologous end joining (NHEJ). HDR is the preferred mechanism for common applications such knock-in, knock-out or precise mutagenesis, but remains inefficient in hPSCs. Here, we demonstrate that synchronizing synchronizing hPSCs in G2/M with ABT phase increases on-target gene editing, defined as correct targeting cassette integration, 3 to 6 fold. We observed improved efficiency using ZFNs, TALENs, two CRISPR/Cas9, and CRISPR/Cas9 nickase to target five genes in three hPSC lines: three human embryonic stem cell lines, neural progenitors and diabetic iPSCs. neural progenitors and diabetic iPSCs. Reversible synchronization has no effect on pluripotency or differentiation. The increase in on-target gene editing is locus-independent and specific to the cell cycle phase as G2/M phase enriched cells show a 6-fold increase in targeting efficiency compared to cells in G1 phase. Concurrently inhibiting NHEJ with SCR7 does not increase HDR or improve gene targeting efficiency further, indicating that HR is the major DNA repair mechanism after G2/M phase arrest. The approach outlined here makes gene editing in hPSCs a more viable tool for disease modeling, regenerative medicine and cell-based therapies.

## INTRODUCTION

Genetic engineering allows for precise manipulation of the genome, facilitating developmental and disease modeling in accessible experimental systems, which are particularly important in regenerative medicine. Human pluripotent stem cells (hPSCs), including induced pluripotent stem cells (iPSCs) and human embryonic stem cells (hESCs), can give rise to any cell type in the body, including cells affected by disease. In order to fully utilize the potential of PSC technology, efficient strategies for gene editing in these cells are essential. Classical gene editing approaches based on homologous recombination (HR) have been fruitfully used in mouse embryonic stem cells for decades; while successful in principal, these same approaches are extremely inefficient in hPSCs.

Recent advances in genetic technology have provided increasingly simpler and more efficient ways to modify the genome based on the generation of double stranded DNA breaks (DSBs) through damage-inducing endonucleases directed by engineered guides to loci of interest. Zinc finger nucleases (ZFNs), transcriptional activator-like effector nucleases (TALENs), and clustered regularly interspaced palindromic repeats (CRISPR) technologies employ modular guides designed by the user to induce DNA damage and increase gene targeting efficiency. With ZFNs, TALENs, and CRISPR, DNA damage can be repaired through non-homologous end joining (NHEJ), leaving an insertion or deletion (indel), or homologous recombination (HR) for homology directed repair (HDR), in which a sister chromatid or template aids in repairing the broken DNA. Both mechanisms of DSB repair — NHEJ and HR — are active in nearly all cell types and species. HR is enriched endogenously during the G2/M phase of the cell cycle. NHEJ is the primary repair mechanism in the G1 phase before DNA synthesis occurs, although it has been detected throughout the cell cycle. When genomic insults such as DSBs occur in hPSCs, damaged cells preferentially undergo apoptosis to limit the replication of compromised DNA and maintain the integrity of the population, leading to a shift away from DNA repair by HR in damaged hPSCs. The result is a decrease in incorporation of homologous template DNA, with effective gene targeting rates oscillating between 0.5–8%17 in hPSCs. HDR allows for precise genome modification and is necessary for many common applications such as knock-in of fluorescent reporters, precise mutations, or selection cassettes that are delivered as exogenous DNA fragments, making HDR crucial and thus gene editing challenging. Therefore, tools directing cells to preferentially undertake one route of DNA repair (HR) over the other (NHEJ) could facilitate the desired targeting events.

Improving the rate of HDR will substantially increase the efficiency of genetic engineering. Recent studies have shown that small molecules like SCR7, BrefeldinA, or L755507 can inhibit NHEJ or manipulate the cell cycle; however, these tools have limitations. For instance, they were tested in carcinoma cell lines or mouse embryos, showed toxicity, have not been thoroughly investigated for various endonuclease or gene targeting strategies, have not been tested bi-directionally to alter the cell cycle, or have not been compared to technologies to influence other phases of the cell cycle. In addition, these studies have only shown the effect of small molecules on targeting efficiencies without delineating the underlying biological mechanism.

Our goal is to find efficient strategies to shift human cells, in particular hPSCs, towards HDR during gene editing using various customizable endonucleases and to improve gene modification efficiency in a locus-independent manner. Here, we systematically determined conditions to increase the effectiveness of precise, template-based repair in genome editing by CRISPR, CRISPR nickase, ZFNs, and TALENs by synchronizing five different hPSC and five different hPSC lines and hPSC-derived cells in the G2/M phase, during which the endogenous repair mechanisms for HR are abundant. We show that the G2/M phase of cell cycle in itself is sufficient to improve targeting efficiencies, with sorted populations of FUCCI-H9 cells showing a robust increase in template repair in G2/M rich populations and a drastic decrease in G1 rich populations. To synchronize cells with small molecules for increased HDR, we sought compounds with low toxicity that arrest cells in the G2/M phase, and have a reversible effect on the cell cycle. In addition, since the most widely targeted cell types are hPSCs, synchronized cells must also maintain their pluripotency and ability to differentiate efficiently into all three germ layers. We identified two compounds that met these requirements: ABT-751 (ABT) and Nocodazole. Both compounds were initially identified as anti-cancer agents and their biological effects on cells are well understood; ABT and Nocodazole inhibit microtubule polymerization, arresting cells in the G2/M phase of the cell cycle. We determined the influence of these compounds on repair mechanisms and the effects on both indel formation and HDR repair.

Further testing was done to asses the combinatorial effect of enhancing HDR with ABT or Nocodazole while concurrently inhibiting NHEJ using SCR7. Using these compounds, we synchronized in the G2/M phase human embryonic stem cell lines H1-WA01 (H1)25, HUES826, and Fucci H9-WA09 (Fucci-H9) and two human diabetic iPSCs (DiPSCs), and hPSC-derived neural progenitor cells (NPCs). Synchronized cells consistently have a higher amount of HR and the efficiency of targeting is increased 3–5 fold compared to unsynchronized cells. After synchronization, cells remain pluripotent, do not prematurely express markers of differentiated germ layers, and successfully differentiate into multiple lineages.

Through increased HR activity, reversible synchronization of cells in the G2/M phase improves the efficiency of donor integration during genome editing. The ability to increase the efficiency of genome editing was confirmed in multiple cell types using several gene-editing technologies and targeting five different genetic loci, showing the broad consequences of this new and useful tool for regenerative medicine and translational research.

# RESULTS

## Nocodazole and ABT Synchronize Human Pluripotent Stem Cells in the G2/M Cell Cycle Phase

As HR contributes to the maintenance of genomic stability in the G2/M cell cycle phase, we hypothesized that the timely arrest of hPSCs in G2/M phase immediately prior gene targeting would result in increased HR efficiency and donor integration. We tested two compounds to arrest cells in the G2/M phase, Nocodazole and ABT, in three different hPSC lines, H1, HUES8 and Fucci-H9 cells, two human diabetic iPSCs (DiPSCs), and hPSC-derived neural progenitor cells (NPCs). ABT and Nocodazole inhibit microtubule polymerization to arrest the cell cycle, and since their mechanism is well studied the risk of unknown targets in the cell is reduced. For H1 and HUES8 cell lines, we tested various synchronization periods and releasing combinations to find the optimal conditions to enrich G2/M phase compared to untreated cells and vehicle control (DMSO) (Fig. 1 and Supplementary Tables 1 and 2). Furthermore, we examined dosage effects and cytotoxicity of ABT and Nocodazole on hPSCs. Detailed optimization of both the concentration of compound and the length of induction, as well as the proper adjustment of cell seeding densities, led to the efficient enrichment in G2/M phase for all tested cell lines. A 16-hour treatment with 1 µg/ml Nocodazole and one hour of releasing in Nocodazole-free medium (Fig. 1a) shows robust G2/M cell phase enrichment with 80% of surviving cells in G2/M phase (Fig. 1b and Supplementary Table 1). At the same time, the majority (58.1%) of untreated cells were in G1 phase (Fig. 1b and Supplementary Table 1). 16 hours of ABT treatment had similar results, with 80% of surviving cells synchronized in the G2/M phase (Fig. 1b and Supplementary Table 2).

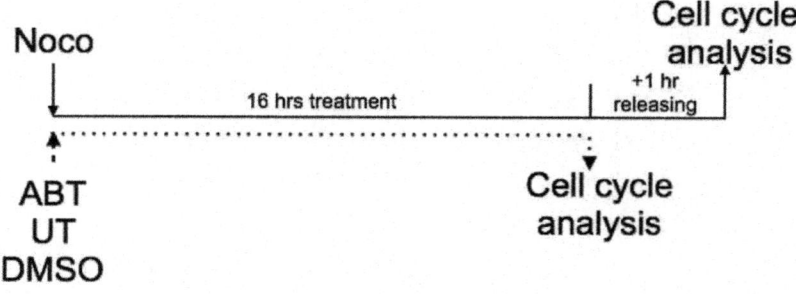

(a)

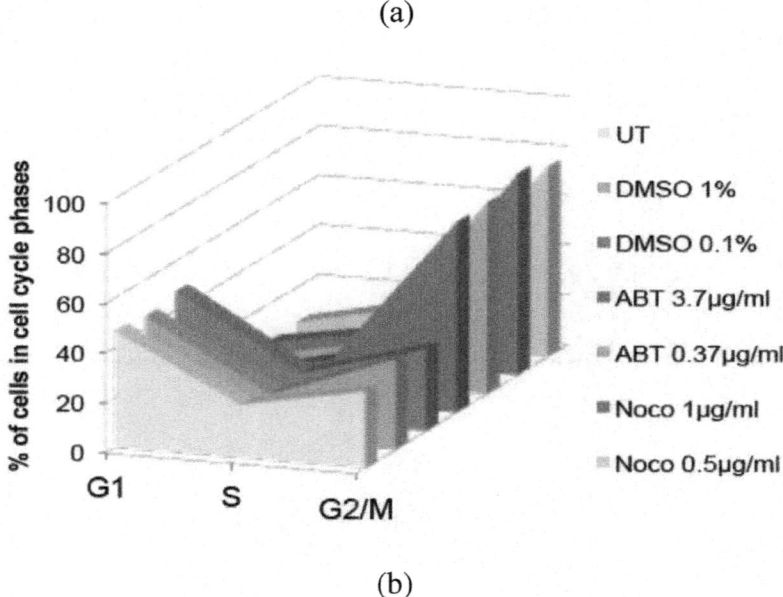

(b)

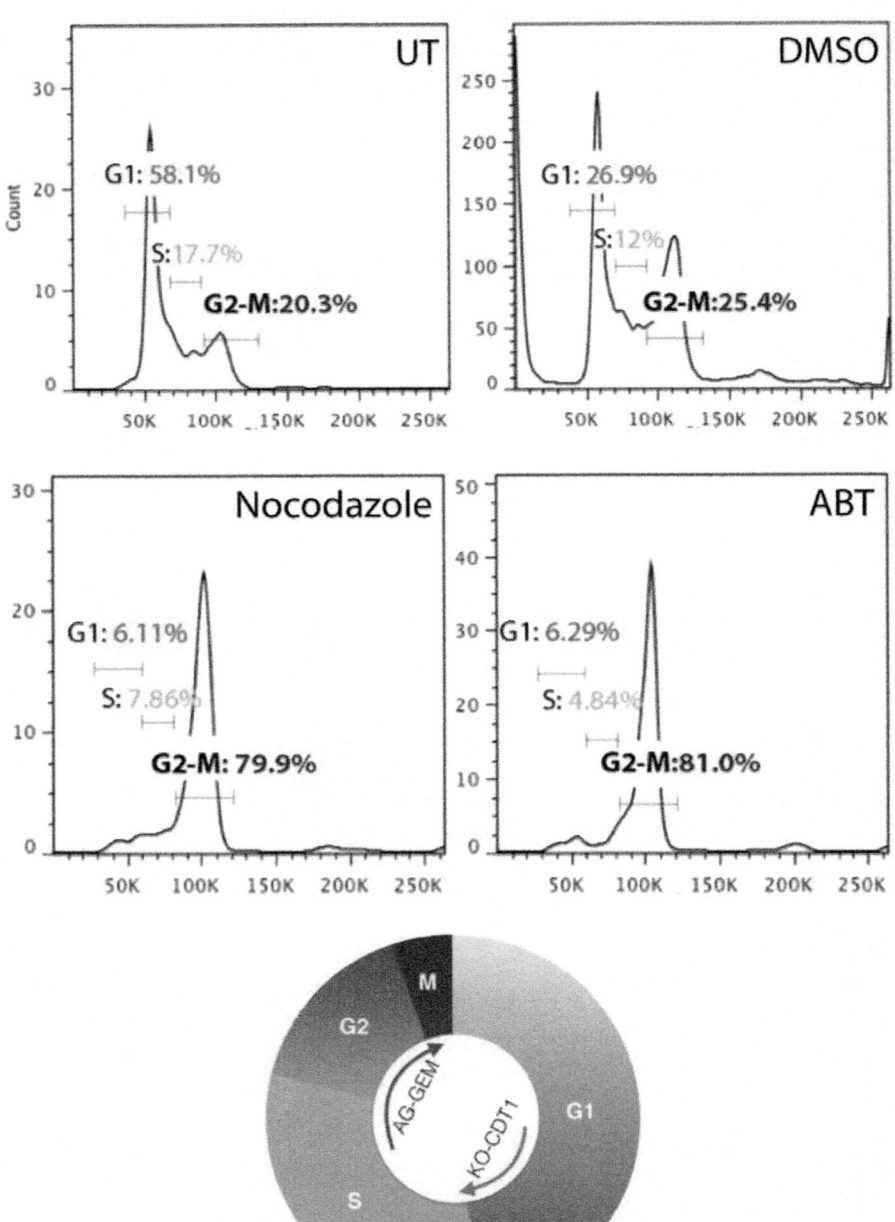

(c)

Enrichment of G2/M Cell Cycle Phase in Human Pluripotent Stem Cells... 223

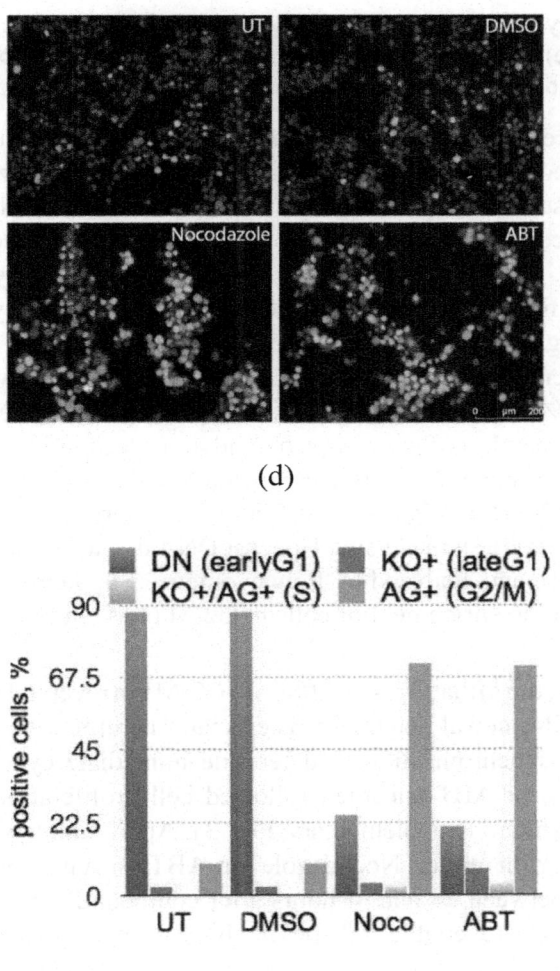

**Figure 1:** Efficient G2/M cell cycle phase synchronization of hPSCs. (a) Treatment timeline and scheme of cell cycle analysis of untreated (UT), vehicle control-DMSO (DMSO), Nocodazole (Noco) and ABT-751 (ABT) treated H1 hESCs. (b) Evaluations of different concentrations of DMSO, Noco and ABT. Percentage of cells in G1, G2/M and S phase were determined by flow cytometry by PI staining (upper right panel). Examples of DNA content captured by flow cytometry of untreated (UT), vehicle control-DMSO (0.1%), Nocodazole (1 µg/ml) and ABT-751 (0.36 µg/ml) treated H1 cell. (c) Schematic diagram of Fucci system. CDT1 are fused to Azami Green-1 (AG) and GEMININ are fused to Kusabira orange-2 (KO). Double negative (DN) fraction represents early G1 cells, the KO (red) fraction represents late G1, the AG (light green) represents S phase, and the AG high (green) represents G2/M. (d) Fucci H9 cells were stained with DNA dye, Hoechst 33342 and analyzed for cell cycle phases by fluores-

cent microscopy in normal culture conditions and after small molecule-mediated synchronization. (e) Quantification of G2/M phase by AG (green), G1 phase by KO (red), S phase by double positive (orange) and early G1 by double negative (blue) cells.

The fluorescent ubiquitination cell cycle indicator, Fucci-H9 hPSCs, encodes two colors, red fluorescence in late G1 phase and Azima green (AG) in G2/M phase, allowing us to follow cell division and further characterize and quantify cell cycle progression in the presence or absence of Nocodazole or ABT (Fig. 1c,d). Fucci-H9 cells were synchronized by the optimized protocol and fluorescence reflecting G2/M and other cell cycle phases was detected using fluorescent microscopy (Fig. 1d) and flow cytometry (Fig. 1e and Supplementary Figs 1 and 2). After synchronization with Nocodazole or ABT, 88.4% and 88.6% of total live cells respectively, expressed AG as detected consistently by flow cytometry and fluorescent microscopy, indicating a 2.5-fold increase in G2/M phase compared to untreated (38.23%) or DMSO-treated cells (34.4%) (Fig. 1d,e, and Supplementary Figs 1 and 2). Lastly, Fucci-H9 cells were stained using Hoechst DNA dye to further determine each cell cycle phase and analyzed by flow cyometry (Supplementary Fig. 1). The dye confirmed the enrichment of cells in G2/M phase induced by Nocodazole or ABT.

Finding factor(s) that arrested hPSCs in G2/M in a reversible manner without any loss in self-renewal potential was crucial. Therefore, we investigated cell survival after synchronization to determine immediate cytotoxicity induced by Nocodazole or ABT and then followed cell proliferation after releasing for 24 and 48 hours (Supplementary Fig. 3). Approximately 40% of hPSCs synchronized with either Nocodazole or ABT, survive and proliferate at normal rates between 24 and 48 hours after compound removal, showing the synchronization is reversible and does not have a long-term deteriorating effect on hPSCs.

## Cell Cycle Synchronization Increases Global Gene Repair, Predominantly HDR

To analyze the extent to which cell synchronization alters the repair system in the cell, we measured the efficiency of HR and NHEJ in a quantitative manner. We used a fluorescent HR reporter (rHR) and NHEJ reporter (rNHEJ), in which a functional *GFP* gene is reconstituted following an HR or NHEJ event (Fig. 2a). H1 cells were treated with ABT or Nocodazole or were transfected with the HR or the NHEJ reporter linearized by digestion with *I-SceI* enzyme (Fig. 2b). After transfection, cells were incubated for 48 hours to allow for the expression of GFP and were analyzed by flow cytometry.

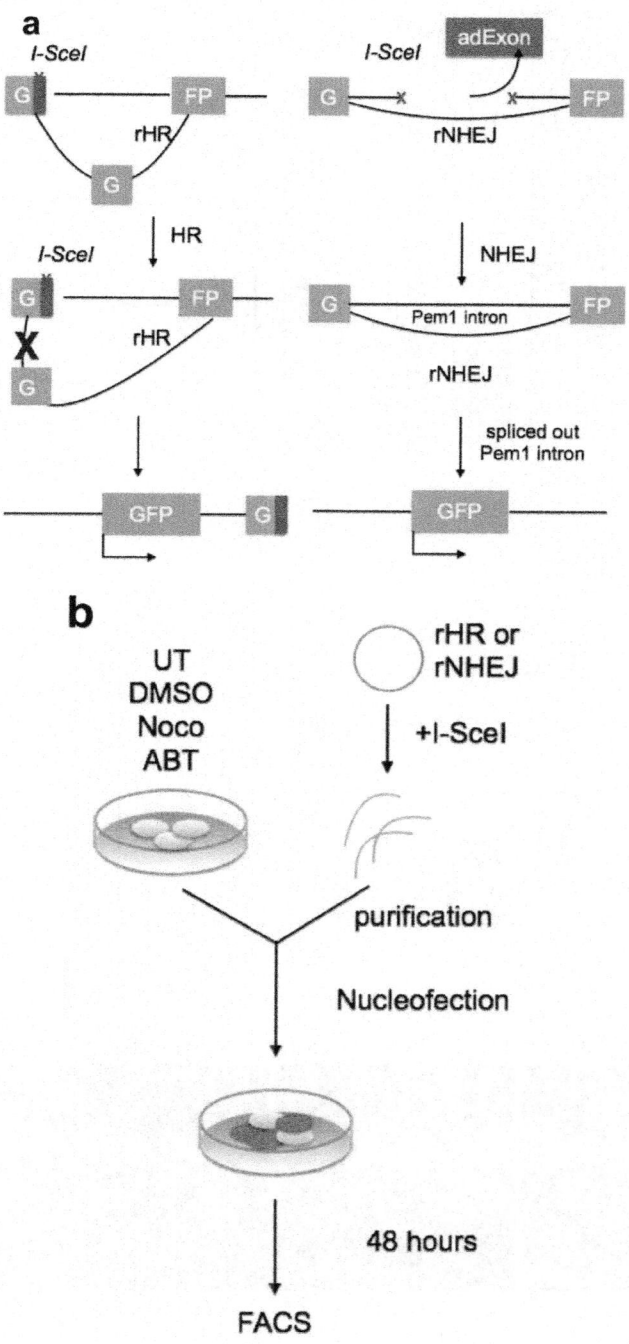

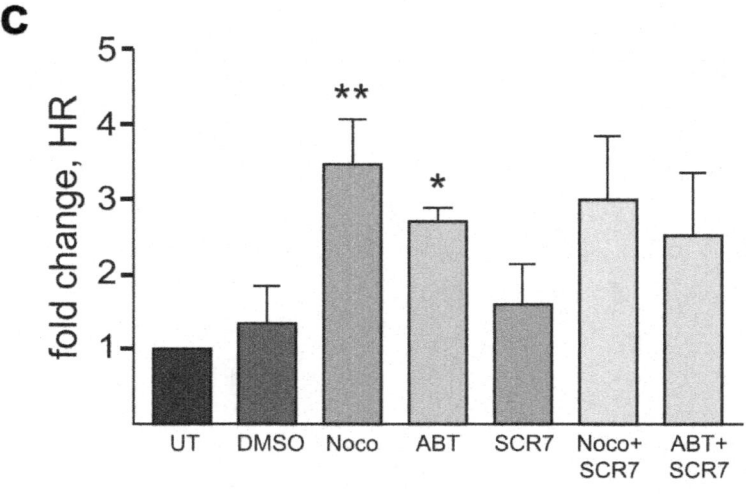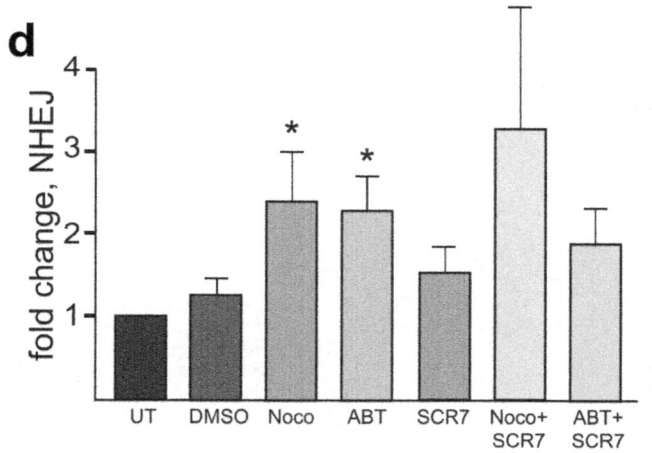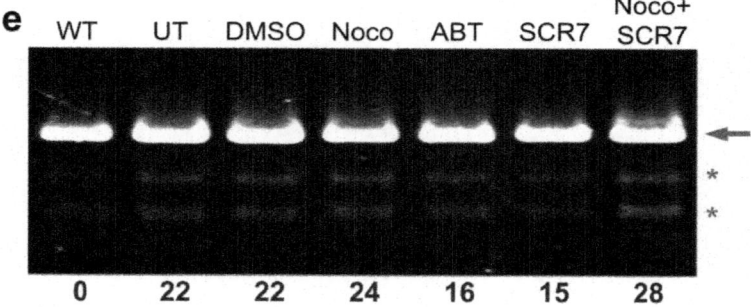

**Figure 2:** HR and NHEJ activity assessment after G2/M synchronization in hPSCs. **(a)** Schematic diagram of the mechanism of rHR (left) and rNHEJ (right) reporter. **(b)** Experimental design of HR and NHEJ analysis after synchronization or treatment of Noco, ABT, SCR7, Noco with SCR7, control DMSO and untreated H1 cells. H1 cells were nucleofected with digested rHR and rNHEJ and the percentage of GFP positive cells was captured by FACS for determining **(c)** HR and **(d)** NHEJ activity in synchronized cells normalized to untreated (n=3 biological replicates, $**P<0.01$ for Noco and $*P<0.05$ or ABT).**(e)** Indel formation after CRISPR/Cas9 targeting at the *OCT4* locus in synchronized and control HEK293t cells using T7 endonuclease assay. Error bars represent SEM.

GFP detection showed a 2–4 fold increase in HR and NHEJ after G2/M phase synchronization compared to untreated cells (0.43%) (Fig. 2c). Synchronization with Nocodazole resulted in a 3.5-fold increase in HR (1.5%), and ABT treatment had a 3.1-fold increase (1.35%) (Fig. 2c). NHEJ activity increases by about 2-fold in Nocodazole (2.75%) and ABT (2.6%) synchronized cells, compared to non-synchronized cells (1.10%) (Fig. 2d). We also used SCR7, an inhibitor of DNA ligase IV necessary for canonical NHEJ pathway, to test whether transiently blocking NHEJ enhances the frequency of HR (Fig. 2c–e). SCR7 treatment alone or in combination with Nocodazole or ABT did not show further increase of HR or change in NHEJ activity (Fig. 2c,d).

Targeting HEK293t cells with CRISPR/Cas9 to create a DSBs at the*OCT4* locus and evaluating NHEJ at the functional level reveals no change in indel formation between Nocodazole synchronized and non-synchronized cells, but 30% decrease in indel for ABT synchronized cells (Fig. 2e and Supplementary Fig. 4). Similarly, a 30% decrease in indel was detected for SCR7 treated cells (Fig. 2e and Supplementary Fig. 4).

To further examine the effects of synchronization and the stress of nucleofection on DNA repair mechanisms, we detected by qPCR acute HR- and NHEJ-related gene expression, including *BRCA1, BRCA2, RAD51* for HR, and *LIG4, XRCC4, XRCC5* for NHEJ, at 3, 6, 12 and 24 hours after transfection. We found no significant differences between Nocodazole-, ABT-treated and untreated cells, indicating that effects on transcriptional machinery did not increase the levels of HR and NHEJ (Supplementary Fig. 5). While the reporter assay shows a 3.5-fold increase in HR (Fig. 2b) and a 2-fold increase in NHEJ (Fig. 2c) with Nocodazole or ABT, functional assays show that synchronizing hPSCs with Nocodazole or ABT increases HR (Figs 3, 4, 5) with little effect on indel formation, or NHEJ at the functional level (Fig. 2d).

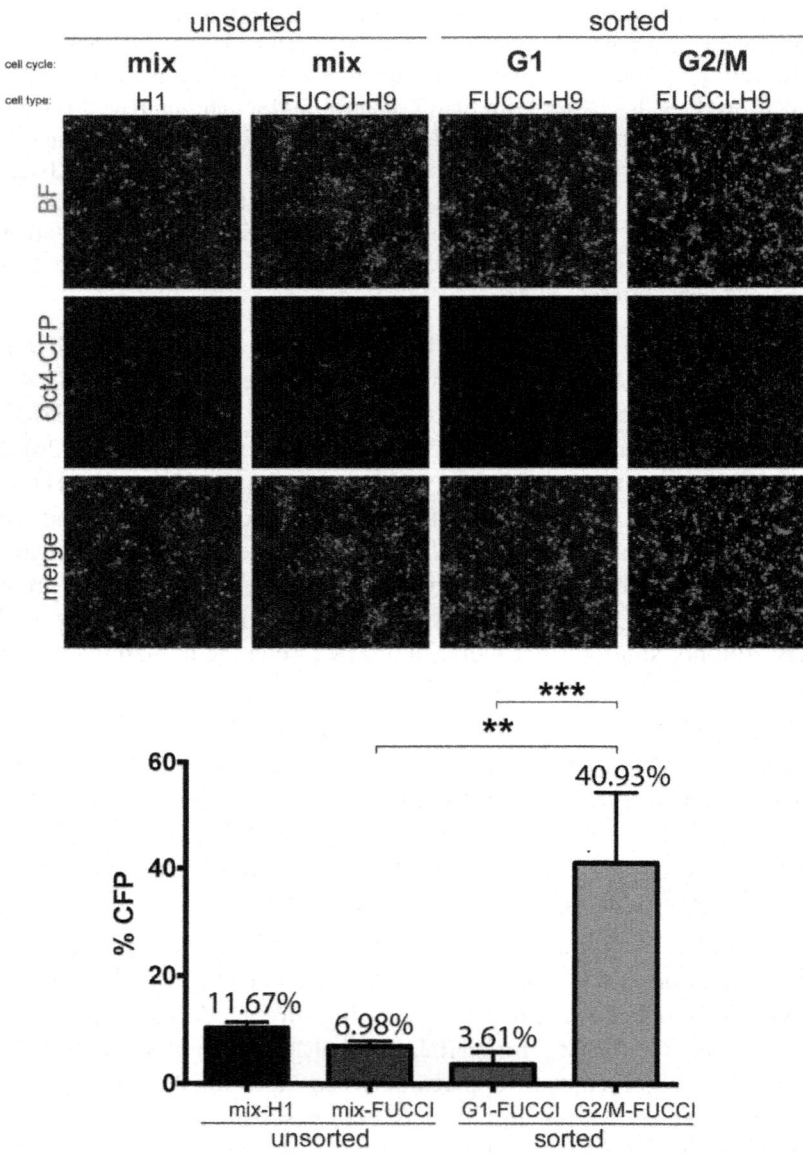

**Figure 3:** Targeting efficiency shifts depending on the cell cycle with an increase in G2/M populations and a decrease in G1 populations. Five days after FACS sorting and CRISPR *S. pyogenes* targeting to insert CFP into the *OCT4*locus, unsorted H1 hPSCs, unsorted FUCCI-H9 cells, sorted FUCCI-H9 in the G1 phase, and sorted FUCCI-H9 in the G2/M phase were imaged by confocal microscopy for CFP expression. Brightfield (BF); Cyan Fluorescent Protein (CFP) (n=3 biological replicates, $**P<0.01$; $***P<0.001$).

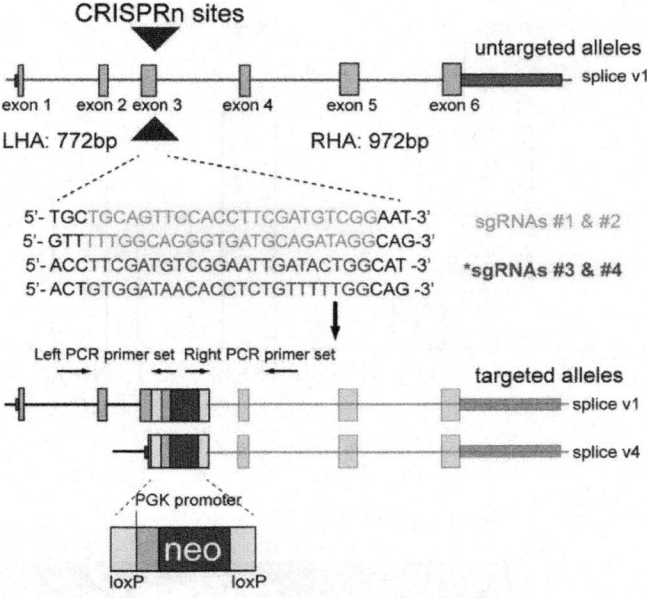

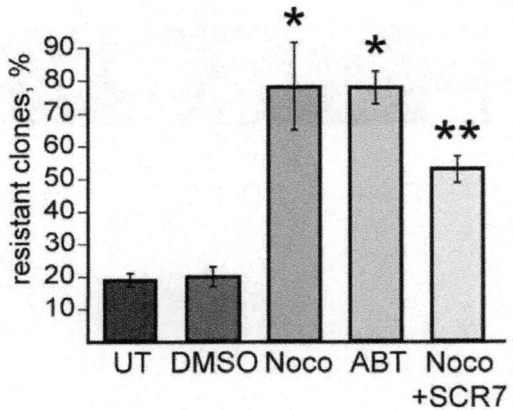

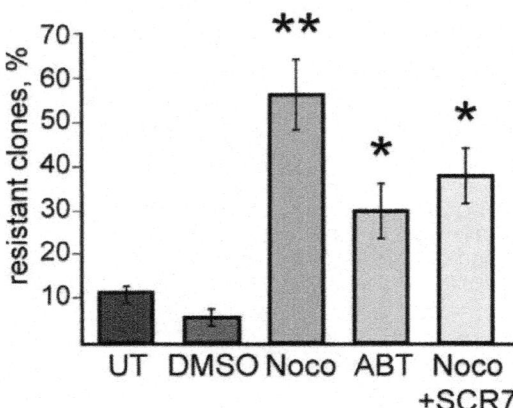
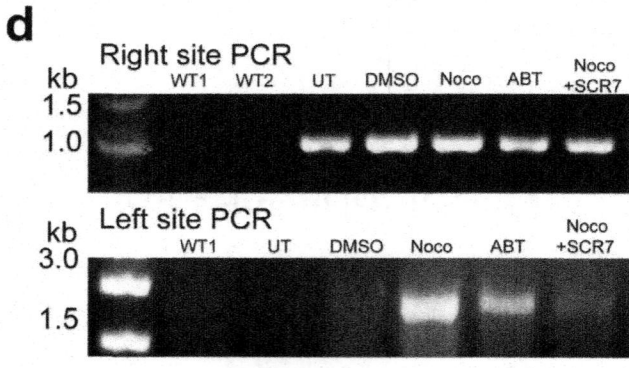
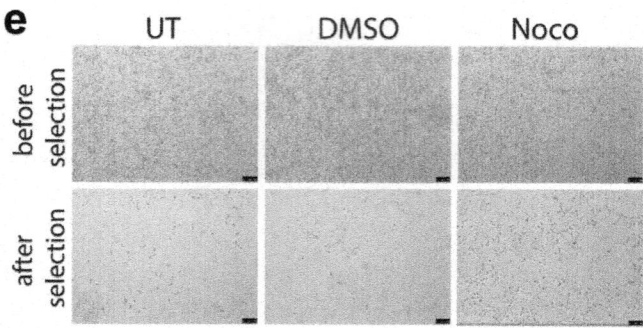

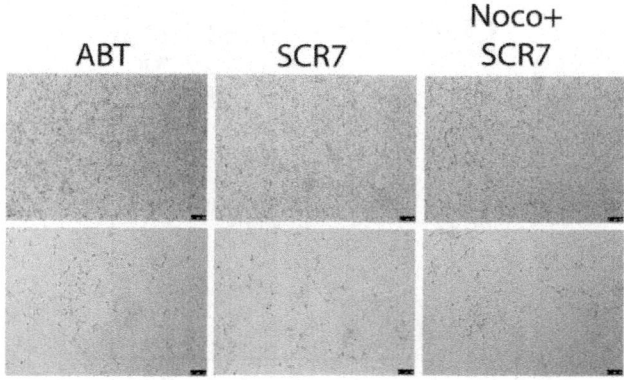

**f** *human NEUROD1*

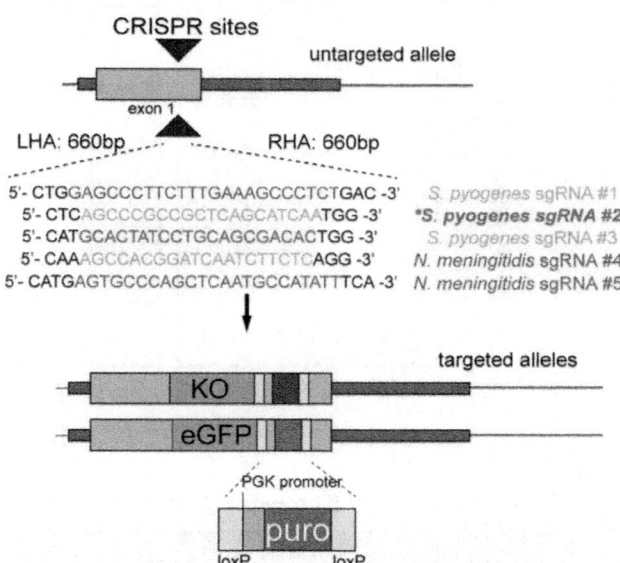

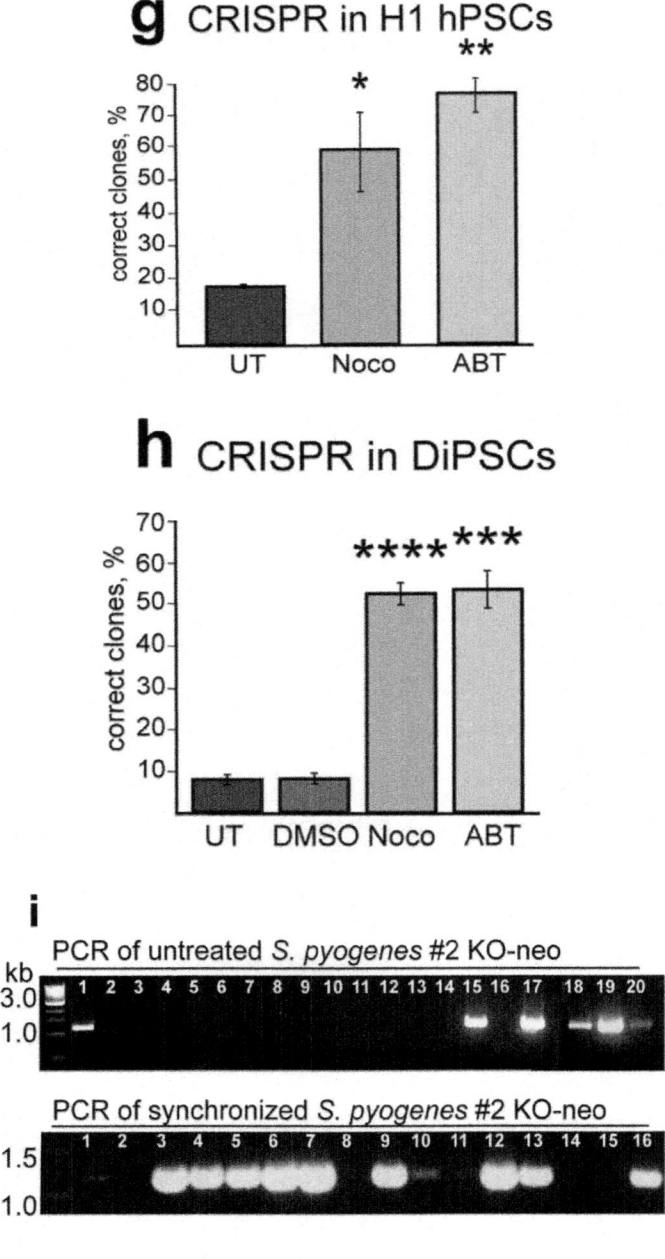

**Figure 4:** G2/M synchronization increases gene targeting efficiency of *WNT5A* by **CRISPR/Cas9** nickase and *NEUROD1* **by two different CRISPR/Cas9 species. (a)** Human *WNT5A* gene structure and targeting scheme. *WNT5A* is targeted for knockout with a floxed neomycin cassette. *indicates sgRNA used for subsequent panels. **(b)** The

efficiency of targeting in H1 cells determined by counting the percent of colonies surviving after antibiotic selection; $n=2$ independent experiments. $*P<0.05$; $**P<0.01$; $****P<0.0001$ (c) The efficiency of targeting in H1-derived NPCs determined by counting the percent of cells surviving after antibiotic selection; $n=2$ biological replicates. $*P<0.05$; $**P<0.01$ (d) PCR confirmation of targeted alleles. (e) Brightfield images of NPCs after targeting and before selection in top row and after targeting and after selection for neomycin resistance in the bottom row. (f) Human gene structure of *NEUROD1* and targeting scheme. *indicates sgRNA used for subsequent panels. (g) The efficiency of targeting in H1 cells by counting percent of correct colonies compared to resistant colonies determined by 3′, 5′ end and internal PCRs; $n=3$ independent experiments. $*P<0.05$; $**P<0.01$. (h) The efficiency of targeting in DiPSCs by counting percent of correct colonies compared to resistant colonies determined by 3′, 5′ end and internal PCRs, $n=4$ independent experiments. $***P<0.001$; $****P<0.0001$ (i) PCR confirmation of the identity of targeted alleles in untreated clones in the top panel or Nocodazole-synchronized cells in bottom panel. Gene structure is annotated including exons (light blue), introns (brown lines), untranslated regions (dark blue), and targeting sites (large navy arrows). Below the untargeted allele(s) the target sequence is shown with left homology arm (LHA) and right homology arm (RHA) sizes. Targeted alleles show the donor cassette integrated into the correct locus. Kusabira orange (KO) and eGFP cassettes have 3′ antibiotic resistance genes, KO with neomycin and eGFP with puromycin. Error bars represent SEM.

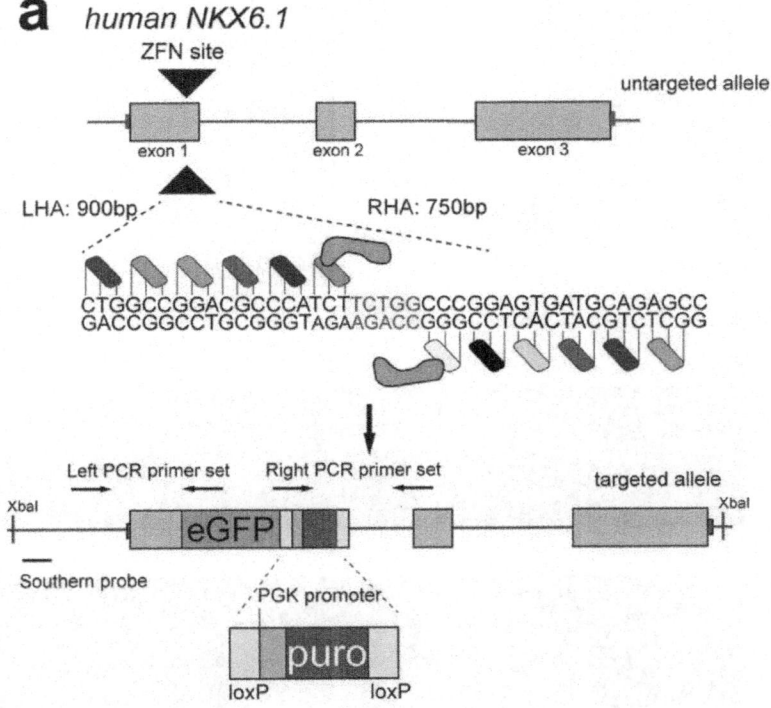

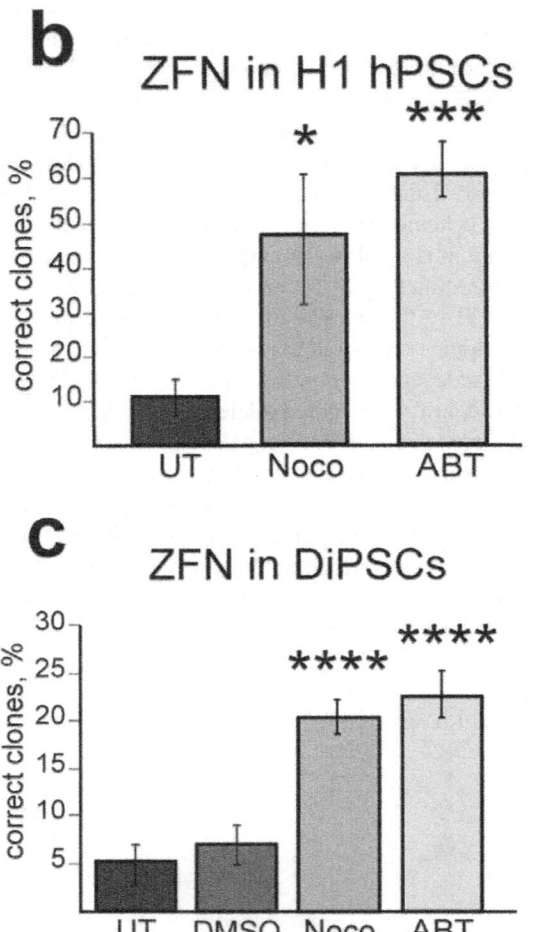
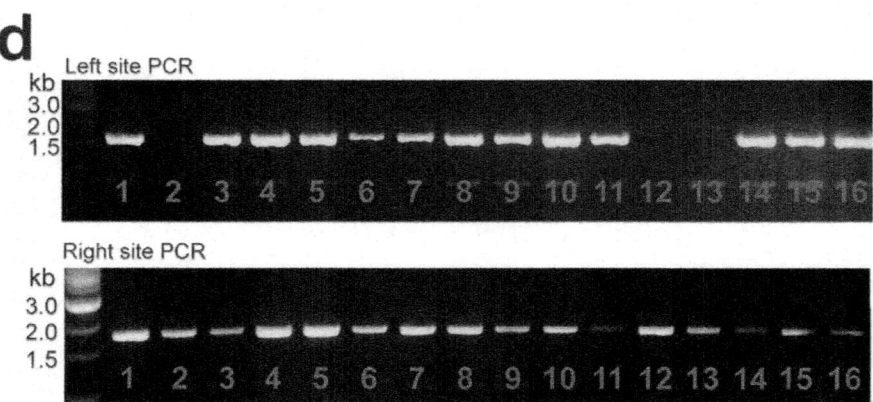

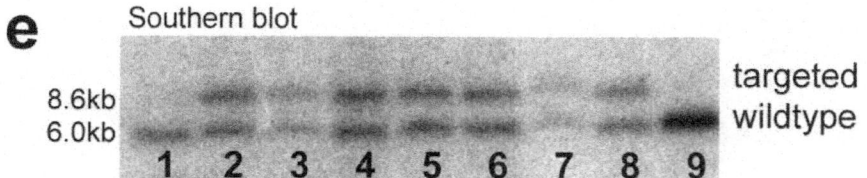
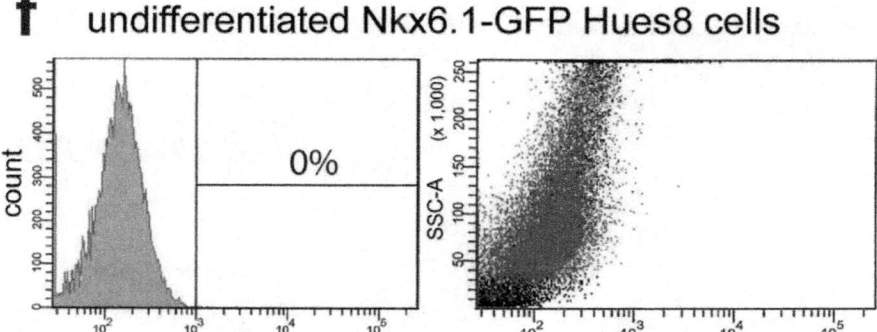
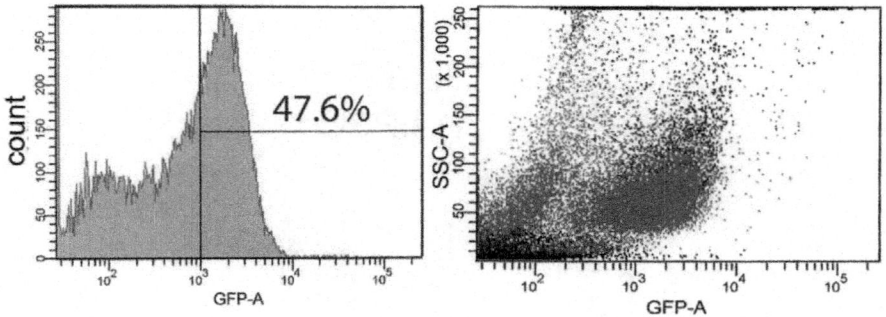

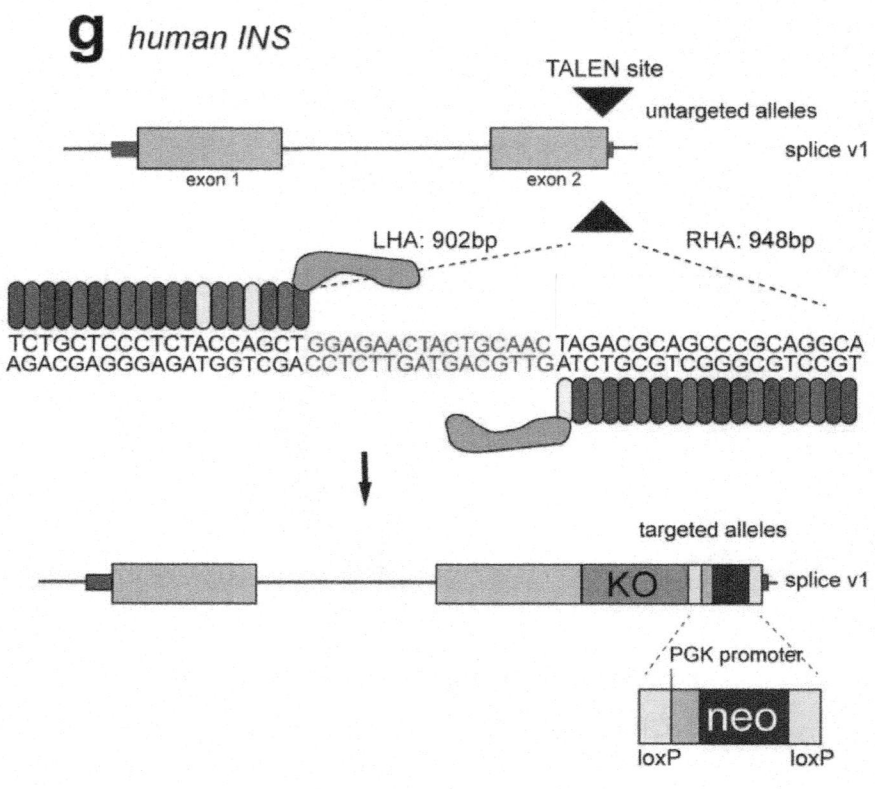
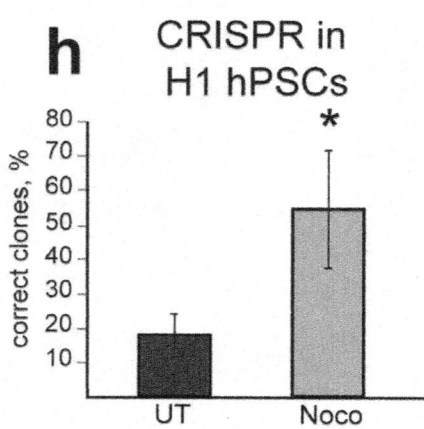

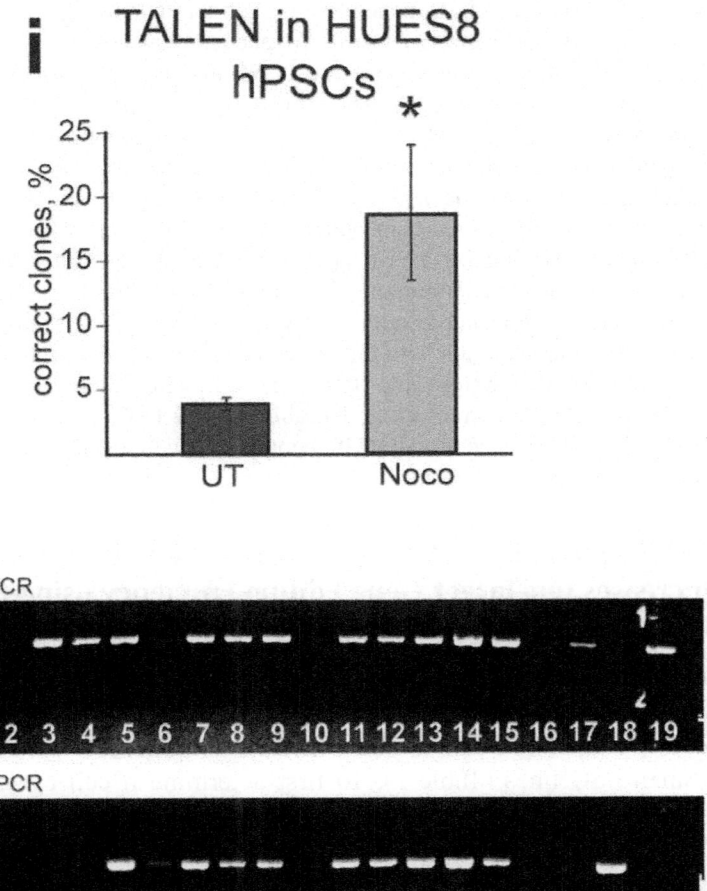

**Figure 5:** G2/M synchronization increases targeting efficiency of *NKX6.1* **with ZFNs and *INS* with TALENs and CRISPR/Cas9.** (a) Human *NKX6.1* gene structure and targeting scheme. ZFNs cleave using FokI nuclease at the 3′ end of the first exon to create DSB for insertion of a fluorescent reporter, eGFP. (b) The efficiency of targeting in H1 cells determined by counting the percentage of correctly targeted colonies compared to resistant colonies after PCR confirmation; $n=2$ independent experiments. $*P<0.05$; $***P<0.001$ (c) The efficiency of targeting in DiPSCs determined by counting the percentage of correctly targeted colonies compared to resistant colonies after PCR confirmation; $n=2$ independent experiments. $****P<0.0001$. (d) PCR confirmation of the identity of targeted individual numbered clones. (e) Southern blot confirmation of targeted alleles after XbaI digestion and 5′ end probe external to the integrating cassette. Correctly targeted clones are 8.6 kb while the wildtype clones are 6.0 kb. (f) GFP is not expressed in undifferentiated NKX6.1-GFP HUES8 cells and is

unregulated specifically in pancreatic progenitors confirming the faithfulness of the reporter. **(g)** Human gene structure of *INS* and targeting scheme. *INS* was targeted using TALENs with FokI nuclease to create a DSB in the 3' end of the gene for insertion of a fluorescent reporter, KO. **(h)** The efficiency of targeting in H1 cells determined by counting the percentage of correctly targeted colonies compared to resistant colonies after PCR confirmation; $n=3$ independent experiments. $*P<0.05$. **(i)** The efficiency of targeting in HUES8 cells determined by counting the percentage of correctly targeted colonies compared to resistant colonies after PCR confirmation; $n=3$ independent experiments. $*P<0.05$. **(j)** PCR confirmation of the identity of targeted individual numbered clones. Gene structure is annotated including exons (light blue), introns (brown lines), untranslated regions (dark blue), and targeting sites (large navy arrows). Below the untargeted allele(s) the target sequence is shown with left homology arm (LHA) and right homology arm (RHA) sizes. Targeted alleles show the donor cassette integrated into the correct locus. Kusabira orange (KO) and eGFP cassettes have 3' antibiotic resistance genes, KO with neomycin and eGFP with puromycin. Error bars represent SEM.

## Synchronization of Multiple Human Cell Lines in G2/M Phase Increases On-Target Gene Editing Efficiency using CRISPR, CRISPRn, ZFNs, and TALENs for Five Different Genetic Loci

To test the efficiency of on-target gene editing after cell synchronization in the G2/M phase, we used customized nuclease gene targeting systems such as, ZFNs, TALENs, CRISPR and CRISPRn to target five genes in five different human PSC lines (Table 1). To first determine if cell cycle phase increases the efficiency of correct targeting, we FACS sorted FUCCI-H9 cells based on their cell cycle indicators to obtain G1 and G2/M enriched populations. These populations were immediately targeted after sorting with CRISPR/Cas9*S. pyogenes* to incorporate Cyan Fluorescent Protein (CFP) at the 3' end of the *OCT4* loci, a gene that is known to be accessible for targeting in hPSCs. Five days post targeting, G2/M populations showed robust CFP expression with 40.93% CFP+ cells out of total cells, compared to 6.98% for unsorted, nonsynchronized, FUCCI-H9 or 11.67% for H1 hPSCs (Fig. 3). G1 populations showed decrease CFP expression (3.61%) compared to unsorted FUCCI-H9. The significant increase in CFP+ cells targeted in G2/M phase and decrease in G1 phase, respectively, indicates that cell cycle arrest influences gene editing efficiency and synchronization of cells with Nocodazole or ABT may be a useful tool to increase the efficiency of these technologies.

**Table 1:** Cell cycle synchronization leads to a marked improvement in genome targeting by HDR regardless of cell line or gene target

|  | Wnt5a | NeuroD1 | Nkx6.1 | Insulin | Oct4 |
|---|---|---|---|---|---|
| Cells targeted | 1. H1 | 1. H1 | 1. H1 2. HUES8 | 1. H1 | 1. FUCCI-H9 |
|  | 2. NPCs | 2. HUES8 | 3. DiPSCs | 2. HUES8 |  |
|  |  | 3. DiPSCs | 3. DiPSCs |  |  |
| Targeting strategy | 1. CRISPR nickase (D10A, S. pyogenes | 1. CRISPR (S. pyogenes) 2. CRISPR (N. meningitides | 1. ZFNs | 1. CRISPR (S. pyogenes)2. TALENs | 1. CRISPR (S. pyogenes |
| Average fold increase in efficiency with synchronization | 3.75 | 4.66 | 5.32 | 4.61 | 5.86 |

Five genes are targeted in five different cell lines using three different targeting strategies, as listed. The effect of synchronization is listed as the difference in percent correct clones (synchronization improvement of efficiency) or the fold-increase in efficiency (fold increase in efficiency with synchronization).

We next used CRISPR *S. pyogenes* D10A nickase to target the *WNT5A* locus, a gene high in heterochromatin content compared to Oct4, to create a knockout by inserting a neomycin resistance gene into the first constitutive exon of *WNT5A* (Fig. 4a). Nocodazole or ABT synchronization again significantly increased the number of clones with integrated donor cassette as measured by % of individual clones before and after antibiotic selection (78.43% and 78.41%, respectively), compared to untreated, non-synchronized cells (18.65%) or cells treated with the vehicle/DMSO (20.14%) (Fig. 4b). While combining Nocodazole with SCR7 had a significant increase in resistant clones compared to control cells (53.0%), this combination did not further improve targeting compared to Nocodazole treatment alone. We tested this approach on H1-derived NPCs, which are known to have a higher rate of NHEJ than hPSCs32, and found that they were also more susceptible to targeting after synchronization in the G2/M phase of the cell cycle. Nocodazole (56.3%) and ABT (29.67%) treatment increased the number of cells resistant to antibiotic treatment compared to untreated cells (11.03%) and DMSO-treated cells (5.57%) (Fig. 4c,e). Again, combining Nocodazole with SCR7 showed increased gene-targeting efficiency (37.86%) when compared to control, but at a lower level than with Nocodazole alone. Donor integration was confirmed by PCR of the right site of the cassette and the left site of the cassette, showing integration of the neomycin cassette in the *WNT5A* locus in targeted H1 hPSCs (Fig. 4d). Targeting of *WNT5A* in hPSCs and NPCs shows that synchronization has a positive effect on the efficiency of genome modification in various human cell types.

To investigate whether synchronization improves the cassette insertion efficiency at any given gene, we chose to target human *NEUROD1* in non-synchronized and Nocodazole-, and ABT-synchronized H1 hPSCs using CRISPR *S. pyogenes* and *N. meningitidis* to create a knock-in of fluorescent genes (Kusabira orange, KO, or eGFP) with an antibiotic resistance gene (Fig. 4f). We tested both the commonly applied CRISPR *S. pyogenes* as well as CRISPR *N. meningitidis*, as CRISPR *N. meningitidis*has been shown to have a higher targeting efficiency in hPSCs than CRISPR *S. pyogenes* and TALENs. Cells synchronized with Nocodazole (59.77%) or ABT (79.91%) showed a 3–5 fold increase in the number of clones with targeting cassette integrated at the 3′ of *NEUROD1* exon1 with compared to untreated cells (17.46%), with CRISPR *S. pyogenes*targeting quantified by % of single clones positive for three independent PCRs confirming the correct targeting out of total analyzed clones (n > 200) (Fig. 4g). These data suggest that improvements in on-target insertion efficiency after elongation of the G2/M phase are not limited to a particular gene. We then targeted DiPSCs, as iPSCs are increasingly used for regenerative medicine studies, using CRISPR *N. meningitides* and again observed a significant increase in on-target donor integration when cells were synchronized with Nocodazole (52.49%) or ABT (53.50%) compared to untreated (7.66%) and DMSO (7.86%) controls (Fig. 4h). Donor integration in each experiment was confirmed by PCRs for individual clones in untreated and synchronized cultures, verifying the integration of the donor cassette in the correct locus and showing a higher efficiency of correctly targeted clones in synchronized cells (Fig. 4i).

Although CRISPR is currently the most popular targeting strategy, we were interested in determining whether the effect of arresting cells in G2/M phase was unique to this technology or if other DSB-based gene editing tools would also benefit from synchronization. We targeted H1 and Hues8 hPSCs and DiPSCs for knock-in of eGFP fluorescent reporter with a floxed puromycin resistance gene in the *NKX6.1* locus with ZFNs (Fig. 5a). Synchronization of H1 and Hues8 hPSCs with Nocodazole (47.19%) or ABT (60.46%) showed 4 to 6-fold increases in efficiency of donor integration at the 3' end of *NKX6.1* exon 1 versus untreated cells (9.55%) (Fig. 5b). Synchronization of DiPSCs with Nocodazole and ABT also increased desired gene editing efficiency with ZFNs, 20.4% and 22.95% respectively, compared to untreated (5.23%) and DMSO-vehicle treated (6.95%) cells (Fig. 5c), as quantified by % of correctly targeted clones out total analyzed clones. PCR was used to confirm the correct insertion of the donor cassette into the *NKX6.1* exon 1 by amplifying the left insertion site and right insertion site, and selected clones were further verified by Southern blot using external 5' end probe (Fig. 5d,e). Since *NKX6.1* is not expressed in the hPSC stage, we additionally confirmed the functionality of the GFP reporter by *in vitro* differentiation of NKX6.1-eGFP Hues8 clones into pancreatic progenitors and analyzed eGFP fluorescence by flow cytometry. The GFP was not expressed in undifferentiated cells but specifically up regulated at pancreatic progenitors, when 47.6% of total cells expressed eGFP (Fig. 5f). Finally, we targeted H1 and Hues8 cells using CRISPR *S. pyogenes* and TALENs to tag human *INSULIN* with a fluorescent KO reporter and a neomycin resistance gene (Fig. 5g). Targeting untreated, non-synchronized H1 cells with CRISPR resulted in 17.77% clones positive for two external PCR confirming the correct integration of KO and selection gene, out of total analyzed clones (n=300), compared to 54.64% for Nocodazole-treated cells (Fig. 5h). Hues8 hPSCs targeted with TALENs to the same locus showed a 5.33-fold increase in *INSULIN* targeting efficiency after synchronization with Nocodazole (Fig. 5i) compared to DMSO treated controls, showing that the increased efficiency of targeting after synchronization is independent of targeting method since both CRISPR and TALEN targeting at the same human *INSULIN* locus was markedly improved. Correct integration was confirmed by 5' and 3' external PCRs as well as cassette specific PCR and Southern Blot (Fig. 5j and data not shown). By targeting different human loci with different donor cassettes in various hPSC lines and hPSC-derived differentiated cells using ZFNs, TALENs, CRISPR *S. pyogenes*, CRISPR *N. meningitides*, and CRISPR *S. pyogenes* D10A nickase in synchronized and unsynchronized cells, we have shown that gene editing is globally improved by synchronization of the cell cycle at the G2/M phase. This improvement is an advance towards high efficiency gene editing in hPSCs and other human cell types.

## hPSCs Retain Pluripotency and are Capable of Differentiating into Different Germ Layers upon G2/M Synchronization

Cell cycle manipulation and G0/G1 phase elongation have been shown to stimulate the differentiation of hPSCs. To explore whether arrest in the G2/M phase stimulates hPSC differentiation, immunofluorescence staining was performed on synchronized H1 hPSCs with pluripotency markers, such as SSEA4, OCT4, and SOX2(Fig. 6a). All cells stimulated with either Nocodazole or ABT exhibited expression of known markers of pluripotency and tri-lineage differentiation potential. Gene expression analysis of pluripotent genes *OCT4, NANOG, REX1, GDF3*, and *LIN28* in Nocodazole- and ABT-synchronized hPSCs showed strong correlation to untreated cells (Fig. 6b and Supplementary Table 11; H1 DMSO $R^2=0.83$, Nocodazole $R^2=0.89$, ABT $R^2=0.93$; Hues8 DMSO $R^2=0.97$, Nocodazole $R^2=0.90$, ABT $R^2=0.97$). Additionally, there was no difference between synchronized and unsynchronized hPSCs in expression of endoderm markers, such as *FOXA2, GATA4, CXCR4,* and *SOX17* (Hues8 DMSO $R^2=0.94$, Nocodazole $R^2=0.88$, ABT $R^2=0.96$; definitive endoderm $R^2=0.25$); mesoderm markers, such as *BRA(T), TPNT,* and *NKX2.5* (H1 DMSO $R^2=0.98$, Nocodazole $R^2=0.98$, ABT $R^2=0.97$; Hues8 DMSO $R^2=0.99$, Nocodazole $R^2=0.99$, ABT $R^2=0.97$; cardiac progenitors $R^2=0.07$), and ectoderm markers, such as *PAX6*48 and *SOX1*49 (Fig. 6b and Supplementary Table 11). Quantitative PCR results confirmed that *in vitro* differentiated iPSCs expressed markers of definitive endoderm, cardiac progenitors, and neural progenitors (Fig. 6b). Genome editing of hPSCs could be applied to model human diseases or to make patient-specific lines to correct patient mutations and must not affect the differentiation potential to generate necessary cell types of any lineage. Therefore, we determined the differentiation potential of synchronized hPSCs into endoderm, pancreatic progenitors, and cardiac progenitors. After synchronization, H1 hPSCs differentiate into endoderm (SOX17), pancreatic progenitors (PDX1), and cardiac progenitors (NKX2.5 and ACTN1) (Fig. 6c). Collectively, these results show that transient G2/M arrest does not affect the pluripotency of hPSCs or their morphology, and they maintain their potential to differentiate into multiple germ layers, making synchronization a useful tool for improving the efficiency in human genome editing.

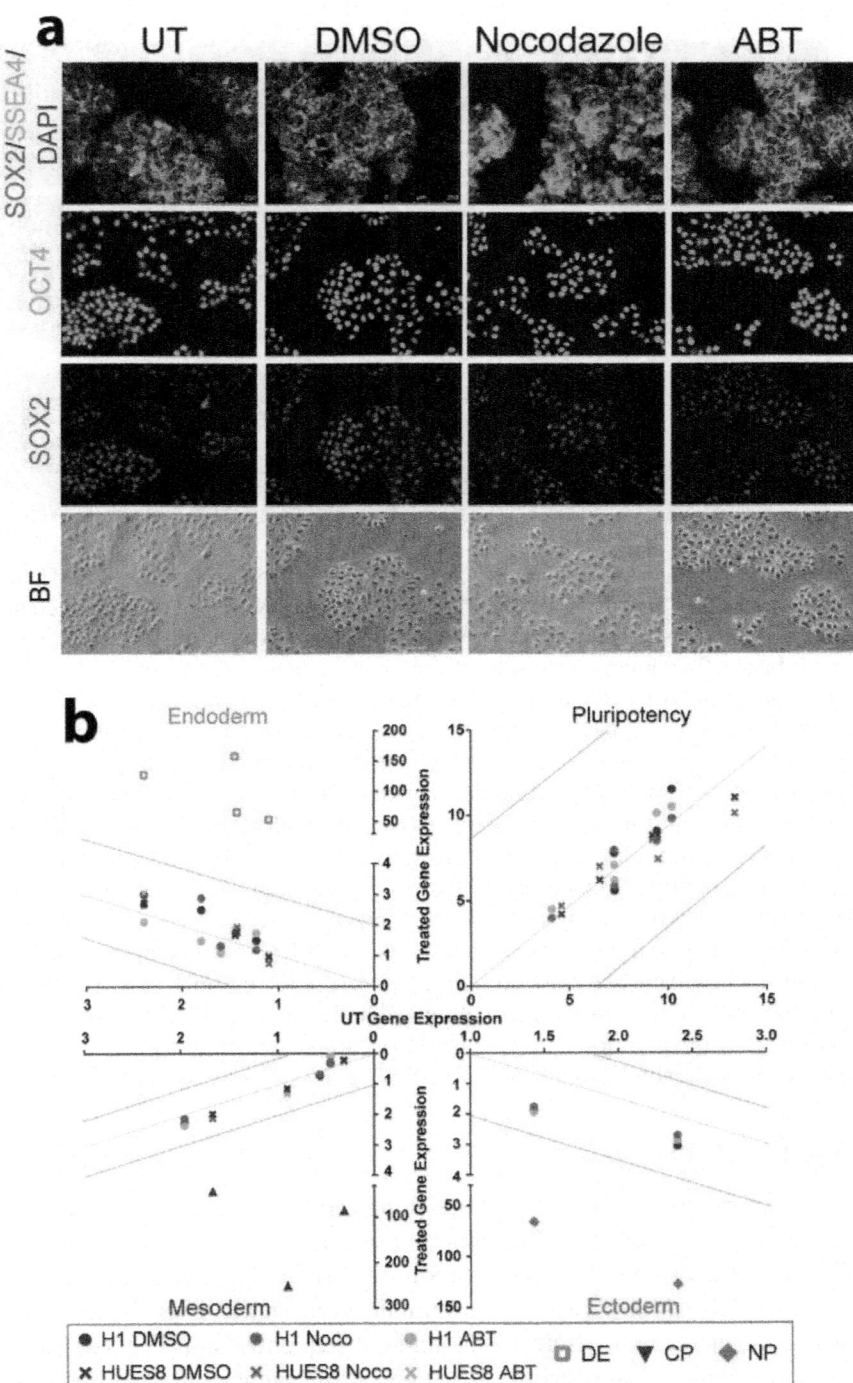

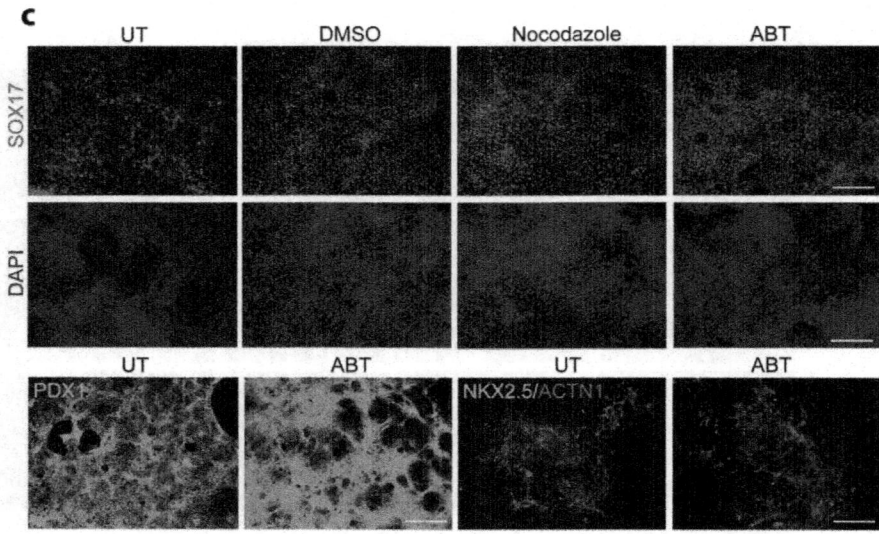

**Figure 6:** Pluripotency is maintained in hPSCs after synchronization. (a) H1 hPSCs were synchronized with Nocodazole, ABT, the vehicle DMSO, or left untreated before staining for markers of pluripotency, including SSEA4, SOX2, and OCT4. The bottom panel is bright-field imaging showing that all cells express OCT4 and SOX2. (b) Gene expression analysis by qPCR of 5 pluripotency markers, 4 markers of endoderm, 3 markers of mesoderm, and 2 markers of ectoderm in Nocodazole- and ABT-synchronized and DMSO-treated H1 and HUES8 hPSCs compared to untreated cells. Correlation of $R=1.0$ shown by a dashed gray line. Dashed red lines annotate the 95% confidence interval from untreated gene expression, with all markers within these two lines not statistically different from untreated cells. H1 and HUES8 cells were differentiated into definitive endoderm (DE), cardiac progenitors (CP), and neural progenitors (NP) to show the fidelity of candidate markers of germ layers. Target genes and expression is shown in Supplementary Table 11(H1 hPSCs marked by circles, HUES8 by X, DMSO by dark blue, Nocodazole by purple, ABT by light blue, DE by green, CP by red, and NP by pink). (c) Synchronized and untreated H1 hPSCs differentiated into pancreatic progenitors (PDX1; upper left), cardiac progenitors (NKX2.5/ACTN1; upper right), and definitive endoderm (SOX17, lower).

## DISCUSSION

We report a simple and robust gene targeting system to increase donor integration by enriching HR events. For the first time, we demonstrate ABT and Nocodazole as separate tools to efficiently synchronize hPSCs in the G2/M cell cycle phase with minimal cytotoxicity. This synchronization leads to an increase in HDR events and subsequently significantly improves on-target genome engineering efficiency. Pluripotency is not affected after

G2/M phase synchronization and cells remain capable of differentiation into multiple lineages, making ABT and Nocodazole great tools for gene editing in hPSCs. Using ZFNs, TALENs, two different species of CRISPR/Cas9, and CRISPR nickase, we successfully targeted five different human cell lines for five different genes. We observed a locus-independent increase in efficiency of on-target gene targeting after synchronization in four different hPSCs as well as differentiated multipotent human cells (such as NPCs). Additionally, synchronization equally improved efficiency of all four targeting tools used to induce DSBs, demonstrating that increased HDR after G2/M phase arrest is a robust and universal property of hPSCs.

During the G2/M phase, cells preferentially repair DNA by HR when sister chromatids are available to act as homologous templates. The principal repair mechanisms of hPSCs at different phases of the cell cycle are not well-established. To understand the mechanism of repair in hPSCs during the G2/M phase, we synchronized cells to test the rate of HR and NHEJ. hPSCs arrested in the G2/M phase had elevated levels of HR compared to untreated hPSCs. In particular, ABT synchronization induces high rates of HR and decreased levels of indel formation compared to untreated cells. Therefore, ABT might be the optimal choice to improve the efficiency of gene targeting.

Other small molecules have been tested to enhance gene targeting either in mouse stem cells or embryos as well as in human carcinoma cell lines, but these studies focused only on CRISPR, showed some toxicity or were technically challenging. To further improve efficiency, with ABT and Nocodazole we also tested SCR7, which acts to inhibit NHEJ through interference of DNA binding in the canonical NHEJ pathway. When we combined Nocodazole or ABT with SCR7, we observed some decrease in NHEJ rates in hPSC, but found no significant increase in HR or gene targeting efficiency. The negligible effect of SCR7 may indicate that HR is the major DNA repair mechanism in the G2/M phase of hPSCs, with the rate of HR unaffected by further inhibition of NHEJ. In differentiated hESC-derived NPCs, in which NHEJ is 1.3× more predominate than in hPSCs, we again found no additional effect on HDR with SCR7. In hPSCs and NPCs, the canonical NHEJ pathway is the primary mechanism of NHEJ repair, but in circumstances when the canonical pathway is blocked, noncanonical NHEJ pathways independent of DNA ligase IV aid in repair. While studies have shown inhibiting NHEJ can increase HDR21, in hPSCs we found that increasing HDR to prevent NHEJ compensation through noncanonical pathways more effectively increases the efficiency of on-target gene editing by customizable endonucleases.

We found that synchronizing hPSCs in the G2/M cell cycle phase increases the efficiency of gene editing on average 4-fold (Table 1). Enriching the

G2/M phase cells to increase gene editing efficiency in hPSCs is not limited to selectable constructs and can also be applied in non-selectable contexts, such as scarless introduction of mutations such as SNPs. The novel tools explored in this report facilitate efficient gene editing of hPSCs, providing the opportunity for human cells to be used for disease modeling, epitope-tagging or fluorescent reporter generation and for patient-derived mutation-corrected tissue for therapeutic application.

## METHODS

### Cell Culture

hPSC lines, H1, Hues8, and FUCCI-H9, were cultured under feeder-free on hESC-qualified Matrigel (BD Biosciences) in E8 media (Stemcell Technologies) with 30% of irradiated mouse embryonic fibroblasts (iMEFs) conditional media. FUCCI-H9 hPSCs were kindly provided by S. Dalton and were maintained under G418 sulfate (Sigma, 200 ug/ml) and Puromycin (0.1 ug/ml) selections. Cells were passaged every 3–5 days at 80% confluent with TrypLE Express (Invitrogen). After dissociation, cells were plated in E8 media with 10 uM Y-27632, (StemGent) for 24 hrs. After 24 hrs media, without Y-27632, was replenished daily. iMEF conditional media was prepared by incubating iMEFs with hPSC media without bFGF for 24 hrs for 7 days. Collected media was filtered, flash frozen and stored at −80 °C.

To maintain neural progenitors, cells were cultured in NPM media: DMEM/F12: Neurobasal media at a 1:1 ratio +0.5xB27 +0.5xN2 +2 mM Glutamax +20 ng/mL bFGF (R&D Systems) +20 ng/mL EGF (R&D Systems) and media was replenished every four days. To split cells, cells were washed with PBS and incubated for 5 min at 37 °C in Accutase (Innovative Cell Technologies). Cells were washed, spun down at 1000 g for 5 min, and replated in NPM media on Geltrex-coated plates.

### Cell Cycle Synchronization

Concentration and time course of different G2/M synchronization and release times was performed as presented in Supplementary Table 2 and for gene targeting and other experiments the 16 hr incubation and 1 hr release for Nocodazole (Sigma) at 1 ug/ml and 16 hr treatment without release for ABT (Selleckchem) at 0.37 µg/ml was selected. As negative control, solvent for small molecules, DMSO was used at 1:100 or 1:1000 dilution.

## Cell Cycle Analysis

For DNA content analysis, hPSCs cells were dissociated by TrypLE Express (Invitrogen) and fixed with 70% EtOH. Propidium iodide was used to for H1 and HUES8 cells and Hoechst 33342 (both Sigma) for FUCCI-H9 cells for 30 min at 37 °C to stain DNA. Samples were then fixed with 2% Paraformaldehyde/PBS (PFA), washed and filtered through 40 um cell strainer (BD Biosciences) before flow cytometry. FACS analysis was performed using LSRII (BD Biosciences) and analyzed using FACSDiva software. Gates were set with reference to negative controls.

## FUCCI Cell Sorting

Cells were sorted on the FACSAria (BD Biosciences) using FACSDiva software.

Gates were set with reference to negative controls. The sorting speed was adjusted to ensure sorting efficiency above 90%. Cells were collected in tubes with Sorting buffer (PBS, 5% FBS and 2 mM EDTA).

## HR and NHEJ Reporter Assay

The HR and NHEJ reporter constructs were kindly provided by V. Gorbunova (University of Rochester). The plasmids were digested overnight with *I-SceI* at 37 °C and confirmed by gel electrophoresis. Digestion products were then purified by phenol/chloroform extraction. Three μg of digested DNA was nucleofected to $1 \times 10^6$ of H1 cells using CA137 program (Lonza). After incubation for 48 hrs at 37 °C, cells were dissociated and GFP (indicating successful HR or NHEJ repair) was detected and quantified by flow cytometry.

## DNA Delivery

hPSCs and NPCs were pre-treated with 10 uM Y-27632, 24 hrs before nucleofection. On the day of nucleofection, hPSCs were disassociated into single cells with TrypLE; Accutase was used for NPCs. 1–2.5 million cells were re-suspended in 100 ul P3 buffer plus Supplement with 10 ug of total DNA and nucleofected using CM113 program using Amaxa 4D-Nuclefector (Lonza). Cells were immediately transferred with 500 ul of appropriate media to 1.5 ml tube and recovered for 10 min at 37 °C before plating on iMEFs at 15,000-cell/cm$^2$ density in media with Y-27632.

For indel analysis 293HT cells were treated with DMSO, Nocodazole, ABT, Nocodazole with SCR7, SCR7, or untreated for 16 hrs. Cells were transfected with Lipofectamine 2000 (Life Technologies) with 1.5 ug of pX330

OCT4 sgRNA. Four days after transfection, cells were collected, spun down at 2000 g for 5 minutes, and DNA was extracted.

## Generation of Gene Targeting and CRISPR-Cas9 CONSTRUCTS

Single stranded oligonucleotides were phosphorylated and annealed (Supplementary Table 5) before digestion and ligation of pX335 D10A Cas9 plasmid (Addgene #42335) for CRISPRn, pX330 plasmid for CRISPR targeting OCT4, pX459 plasmid for *S. pyogenes* Cas9 targeting of*NEUROD1* (Addgene #48139), or pSimpleII backbone for *N. meningitidis*(Addgene #47868) with paired oligos performed as described51 with digestion/ligation with BbsI and T7 ligase by PlasmidSafe exonuclease digestion (EpiCentre).

*WNT5A* was targeted within the first constitutive exon (Supplementary Table 6) using plasmids shown in Supplementary Table 4, with homology arms as shown in Supplementary Table 7. *NEUROD1* was targeted with two different Cas9 species (Supplementary Table 4) with homology arms for the same genomic region (Supplementary Tables 5–7). *NKX6.1* was targeted in the 3' end using Zinc finger nucleases with FokI endonuclease (Supplementary Tables 4–6). ZFNs were designed to bind to 18 bp flanking a 5 bp spacer sequence using CompoZr™ Custom Zinc Finger Nucleases (Sigma Aldrich) to construct PZFN plasmid DNA as well as *NKX6.1* ZFN mRNA. Activity of *NKX6.1* ZFN plasmid DNA and mRNA was assayed by PCR amplification of the target region with AccuPrimer GC Rich DNA Polymerase (Sigma Aldrich) followed by CelI nucleotide mismatch assay. *INSULIN* was targeted at the 3' end of the gene with TALENs with FokI (Supplementary Tables 4–7). TALENs were designed using GeneArt® Precision TALs (Life Technologies) and cloned into aTAL entry clones before recombination into the destination vector pcDNA™-DEST40 by Gateway® cloning catalyzed by LR Clonase® II enzyme (Life Technologies). PCR of eCFP using high fidelity Q5 polymerase with primers with a 5' NheI overhang and 3' AscI overhang was used to create OCT4-eCFP donor (Supplementary Tables 4–8). All oligonucleotides and homology arms are from IdtDNA. All enzymes unless otherwise noted are from NEB.

## Indel Analysis

PCR was performed on DNA using Q5 High Fidelity Polymerase (NEB) to amplify the targeted region (primers in Supplementary Table 8) before isolating by gel extraction. A hybridization reaction with NEBuffer 2, 200 ng PCR product, and water was run at 95 °C for 5 minutes, ramping down to 85 °C at 2 °C/s, ramping down to 25 °C at 0.1 °C/s. Hybridized PCR products are incubated at 37 °C with T7 endonuclease I for 15 minutes and then EDTA was used to stop the reaction. Products are run on 6% polyacrylamide gel to distinguish cut and uncut bands.

## Drug Selection and Clonal Expansion

hPSC clones that have stably integrated the targeting vectors were identified through the selection with 50 ug/ml G418 or 0.1 ug/ml puromycin. Selection was started at day 2 post-nucleofection and at day 4 the G418 concentration was increased to 100 ug/ml and puromycin to 0.5 ug/ml. After 7 days of selection, drug-resistant colonies were individually picked into 96-well plates with iMEFs and expanded for cell banking and DNA extraction. Cells for genomic DNA extraction were cultured on 0.1% gelatin coated plates until >90% confluent. A similar procedure was used to identify drug-resistant NPCs.

## Identification of Targeted Clones

Genomic DNA was isolated using DNaesy Blood and Cell Culture kit (Qiagen) or KAPA direct mouse genotyping kit (KAPA Biosystems). For each targeted locus, at least 3 sets of genotyping primers were designed, one set internal to detect the cassette, and two pairs spanning the junction of genomic sequences and targeting vector. Sequences of all primers are listed in Supplementary Table 8.

## Southern Blot

External and internal probes were generated by PCR using the PCR DIG Probe Synthesis Kit (Roche). For the external probes we used HUES8 genomic DNA as the template and for internal probe generation the targeting plasmids. Genomic DNA was isolated from hPSCs by proteinase K digest and phenol/chloroform extraction. 10 ug of DNA was digested overnight with appropriate restriction enzymes, separated on a 0.8% agarose gel, denatured, neutralized and transferred by capillarity onto a Hybond-N membrane (GE Healthcare) using 10×SSC buffer. The hybridization was carried out overnight at 65 °C with rotation with a 5 ul of DIG labeled probe in 15 ml of hybridization buffer. Blots were prepared and washed using commercial buffers (DIG block and wash buffers, Roche) following manufacturer's instructions. The DIG probe was detected with an AP-conjugated anti-DIG antibody followed by chemiluminescent detection (CDP-Star, Roche) and developed on film.

## Directed Differentiation

To initiate pancreatic differentiation, the cells were dissociated using TrypLE Express to single cells and seeded at 150,000 cell/cm$^2$ onto 1:30 dilution of growth factor reduced Matrigel (BD Biosciences) in DMEM/F12 in E8-MEF conditional media with 10 uM Y27632. Two days following seeding the differentiation was started. Day 1 cells were exposed to RPMI +3 uM CHIR-99021 (Stemgent) +100 ng/ml rhActivinA (R&D Systems). Day 2–3: +100 ng/ml rhActivinA. Day 4–5: +50 ng/ml FGF7 (Peprotech). Day 6–9: DMEM +50 ng/ml FGF7 +2 µM RA (Sigma) +0.25 µM SANT-1 (Sigma) +100 ng/ml rhNoggin (R&D Systems).

To generate neural progenitor cells, H1 hPSCs were cultured on Matrigel coated plates in E8 media for two days before induction to the neural program for five days DMEM/F12:Neurobasal media at 1:1 ratio +1xB27 + 1×N2 +2mM Glutamax (all Invitrogen).

Prior to cardiac differentiation, hPSCs were passaged onto Matrigel coated plates. At 70–95% confluence cells were exposed to rhActivinA and WNT3A for 1 day (R&D Systems) in Advanced RPMI (Invitrogen) supplemented with 2% KOSR (Gibco), ascorbic acid (Sigma), NEAA (Gibco), BSA (Gibco) and thioglycerol (Sigma). For the next 2 days, cells were treated with ActivinA, WNT3A, BMP4, and transferrin; at day 4 with BMP4 and transferrin; from day 6 onwards, with basal differentiation medium only.

## Antibody Staining

For immunofluorescent analysis, cells were briefly washed with PBS and then fixed with 4% PFA in PBS for 20 min at RT followed by washing three times with PBST (PBS +0.1% Triton-X). The unspecific binging of antibodies was blocked by 30 min incubation with blocking solution (10% donkey serum in PBST) at RT. The primary antibodies were in blocking solution for 16 h at 4 °C with shaking and then cells were washed three times with PBST for 10 min. The secondary antibodies were conjugated with appropriated Alexa Fluor Dye (Jackson ImmunoResearch Laboratories), diluted with blocking solution and incubated with cells for 1 h at RT. Cells were then washed three times with PBST and nuclei were stained with DAPI (Invitrogen). All primary antibodies and dilutions are listed in Supplementary Table 10.

## Gene Expression Analysis

For quantitative RT-PCR, total RNA was isolated using Trizol (Invitrogen). The RNase-free DNAse treatment was used to remove any traces of genomic DNA according to the manufacturer's protocol (Qiagen). One ug of RNA was used for reverse transcription using iScript (Biorad). 1/10 of cDNA was used for PCR using SYBR Green (KAPA Biosystems) and a Connect CFX light cycler (Biorad) ($\leq$40 cycles). Primers were designed to span or amplify across the exon junctions using Primer3 software. The specificity of PCR products was verified by melt curve analysis using Precision Melt Analysis Software (Biorad) and gel electrophoresis, followed by TOPO cloning (Invitrogen) and Sanger sequencing. Threshold data were analyzed in CFX Manager Software v3.1 (Applied Biosystems) using the Comparative Ct relative quantitation method, with beta-actin and TBP as the endogenous controls. Sequences of primers are shown in Supplementary Table 9. All primers were purchased from IdtDNA.

## Statistical Analysis

P values were calculated using t-test and PRISM6 software. For multiple comparison tests Bonferroni correction was used. Unless otherwise stated in figure legends, data are presented as mean±SD and the following symbols are used to represent P values, $*P<0.05$, $**P<0.01$, $***P<0.001$, and $****P<0.0001$. N represents number of independent experiments.

# SUPPLEMENTARY INFORMATION

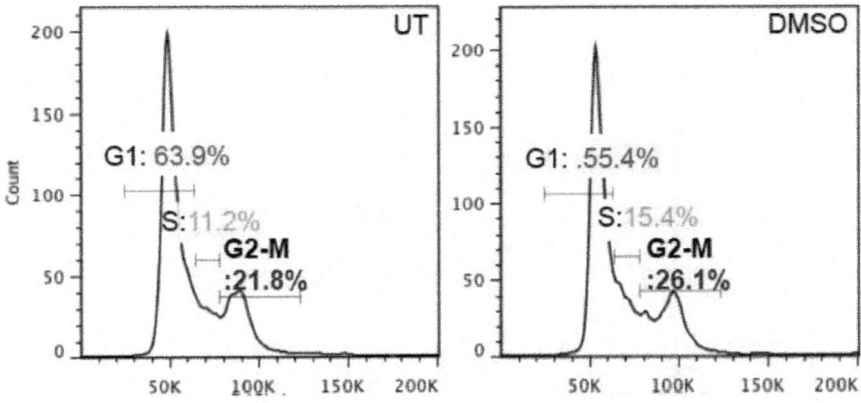

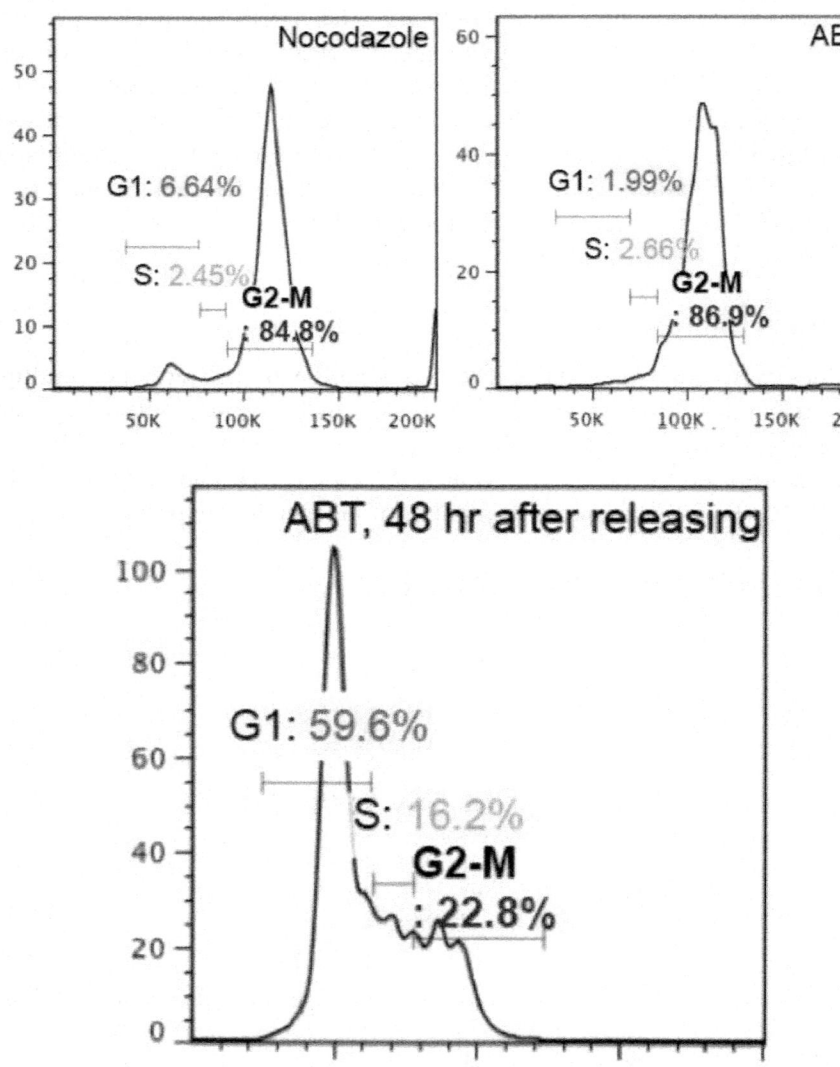

|  | UT | DMSO | Nocodazole | ABT |
|---|---|---|---|---|
| G1 (%) | 59.13±4.88 | 63.83±9.08 | 4.27±1.89 | 4.37±3.11 |
| S (%) | 11.67±1.15 | 10.87±1.97 | 1.6±0.36 | 1.57±0.35 |
| G2/M (%) | 28.8±4.45 | 24.7±8.22 | 90.73±1.56 | 91.43±3.29 |

**Supplementary Figure 1:** Synchronized Fucci H9 cells detected measured by DNA content and flow cytometry. Representative FACS plots of synchronized Fucci H9 cells stained with Hoechst show efficient G2/M arrest in ABT or Nocodazole treated cells as compared to untreated and DMSO vehicle control cells (Top panel). ABT treatment shows a reversible effect of cell cycle after 48 hours of releasing. Lower panel shows Mean ± standard deviation for percentage of each cell cycle phase (n=3 biological replicates).

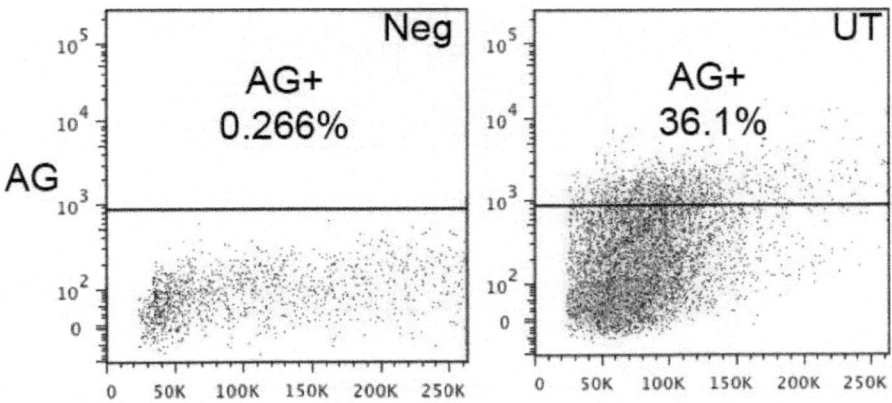

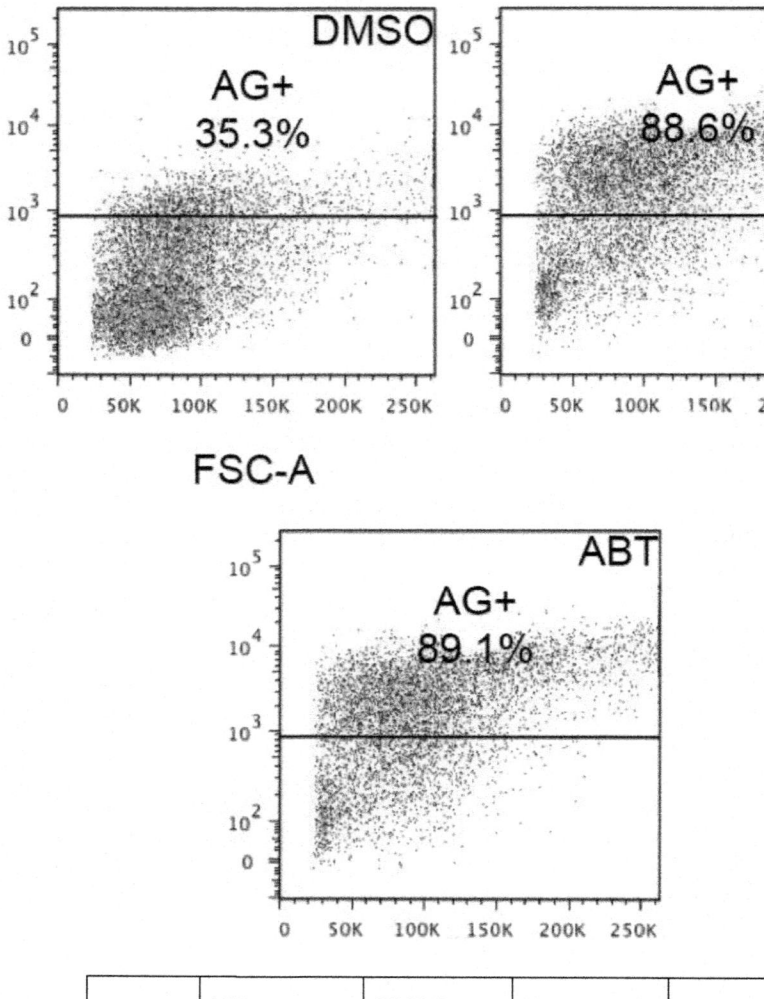

|  | UT | DMSO | Nocodazole | ABT |
|---|---|---|---|---|
| DN (%) | 45.07±27.43 | 59.93±1.01 | 7.33±0.67 | 7.4±0.61 |
| KO+(%) | 9.4±15.07 | 0.67±0.06 | 3.37±0.67 | 2.03±1.59 |
| AG+(%) | 38.23±8.41 | 34.4±1.31 | 88.43±0.38 | 88.6±0.7 |

**Supplementary Figure 2:** G2/M cell cycle phase of Fucci H9 cells measured by AG fluorescence with and without synchronization using flow cytometry. Representative FACS plots of synchronized Fucci H9 cells show efficient G2/M arrest by ABT or Nocodazole treated cells as compared to untreated and DMSO vehicle control cells

(Top panel). Lower panel shows Mean ± standard deviation for percentage of each cell cycle phase (n=3 biological replicates). DN, double negative. KO, Kusabira orange. AG, Azami Green.

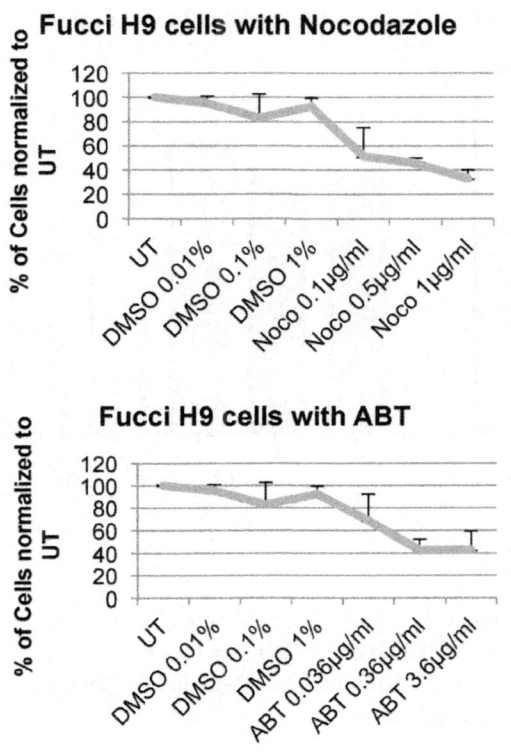

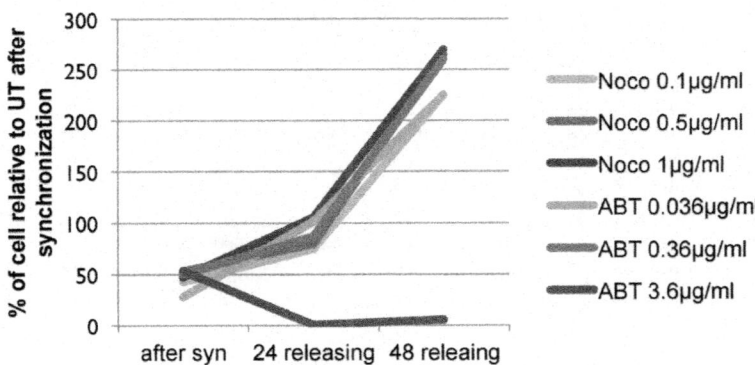

**Supplementary Figure 3:** Cell survival rate after synchronization with Nocodazole and ABT. Fucci H9 cell counts immediately after synchronization normalized to untreated cells show a decrease after treatment (Top panel, n=3 biological replicates). Nocodazole and ABT induce reversible growth arrest and cells recover their number around 30hrs post synchronization as shown by survival and recover cure normalized to untreated cell number at 0 hours post synchronization (Lower panel). The high concentration of ABT (3.6 µg/ml) inhibits cell number recovery.

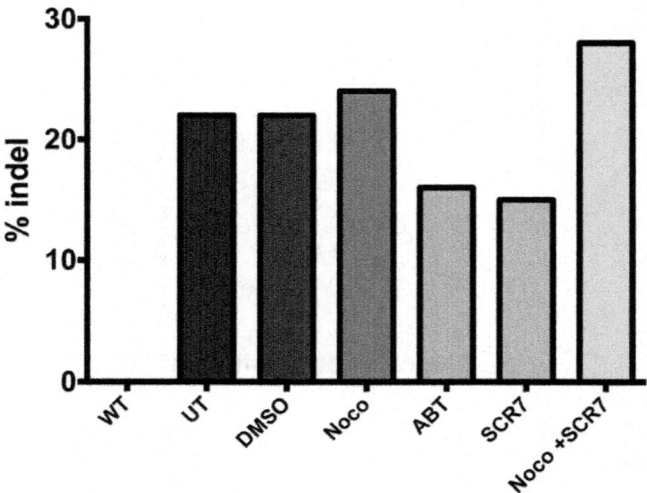

**Supplementary Figure 4:** Quantification of indel formation after targeting of Oct4 using CRISPR/Cas9 in HEK293t cells. CRISPR/Cas9 targeted region was amplified and indels were analyzed by T7 endonuclease assay in HEK293t untreated, DMSO vehicle treated, synchronized, and control untargeted cells.

| Normalized to UT | | BRCA1 | BRCA2 | RAD51 | RAD52 | LIG4 | XRCC4 | XRCC5 | XRCC6 |
|---|---|---|---|---|---|---|---|---|---|
| 3hr | UT | 1.00 | 1.00 | 1.00 | 1.00 | 1.000 | 1.00 | 1.00 | 1.00 |
| | Noco | 2.01 | 29.70 | 10.43 | 93.11 | 0.274 | 0.14 | 0.51 | 0.61 |
| | ABT | 0.22 | 0.74 | 0.27 | 1.61 | 1.407 | 1.34 | 1.00 | 1.38 |
| 6hr | UT | 1.00 | 1.00 | 1.00 | 1.00 | 1.001 | 1.00 | 1.00 | 1.00 |
| | Noco | 1.28 | 0.77 | 1.51 | 7.55 | 1.360 | 0.66 | 0.58 | 0.70 |
| | ABT | 1.58 | 0.56 | 0.45 | 0.83 | 1.461 | 0.66 | 0.56 | 0.64 |
| 12hr | UT | 1.00 | 1.00 | | | 1.000 | 1.00 | 1.00 | 1.00 |
| | Noco | 0.50 | 2.11 | 1.53 | | 0.483 | 0.68 | 0.57 | 0.76 |
| | ABT | 2.43 | 3.25 | 2.62 | | 1.665 | 1.06 | 0.53 | 0.49 |
| 24hr | UT | 1.00 | 1.00 | 1.00 | 1.00 | 1.000 | 1.00 | 1.00 | 1.00 |
| | Noco | 1.33 | 1.52 | 2.00 | 13.20 | 3.305 | 0.79 | 1.04 | 1.41 |
| | ABT | 5.86 | 1.48 | 0.43 | 9.71 | 2.424 | 1.02 | 1.15 | 0.49 |

Low    High

**Supplementary Figure 5:** HR related (BRCA1, BRCA2, RAD51, RAD52) and NHEJ related (LIG4 XRCC4, XRCC5 and XRCC6) gene expression detected by qPCR. H1 synchronized cell gene expression analyzed by qPCR at 3, 6, 12, and 24 hours post nucleofection normalized to untreated cells (n=2 technical replicates).

**Supplementary Table :** H1 hPSCs synchronized using different concentrations of Nocodazole and ABT (n=3 biological replicates). Data presented as Mean ± Standard deviation

| Treatment | G0/G1 (%) | S (%) | G2/M (%) |
|---|---|---|---|
| UT | 46.73±5.66 | 21.4±6.95 | 30.57±4.92 |
| DMSO 1% | 45.2±7.13 | 18±0.79 | 35.97±6.74 |
| DMSO 0.1% | 47.77±6.56 | 18.9±0.7 | 32.67±6.82 |
| ABT 3.6µg/ml | 5.87±2.57 | 16.93±6.45 | 76.6±9.4 |
| ABT 0.36µg/ml | 6.5±4.16 | 15.97±8.08 | 77±12.99 |
| Noco 1µg/ml | 5.07±2.23 | 13.5±2.65 | 81.43±5.25 |
| Noco 0.5µg/ml | 5.77±4.41 | 16.5±9.23 | 76.77±14.01 |

**Supplementary Table 2:** H1 hPSCs with different synchronization and releasing time. Treatment and releasing time are annotated as Treatment/Releasing time (hr) (n=3 biological replicates). Data presented as Mean ± Standard deviation

| Treatment /Releasing time (hr) | Treatment | G1 (%) | S (%) | G2/M (%) |
|---|---|---|---|---|
| | UT | 52±8.66 | 27.6±1.67 | 18.07±7.04 |
| | DMSO | 47.23±3.9 | 26.27±2.6 | 25.9±4.84 |
| 4/3 | Nocodazole | 25.53±1.1 | 26.83±1.15 | 48.83±1.81 |
| | ABT | 24.37±1.63 | 27.47±1.35 | 46.57±0.21 |
| 4/1 | Nocodazole | 33.63±1.79 | 23.47±1.98 | 41.36±0.35 |
| | ABT | 32.27±3.26 | 25.23±2.63 | 40.7±2.67 |
| 4/0 | Nocodazole | 34.6±2.05 | 23.2±0.87 | 40.13±2.05 |
| | ABT | 38.07±0.86 | 22.8±0.72 | 36.73±1.18 |
| 16/3 | Nocodazole | 23.9±15.25 | 6.41±1.36 | 68.37±17.55 |
| | ABT | 30.77±26.72 | 5.05±2.21 | 62.47±25.61 |
| **16/1** | **Nocodazole** | **5.7±1.18** | **2.17±0.46** | **91±1.25** |
| | ABT | 14.76±9.87 | 3.55±2.36 | 80.13±11.18 |
| **16/0** | Nocodazole | 6.41±1.59 | 2.77±0.59 | 89.8±2.88 |
| | **ABT** | **6.51±3.2** | **2.86±0.99** | **89.23±4.42** |
| 12/3 | Nocodazole | 11.64±8.83 | 2.75±0.66 | 84.43±9.79 |
| | ABT | 21.82±24.49 | 2.32±0.66 | 74.73±25.69 |
| 12/1 | Nocodazole | 5.54±0.23 | 2.6±1.3 | 61.72±48.61 |
| | ABT | 10.93±6.5 | 2.74±0.58 | 84.77±7.62 |
| 12/0 | Nocodazole | 8.34±2.41 | 3.27±1.58 | 84.53±2.54 |
| | ABT | 6.23±2.94 | 3.23±1.38 | 89.13±4.52 |
| 18/3 | Nocodazole | 9.7± 6.29 | 3.4±1 | 86.1±7.31 |
| | ABT | 14.13±5.23 | 5.72±2.45 | 79.47±7.48 |
| 18/1 | Nocodazole | 5.95±2.19 | 1.43±0.27 | 91.87±2.57 |
| | ABT | 5.95±1.35 | 1.6±0.48 | 91.47±1.52 |
| 18/0 | Nocodazole | 6.83±3.85 | 2.04±0.82 | 90.37±4.7 |
| | ABT | 8.25±3.01 | 1.74±0.23 | 89.27±2.68 |

**Supplementary Table 3:** Various nucleofection programs were tested using 10ug pMAXGFP of 2.5*10^5 of H1. GFP and cell viability was measured by FACS after 48 hours of nucleofection

| Program | GFP+(%) | Viability |
|---|---|---|
| CA137 | 67.6 | ~60% |
| CB150 | 37.4 | ~60% |
| CD118 | 29.2 | ~80% |
| CE118 | 28.6 | ~80% |
| CM113 | 38.6 | ~80% |
| DC100 | 66.3 | ~30% |
| DN100 | 84.4 | ~20% |

**Supplementary Table 4:** Plasmids for gene targeting method and donor plasmids for HDR. NEUROD1 CRISPR/Cas9 are separated by species with sgRNA numbers corresponding to Supplementary Table 5

| Gene target | Targeting method | Plasmid |
|---|---|---|
| WNT5A | CRISPR; D10A Cas9; *S. pyogenes* | Targeting: pX335<br>Donor(s): PL452 |
| NEUROD1 1-3 | CRISPR; Cas9; *S. pyogenes* | Targeting: pX459<br>Donor(s): OCT4-2A-eGFP-PGK-Puro;<br>Kume_mKO1_pPL452loxP_EM7_PGK_neo |
| NEUROD1 4-5 | CRISPR; Cas9; *N. meningitides* | Targeting: pSimpleII-U6-tracr-U6-BsmBI-NLS-NmCas9-HA-NLS<br>Donor(s): OCT4-2A-eGFP-PGK-Puro;<br>Kume_mKO1_pPL452loxP_EM7_PGK_neo |
| NKX6.1 | ZFNs | Targeting: PZFN1_Nkx6.1 and PZFN2_Nkx6.1<br>Donor(s): OCT4-2A-eGFP-PGK-Puro |
| INS | CRISPR; Cas9 TALENs | Targeting: 13ABPZVC_INS_TAL_3R_TALtrunc_FokI<br>Donor(s): Kume_mKO1_pPL452loxP_EM7_PGK_neo |
| OCT4 | CRISPR; Cas9; *S. pyogenes* | Targeting: pX330<br>Donor(s): CMV-Brainbow-1.1M and Oct4-eGFP-PGK-Puro |

**Supplementary Table 5:** Guide sequences for gene targeting

| Gene target | Targeting strategy | Guide set | Direction | Sequence (5' to 3') |
|---|---|---|---|---|
| | CRISPR nickase | 1 | F | CACCGTGCAGTTCCACCTTCGATGTCGG |
| | CRISPR nickase | 1 | R | AAACCCCGACATCGAAGGTGGAACTGCAC |

| Gene | Type | # | Dir | Sequence |
|---|---|---|---|---|
| WNT5A | CRISPR nickase | 2 | F | CACCGTTTGGCAGGGTGATGCAGATAGG |
| | CRISPR nickase | 2 | R | AAACCCTATCTGCATCACCCTGCCAAAC |
| | CRISPR nickase | 3 | F | CACCGTTCGATGTCGGAATTGATACTGG |
| | CRISPR nickase | 3 | R | AAACCAGTATCAATTCCGACATCGAAC |
| | CRISPR nickase | 4 | F | CACCGGTGGATAACACCTCTGTTTTTGG |
| | CRISPR nickase | 4 | R | AAACCCAAAAACAGAGGTGTTATCCACC |
| NEUROD1 | CRISPR | 1 | F | CACCGAGCCACGGATCAATCTTCTC |
| | CRISPR | 1 | R | AAACGAGAAGATTGATCCGTGGTCC |
| | CRISPR | 2 | F | CACCGGCACTATCCTGCAGCGACAC |
| | CRISPR | 2 | R | AAACGTGTCGCTGCAGGATAGTGCC |
| | CRISPR | 3 | F | CACCGGCACTATCCTGCAGCGACAC |
| | CRISPR | 3 | R | AAACGTGTGGCTGCAGGATAGTGCC |
| | CRISPR | 4 | F | GAGCCCTTCTTTGAAAGCCCTCT |
| | CRISPR | 4 | R | AGAGGGCTTTCAAAGAAGGGCTC |
| | CRISPR | 5 | F | AGTGCCCAGCTCAATGCCATATT |
| | CRISPR | 5 | R | AATATGGCATTGAGCTGGGCACT |
| NKX6.1 | ZFN | 1 | - | CTCTACTTCAGCCCCAGnnCCGCGGCCGTGGCCGCCGG |
| | ZFN | 2 | - | GGACGCCCATCTTCTGGnCCGGAGTGATGCAGAGCC |
| | ZFN | 3 | - | **CTGGCCGGACGCCCATCTTCnGGCCCGGAGTGATGCAGAGCC** |
| | ZFN | 4 | - | CCGGTACCCCAAGCCGCnnGCTGAGCTGCCTGGCCGGC |
| | ZFN | 5 | - | CGGACGCCCATCTTCTGnnCCGGAGTGATGCAGAGCC |
| | ZFN | 6 | - | TACTTCAGCCCCAGCGCCGCnnCCGTGGCCGCCGTGGGCCGG |
| | ZFN | 7 | - | CTGGCCGGACGCCCATCTTCTnGCCCGGAGTGATGCA |
| | ZFN | 8 | - | CGCCGTGGGCCGGTACCCCAAnnCGCTGGCTGAGCTG |
| | ZFN | 9 | - | CCGGTACCCCAAGCCGCTGGnnGAGCTGCCTGGCCGGC |
| | ZFN | 10 | - | CGTGGGCCGGTACCCCAAGCnnCTGGCTGAGCTGCCTG |
| | ZFN | 11 | - | GGCCGCCGTGGGCCGGTACCnnAAGCCGCTGGCTGAGT |

|  |  |  |  |  |
|---|---|---|---|---|
|  | ZFN | 12 | - | CCTGGCCGGACGCCCATCTTnT GGCCCGGAGTGATGCAGAG |
|  | ZFN | 13 | - | GGCCGCCGTGGGCCGGTACCCn AAGCCGCTGGCTGAGT |
|  | ZFN | 14 | - | GACGCCCATCTTCTGGCCCGnn GTGATGCAGAGCCCGCC |
|  | ZFN | 15 | - | CTGGCCGGACGCCCATCnnCTG GCCCGGAGTGATGCAG |
|  | ZFN | 16 | - | CGGACGCCCATCTTCTGGCCnn GAGTGATGCAGAGCCCGCCCT |
| INS | TALENs | 1 | F | TCTGCTCCCTCTACCAGCT |
|  | TALENs | 1 | R | TAGACGCAGCCCGCAGGCA |
| OCT4 | CRISPR | 1 | F | CACCGCTCTCCCATGCATTCAAA CTGAGG |
|  | CRISPR | 1 | R | AAACCCTCAGTTTGAATGCATGG GAGAGC |

**Supplementary Table 6:** Genomic location of targeting. Red sequence shows complementary sequence binding targeting guides

| Gene target | Chromosomal location | Sequence |
|---|---|---|
| WNT5A first constitutive exon | chr3:55,513,341-55,513,466 | atgcagtacatcggagaaggcgcgaagacag gcatcaaagaatgccagtatcaattccgacatcg aaggtggaactgcagcactgtggataacacctct gttttggcagggtgatgcagataggcagccgcg agacggccttcacatacgcggtgagcgcagca ggggtggtgaacgccatgagccgggcgtgccg cgagggcgagctgtccacctgcggctg |
| NEUROD1, exon 1 | chr2:182,540,833-182,545,392 | acgcctacagcgcagcgctggagcccttctt tgaaagccctctgactgattgcaccagcccctt cctttgatggacccctcagcccgccgctcag catcaatggcaacttctcttcaaacacgaac cgtccgccgagtttgagaaaaattatgcctta ccatgcactatcctgcagcgacactggcagg ggcccaaagccacggatcaatcttctcaggc accgctgccctcgctgcgagatccccatag acaatattatgtccttcgatagccattcacatc atgagcgagtcatgagtgcccagctcaatgc catatttcatgattag |
| NKX6.1, 3' of exon 1 | chr4:85,403,140-85,429,603 | gcggccgtggccgccgtgggccggtaccccaa gccgctggctgagctgcctggtggccggacgcc catcttctggcccggagtgatgcagagcccgccc tggagggacgcacgcctggcctgtacccctcgtg agtac |
| INS | chr11:2,171,009-2,192,571 | tggggcaggtggagctgggcgggggccctggt gcaggcagcctgcagcccttggccctggaggg gtccctgcagaagcgtggcattgtggaacaatgc tgtaccagcatctgctccctctaccagctggagaa ctactgcaactagacgcagcccgcaggcagcc ccacacccgccgcctcctgcaccgagagagat ggaataaagcccttgaaccagc |

| | | |
|---|---|---|
| OCT4 | chr6_ssto_hap7:2,467,051-2,473,404 | cctcggtccctttccctgaggggggaagcctttccc cctgtctctgtcaccactctgggctctcccatgcatt caaactgaggtgcctgcccttctaggaatggggg acaggggggaggggaggagctagggaaag |

**Supplementary Table 7:** Homology arm sequences for donor integration. Guide sequences are mutated or interrupted by the donor cassette to prevent binding to homologous donors after integration. Mutated sequences and interrupted sequences are shown in lower case

| Homologous arms of donor | Sequence | Mutated sgRNA binding site |
|---|---|---|
| WNT5a left homology arm, guides 1 & 2; 3 & 4 | CGTCATGGTACCCCTCACAATCACATAAGGAGCCCTTTTAAAATGTCAAGAAACATGGTGAATTTTCTCCAGTCTGTATACTTGTCCCTCTGTTCTTCCATCAGTCAAAAGGAACTGCTATCTCCTACCCGTATCTAATTGTGTCTACTTTCAGCATTCATAAGTTTTCAAGGCAGATTGGTCCAATTTGTGGGGCAAAGTGGGTCACTTTCTCTAATGTCAGAGGCCTGTCCCTCAAACTGCTGTTTCCTCCTTTTCTGAGTGGATTTTGGAAATACTGGTTCAGGTATATGGTGACAGGGGCTAGGAAGGGAGTGAAAATACTAACTGAGGAGCCAGATGGCAGGATGGGGACTGCACAATACCTCCGTTTTCTGAGTGCAGTTTAAACAGAGGAGTTTTATGAAGCCCTTCTCATACTTTGTCATGAGGACAAGCAGGAGAGAAAAAAGCATCTGTGTGCTTAAAACTATGTATTGCTGACATGCTTTCAGGGGGCCCACCTCGAGATTTCACGTAAATCTTGAATGCACATCTCCCCCCCTTGCAGGTCGCTAGGTATGAATAACCCTGTTCAGATGTCAGAAGTATATATTATAGGAGCACAGCCTCTCTGCAGCCAACTGGCAGGACTTTCTCAAGGACAGAAGAAACTGTGCCACTTGTATCAGGACCACATGCAGTACATCGGAGAAGGCGCGAAGACAGGCATCAAAGAATGAATGTTTCAATTATCACATCCTTGGTGGAACTGCAGCAGTCGACCTAGCTG | ATCACATCCTTGGTGGAACTGCA |
| WNT5a right homology arm, guides 1 & 2; 3 & 4 | GTCAGTAAGCGGCCGCCTGTGGATAACACCGTGGTTTTGAACAGGGTGATGCTTATTAATAGGAAAACATTTAAAATATTTTTTAAAACATTAACTTTTTTTCCTAGAAGAAATCCTTAGCATTTACTTCATGCTTGATAGAAAGCATGAAGGAGGATGCCTTCCAGCTCGGGTGGTGGCTTGGGCATTTTTGTCATTATAGCAAATGATGTTAATTCCTGTCCCTACACCTGATGATAAGTTTTGGATTTTAATCAAATCAGGAGGCAAGTATCACAGACACCTTCATTTTTGCAAACACTTAATTTTTCTATAAAGCATTTGTTTAAAATAGAAGCAGAACACTTATTTTGGTTTCTTAAGAGTAGGTGGGGGAAGCAAGTCATGTATATGCTTTTAAACGTCGAAACTCTCTTGCCTGTTAGTTTTTTAAGCCACAATTGAAACGGCCCGAGCATCGAACTAAAAGTTGCCG | TTGAACAGGGTGATGCTTATTAA |

| | | |
|---|---|---|
| | TCTTTTAGAATCCTGGATATATTGAAAGTGAA<br>CAAAGTTTTAATTTTCCTAATAGAACTATTTCC<br>TGAGGCTATAAAGATATAGACGTTAAAAGCTT<br>TATTTTTAGTCCCAGAAAGAATTTTAATTAAAA<br>GTGATCCAATTTATGTTTGTGATGTCAGCAAT<br>TGATATAATAAAAGCACATGCTTGGGATGTCA<br>ACCTAGGAACTTGAAAAAAAAATTAAGAGTAT<br>TGCTCACAGTCCGATTCTTCATTTTAAATACG<br>TTGTTGAATGTCTTACTTCCCCATTCAAAGAG<br>ATAGCATTTCAAATGGGCTAAAGGCAAAGATT<br>TTAAAAAATCTAGTAATGATAGTGAGATCCAG<br>ACATGATAAGATACATTGATGAGTTTGGACAA<br>ACCACAACTAGAATGCAGTGAAAAAAATGCTT<br>TATTTGTGAAATTTGTGATGCTATTGCTTTATT<br>TGTAACCATTATAAGCTGCAATAAACAAGTTC<br>CGCGGCTATGAC | |
| *NEUROD1*<br>left homology<br>arm, guide 1 | TTAATTggatccATGAAGGCTAACGCCCGGGAG<br>CGGAACCGCATGCACGGACTGAACGCGGCG<br>CTAGACAACCTGCGCAAGGTGGTGCCTTGCT<br>ATTCTAAGACGCAGAAGCTGTCCAAAATCGA<br>GACTCTGCGCTTGGCCAAGAACTACATCTGG<br>GCTCTGTCGGAGATCCTGCGCTCAGGCAAAA<br>GCCCAGACCTGGTCTCCTTCGTTCAGACGCT<br>TTGCAAGGGCTTATCCCAACCCACCACCAAC<br>CTGGTTGCGGGCTGCCTGCAACTCAATCCTC<br>GGACTTTTCTGCCTGAGCAGAACCAGGACAT<br>GCCCCCCCACCTGCCGACGGCCAGCGCTTC<br>CTTCCCTGTACACCCCTACTCCTACCAGTCG<br>CCTGGGCTGCCCAGTCCGCCTTACGGTACCA<br>TGGACAGCTCCCATGTCTTCCACGTTAAGCC<br>TCCGCCGCACGCCTACAGCGCAGCGCTGGA<br>GCCCTTCTTTGAAAGCCCTCTGACTGATTGC<br>ACCAGCCCTTCCTTTGATGGACCCCTCAGCC<br>CGCCGCTCAGCATCAATGGCAACTTCTCTTT<br>CAAACACGAACCGTCCGCCGAGTTTGAGAAA<br>AATTATGCCTTTACCATGCACTATCCTGCAGC<br>GACACTGGCAGGGGCCCAAAGCCACGGATC<br>AATCTTtgctagcTTAATT | AGCCACGG<br>ATCAATCTT<br>tgc |
| *NEUROD1*<br>right<br>homology<br>arm, guide 1 | TTAATTggcgcgccCTCAGGCACCGCTGCCCCT<br>CGCTGCGAGATCCCCATAGACAATATTATGT<br>CCTTCGATAGCCATTCACATCATGAGCGAGT<br>CATGAGTGCCCAGCTCAATGCCATATTTCAT<br>GATTAGAGGCACGCCAGTTTCACCATTTCCG<br>GGAAACGAACCCACTGTGCTTACAGTGACTG<br>TCGTGTTTACAAAAGGCAGCCCTTTGGGTAC<br>TACTGCTGCAAAGTGCAAATACTCCAAGCTTC<br>AAGTGATATATGTATTTATTGTCATTACTGCCT<br>TTGGAAGAAACAGGGGATCAAAGTTCCTGTT<br>CACCTTATGTATTATTTTCTATAGCTCTTCTAT<br>TTAAAAAATAAAAAAATACAGTAAAGTTTAAAA<br>AATACACCACGAATTTGGTGTGGCTGTATTCA<br>GATCGTATTAATTATCTGATCGGGATAACAAA | ttaattggcgcg<br>ccCTC |

| | | |
|---|---|---|
| | ATCACAAGCAATAATTAGGATCTATGCAATTT TTAAACTAGTAATGGGCCAATTAAAATATATA TAAATATATATTTTTCAACCAGCATTTTACTAC TTGTTACCTTTCCCATGCTGAATTATTTTGTTG TGATTTTGTACAGAATTTTTAATGACTTTTTAT AATGTGGATTTCCTATTTTAAAACCATGCAGC TTCATCAATTTTTATACATATCgcggccgcTTAAT T | |
| NEUROD1 left homology arm, guide 2 | TTAATTggatccACTAAGGCTCGCCTGGAGCGT TTTAAATTGAGACGCATGAAGGCTAACGCCC GGGAGCGGAACCGCATGCACGGACTGAACG CGGCGCTAGACAACCTGCGCAAGGTGGTGC CTTGCTATTCTAAGACGCAGAAGCTGTCCAA AATCGAGACTCTGCGCTTGGCCAAGAACTAC ATCTGGGCTCTGTCGGAGATCCTGCGCTCAG GCAAAAGCCCAGACCTGGTCTCCTTCGTTCA GACGCTTTGCAAGGGCTTATCCCAACCCACC ACCAACCTGGTTGCGGGCTGCCTGCAACTCA ATCCTCGGACTTTTCTGCCTGAGCAGAACCA GGACATGCCCCCCCACCTGCCGACGGCCAG CGCTTCCTTCCCTGTACACCCCTACTCCTAC CAGTCGCCTGGGCTGCCCAGTCCGCCTTAC GGTACCATGGACAGCTCCCATGTCTTCCACG TTAAGCCTCCGCCGCACGCCTACAGCGCAG CGCTGGAGCCCTTCTTTGAAAGCCCTCTGAC TGATTGCACCAGCCCTTCCTTTGATGGACCC CTCAGCCCGCCGCTCAGCATCAATGGCAACT TCTCTTTCAAACACGAACCGTCCGCCGAGTT TGAGAAAAATTATGCCTTTACCATGCACTATC CTGCAGCGAtggctagcTTAATT | GCACTATC CTGCAGCG Atgg |
| NEUROD1 right homology arm, guide 2 | TTAATTggcgcgccCACTGGCAGGGGCCCAAAG CCACGGATCAATCTTCTCAGGCACCGCTGCC CCTCGCTGCGAGATCCCCATAGACAATATTA TGTCCTTCGATAGCCATTCACATCATGAGCG AGTCATGAGTGCCCAGCTCAATGCCATATTT CATGATTAGAGGCACGCCAGTTTCACCATTT CCGGGAAACGAACCCACTGTGCTTACAGTGA CTGTCGTGTTTACAAAAGGCAGCCCTTTGGG TACTACTGCTGCAAAGTGCAAATACTCCAAG CTTCAAGTGATATATGTATTTATTGTCATTACT GCCTTTGGAAGAAACAGGGGATCAAAGTTCC TGTTCACCTTATGTATTATTTTCTATAGCTCTT CTATTTAAAAAATAAAAAAATACAGTAAAGTTT AAAAAATACACCACGAATTTGGTGTGGCTGTA TTCAGATCGTATTAATTATCTGATCGGGATAA CAAAATCACAAGCAATAATTAGGATCTATGCA ATTTTTAAACTAGTAATGGGCCAATTAAAATAT ATATAAATATATATTTTTCAACCAGCATTTTAC TACTTGTTACCTTTCCCATGCTGAATTATTTTG TTGTGATTTTGTACAGAATTTTTAATGACTTTT TATAATGTGGATTTCCTATTTTAgcggccgcTTAA TT | ttaattggcgcgc ccCAC |

| | | |
|---|---|---|
| NEUROD1 left homology arm, guide 3 | TTAATTggatccGATGAGGACCTGGAAGAGGAG GAAGAAGAGGAAGAGGAGGATGACGATCAA AAGCCCAAGAGACGCGGCCCCAAAAAGAAG AAGATGACTAAGGCTCGCCTGGAGCGTTTTA AATTGAGACGCATGAAGGCTAACGCCCGGGA GCGGAACCGCATGCACGGACTGAACGCGGC GCTAGACAACCTGCGCAAGGTGGTGCCTTGC TATTCTAAGACGCAGAAGCTGTCCAAAATCG AGACTCTGCGCTTGGCCAAGAACTACATCTG GGCTCTGTCGGAGATCCTGCGCTCAGGCAAA AGCCCAGACCTGGTCTCCTTCGTTCAGACGC TTTGCAAGGGCTTATCCCAACCCACCACCAA CCTGGTTGCGGGCTGCCTGCAACTCAATCCT CGGACTTTTCTGCCTGAGCAGAACCAGGACA TGCCCCCCCACCTGCCGACGGCCAGCGCTT CCTTCCCTGTACACCCCTACTCCTACCAGTC GCCTGGGCTGCCCAGTCCGCCTTACGGTAC CATGGACAGCTCCCATGTCTTCCACGTTAAG CCTCCGCCGCACGCCTACAGCGCAGCGCTG GAGCCCTTCTTTGAAAGCCCTCTGACTGATT GCACCAGCCCTTCCTTTGATGGACCCCTCAG CCCGCCGCTCAGCATggctagcTTAATT | AGCCCGCC GCTCAGCA Tggc |
| NEUROD1 right homology arm, guide 3 | TTAATTggcgcgccCAATGGCAACTTCTCTTTCA AACACGAACCGTCCGCCGAGTTTGAGAAAAA TTATGCCTTTACCATGCACTATCCTGCAGCGA CACTGGCAGGGCCCAAAGCCACGGATCAA TCTTCTCAGGCACCGCTGCCCCTCGCTGCGA GATCCCCATAGACAATATTATGTCCTTCGATA GCCATTCACATCATGAGCGAGTCATGAGTGC CCAGCTCAATGCCATATTTCATGATTAGAGGC ACGCCAGTTTCACCATTTCCGGGAAACGAAC CCACTGTGCTTACAGTGACTGTCGTGTTTACA AAAGGCAGCCCTTTGGGTACTACTGCTGCAA AGTGCAAATACTCCAAGCTTCAAGTGATATAT GTATTTATTGTCATTACTGCCTTTGGAAGAAA CAGGGGATCAAAGTTCCTGTTCACCTTATGTA TTATTTTCTATAGCTCTTCTATTTAAAAAATAA AAAAATACAGTAAAGTTTAAAAAAATACACCAC GAATTTGGTGTGGCTGTATTCAGATCGTATTA ATTATCTGATCGGGATAACAAAATCACAAGCA ATAATTAGGATCTATGCAATTTTTAAACTAGTA ATGGGCCAATTAAAATATATATAAATATATATT TTTCAACCAGCATTTTACTACTTGgcggccgcTT AATT | ttaattggcgcg ccCAA |
| NEUROD1 left homology arm, guide 4 | TTAATTggatccGAGGACGACCTCGAAACCATG AACGCAGAGGAGGACTCACTGAGGAACGGG GGAGAGGAGGAGGACGAAGATGAGGACCTG GAAGAGGAGGAAGAAGAGGAAGAGGAGGAT GACGATCAAAAGCCCAAGAGACGCGGCCCC AAAAAGAAGAAGATGACTAAGGCTCGCCTGG AGCGTTTTAAATTGAGACGCATGAAGGCTAA CGCCCGGGAGCGGAACCGCATGCACGGACT | GAGCCCTT CTTTGAAA GCCCggc |

| | | |
|---|---|---|
| | GAACGCGGCGCTAGACAACCTGCGCAAGGT GGTGCCTTGCTATTCTAAGACGCAGAAGCTG TCCAAAATCGAGACTCTGCGCTTGGCCAAGA ACTACATCTGGGCTCTGTCGGAGATCCTGCG CTCAGGCAAAAGCCCAGACCTGGTCTCCTTC GTTCAGACGCTTTGCAAGGGCTTATCCCAAC CCACCACCAACCTGGTTGCGGGCTGCCTGC AACTCAATCCTCGGACTTTTCTGCCTGAGCA GAACCAGGACATGCCCCCCCACCTGCCGAC GGCCAGCGCTTCCTTCCCTGTACACCCCTAC TCCTACCAGTCGCCTGGGCTGCCCAGTCCG CCTTACGGTACCATGGACAGCTCCCATGTCT TCCACGTTAAGCCTCCGCCGCACGCCTACAG CGCAGCGCTGGAGCCCTTCTTTGAAAGCCCg gctagcTTAATT | |
| NEUROD1 right homology arm, guide 4 | TTAATTggcgcgccTCTGACTGATTGCACCAGCC CTTCCTTTGATGGACCCCTCAGCCCGCCGCT CAGCATCAATGGCAACTTCTCTTTCAAACACG AACCGTCCGCCGAGTTTGAGAAAAATTATGC CTTTACCATGCACTATCCTGCAGCGACACTG GCAGGGCCCAAAGCCACGGATCAATCTTCT CAGGCACCGCTGCCCCTCGCTGCGAGATCC CCATAGACAATATTATGTCCTTCGATAGCCAT TCACATCATGAGCGAGTCATGAGTGCCCAGC TCAATGCCATATTTCATGATTAGAGGCACGCC AGTTTCACCATTTCCGGGAAACGAACCCACT GTGCTTACAGTGACTGTCGTGTTTACAAAAG GCAGCCCTTTGGGTACTACTGCTGCAAAGTG CAAATACTCCAAGCTTCAAGTGATATATGTAT TTATTGTCATTACTGCCTTTGGAAGAAACAGG GGATCAAAGTTCCTGTTCACCTTATGTATTAT TTTCTATAGCTCTTCTATTTAAAAAATAAAAAA ATACAGTAAAGTTTAAAAAAATACACCACGAAT TTGGTGTGGCTGTATTCAGATCGTATTAATTA TCTGATCGGGATAACAAAATCACAAGCAATAA TTAGGATCTATGCAATTTTTAAACTAGTAATG GGCCAATTAAAATgcggccgcTTAATT | ttaattggcgcg ccTCT |
| NEUROD1 left homology arm, guide 5 | TTAATTggatccAGACGCAGAAGCTGTCCAAAA TCGAGACTCTGCGCTTGGCCAAGAACTACAT CTGGGCTCTGTCGGAGATCCTGCGCTCAGG CAAAAGCCCAGACCTGGTCTCCTTCGTTCAG ACGCTTTGCAAGGGCTTATCCCAACCCACCA CCAACCTGGTTGCGGGCTGCCTGCAACTCAA TCCTCGGACTTTTCTGCCTGAGCAGAACCAG GACATGCCCCCCACCTGCCGACGGCCAGC GCTTCCTTCCCTGTACACCCCTACTCCTACCA GTCGCCTGGGCTGCCCAGTCCGCCTTACGG TACCATGGACAGCTCCCATGTCTTCCACGTT AAGCCTCCGCCGCACGCCTACAGCGCAGCG CTGGAGCCCTTCTTTGAAAGCCCTCTGACTG ATTGCACCAGCCCTTCCTTTGATGGACCCCT CAGCCCGCCGCTCAGCATCAATGGCAACTTC | AGTGCCCA GCTCAATG CCATagc |

| | | |
|---|---|---|
| | TCTTTCAAACACGAACCGTCCGCCGAGTTTG<br>AGAAAAATTATGCCTTTACCATGCACTATCCT<br>GCAGCGACACTGGCAGGGGCCCAAAGCCAC<br>GGATCAATCTTCTCAGGCACCGCTGCCCCTC<br>GCTGCGAGATCCCCATAGACAATATTATGTC<br>CTTCGATAGCCATTCACATCATGAGCGAGTC<br>ATGAGTGCCCAGCTCAATGCCATagctagcTTA<br>ATT | |
| NEUROD1 right homology arm, guide 5 | TTAATTggcgcgccATTTCATGATTAGAGGCACG<br>CCAGTTTCACCATTTCCGGGAAACGAACCCA<br>CTGTGCTTACAGTGACTGTCGTGTTTACAAAA<br>GGCAGCCCTTTGGGTACTACTGCTGCAAAGT<br>GCAAATACTCCAAGCTTCAAGTGATATATGTA<br>TTTATTGTCATTACTGCCTTTGGAAGAAACAG<br>GGGATCAAAGTTCCTGTTCACCTTATGTATTA<br>TTTTCTATAGCTCTTCTATTTAAAAAATAAAAA<br>AATACAGTAAAGTTTAAAAAATACACCACGAA<br>TTTGGTGTGGCTGTATTCAGATCGTATTAATT<br>ATCTGATCGGGATAACAAAATCACAAGCAATA<br>ATTAGGATCTATGCAATTTTTAAACTAGTAAT<br>GGGCCAATTAAAATATATATAAATATATATTTT<br>TCAACCAGCATTTTACTACTTGTTACCTTTCC<br>CATGCTGAATTATTTTGTTGTGATTTTGTACA<br>GAATTTTTAATGACTTTTTATAATGTGGATTTC<br>CTATTTTAAAACCATGCAGCTTCATCAATTTTT<br>ATACATATCAGAAAAGTAGAATTATATCTAATT<br>TATACAAAATAATTTAACTAATTTAAACCAGCA<br>GAAAAGTGCTTAGAAAGTTATTGTGTTGCCTT<br>AGCACTTCTTTCCTCTCCAATTGTAAAAAAAA<br>gcggccgcTTAATT | ttaattggcgcg<br>ccATT |
| NKX6.1, left homology arm, ZFN 3 | AGCCAGGGATCGAATCTAGGACTCGCGGAA<br>CGAAAGGACTGCCTAGCCCGCCGGGACGCC<br>TGCTTTTCTCGGCGAGCTGCCGCCTCCCGCG<br>TGGAGGGTTTGGACATCTCTGCTGCGCAGCT<br>AGGCGAGCAACTCCCGGCAGCGGCATTTTG<br>GTTCAGTTGGCAGCTCGCCTCCGGGCGCGC<br>CGAGTGCCTCTCCGCTCGCGCCCTCGGCGC<br>TTCCGGCTCCTCTGAGCCCCGCGGGGGGCA<br>CCAGCCAGCGCCCTCGCTGCAAGGCTACGG<br>TCTCCGGCGTGGCCGTGGGATGTTAGCGGT<br>GGGGGCAATGGAGGGCACCCGGCAGAGCG<br>CATTCCTGCTCAGCAGCCCTCCCCTGGCCGC<br>CCTGCACAGCATGGCCGAGATGAAGACCCC<br>GCTGTACCCTGCCGCGTATCCCCCGCTGCCT<br>GCCGGCCCCCCTCCTCCTCGTCCTCGTCGT<br>CGTCCTCCTCGTCGCCCTCCCCGCCTCTGG<br>GCACCCACAACCCAGGCGGCCTGAAGCCCC<br>CGGCCACGGGGGGCTCTCATCCCTCGGCA<br>GCCCCCGCAGCAGCTCTCGGCCGCCACCC<br>CACACGGCATCAACGATATCCTGAGCCGGCC<br>CTCCATGCCCGTGGCCTCGGGGGCCGCCCT<br>GCCCTCCGCCTCGCCCTCCGGTTCCTCCTCC | CTGGCCGG<br>ACGCCCAT<br>CTtctggCCC<br>GGAGTGAT<br>GCAGAGCC |

| | | |
|---|---|---|
| | TCCTCTTCCTCGTCCGCCTCTGCCTCCTCCG CCTCTGCCGCCGCCGCGGCTGCTGCCGCGG CCGCAGCCGCCGCCTCATCCCGGCGGGGC TGCTGGCCGGACTGCCACGCTTTAGCAGCCT GAGCCCGCCGCCGCCGCCCGGGCTCTA CTTCAGCCCCAGCGCCGCGGCCGTGGCCGC CGTGGGCCGGTACCCCAAGCCGCTGGCTGA GCTGCCTGGCCGGACGCCCATCTtctggGCTA GC | |
| *NKX6.1*, right homology arm, ZFN 3 | CCCGGAGTGATGCAGAGCCCGCCCTGGAGG GACGCACGCCTGGCCTGTACCCCTCGTGAG TACTACCACCCGCGCCCCGATGCCTGCCTGC CGTGCCTGTTCTGCCCACTCCCGGGTCGCG GCCCTGGTGTGCATGCCGCTCAGTCCATCC TGTGGCCCGCCCCAGTAGTTTGACAGCGCC GGAATCACACCAGTTAGGTTGGGCCAGGGCT TCCAGAAAGGGATTCTCGCTTTTCTGAGCTC GCGTGGCTGTATCCAAGCGCCTCCCTCTCTC TAAGTCTTGAGTACGCCTGGGCTGAGCCGGT GTGTGTGTTCGTGTGGACCACAACGCGGT GAGGGACGTGTGTCTCGTCCGAGGCCCTGG CTGTTCCTGAAACGGCCTGGCCAGGACCGC AGGCCTGGCCTGGATTAGAAAGATGGGGAG GAGAGGGGAGGAGAGGCCACTGCATCTCTG CTTTCCCGGCTCCTCCCTGTCGCCAGCGCCC CATTCCGCCTTCGCCAGCCTAACCGCTGCCT CTCTTCGAAGCAGTGAGGCCCGGCAGCGGT GAGCCCTCCACCGCCCGGGAAGTGAGTGTA TTTGCATGCACGTCTCCGCCTGGGGTGCCCC ACCTCACCTCTTCGCCCTCGCCGCCCGAGTA CACTCGGCCTGGGTGAGCCCATTGAAGACA GTCCAGCGTCCCAGGGCAGGGCGAGCGCT TCCAGCCGCTCACCAGCGCTGGGCCCGCGC TTTCCCACCTGCCTCCACTCTGGGGGCTGGC ACACGCTGCGTCCGGGCA | CTGGCCGG ACGCCCAT CTtctggCCC GGAGTGAT GCAGAGCC |
| *INS* left homology arm | GCCGCCCCAGCCACCCCTGCTCCTGGCG CTCCCACCCAGCATGGGCAGAAGGGGGCAG GAGGCTGCCACCCAGCAGGGGTCAGGTGC ACTTTTTAAAAAGAAGTTCTCTTGGTCACGT CCTAAAAGTGACCAGCTCCCTGTGGCCCAGT CAGAATCTCAGCCTGAGGACGGTGTTGGCTT CGGCAGCCCCGAGATACATCAGAGGGTGGG CACGCTCCTCCCTCCACTCGCCCCTCAAACA AATGCCCCGCAGCCCATTTCTCCACCCTCAT TTGATGACCGCAGATTCAAGTGTTTTGTTAAG TAAAGTCCTGGGTGACCTGGGGTCACAGGGT GCCCCACGCTGCCTGCCTCTGGGCGAACAC CCCATCACGCCCGGAGGAGGGCGTGGCTGC CTGCCTGAGTGGGCCAGACCCCTGTCGCCA GGCCTCACGGCAGCTCCATAGTCAGGAGAT GGGGAAGATGCTGGGGACAGGCCCTGGGGA GAAGTACTGGGATCACCTGTTCAGGCTCCCA | - |

| | | |
|---|---|---|
| | CTGTGACGCTGCCCCGGGGCGGGGGAAGGA GGTGGGACATGTGGGCGTTGGGGCCTGTAG GTCCACACCCAGTGTGGGTGACCCTCCCTCT AACCTGGGTCCAGCCCGGCTGGAGATGGGT GGGAGTGCGACCTAGGGCTGGCGGGCAGG CGGGCACTGTGTCTCCCTGACTGTGTCCTCC TGTGTCCCTCTGCCTCGCCGCTGTTCCGGAA CCTGCTCTGCGCGGCACGTCCTGGCAGTGG GGCAGGTGGAGCTGGGCGGGGGCCCTGGT GCAGGCAGCCTGCAGCCCTTGGCCCTGGAG GGGTCCCTGCAGAAGCGTGGCATTGTGGAA CAATGCTGTACCAGCATCTGCTCCCTCTACC AGCTGGAGAACTACTGCAAC | |
| *INS* right homology arm | TAGACGCAGCCCGCAGGCAGCCCCACACCC GCCGCCTCCTGCACCGAGAGAGATGGAATAA AGCCCTTGAACCAGCCCTGCTGTGCCGTCTG TGTGTCTTGGGGGCCCTGGGCCAAGCCCCA CTTCCCGGCACTGTTGTGAGCCCCTCCCAGC TCTCTCCACGCTCTCTGGGTGCCCACAGGTG CCAACGCCGGCCAGGCCCAGCATGCAGTGG CTCTCCCCAAAGCGGCCATGCCTGTCGGCTG CCTGCTGCCCCCACCCTGTGGCTCAGGGTC CAGTATGGGAGCTGCGGGGGTCTCTGAGGG GCCAGGGGTGGTGGGGCCACTGAGAAGTGA CTTCTTGTTCAGTAGCTCTGGACTCTTGGAGT CCCCAGAGACCTTGTTCAGGAAAGGGAATGA GAACATTCCAGCAATTTTCCCCCCACCTAGC CCTCCCAGGTTCTATTTTTAGAGTTATTTCTG ATGGAGTCCCTGTGGAGGGAGGAGGCTGGG CTGAGGGAGGGGGTCCTGCAGGGCGGGGG GCTGGGAAGGTGGGGAGAGGCTGCCGAGAG CCACCCGCTATCCCCAGCTCTGGGCAGCCC CGGGACAGTCACACACCCTGGCCTCGCGGC CCAAGCTGGCAGCCGTCTGCAGCCACAGCTT ATGCCAGCCCAGGTCCAGCCAGACACCTGA GGGACCCACTGGTGCCTTGGAGGAAGCAGG AGAGGTCAGATGGCACCATGAGCTGGGGCA GGTGCAGGGACCGTGGCAGCACCTGGCAGG GCCTCAGAACCCATGCCTTGGGCACCCCGG CCATGAGGCCCTGAGGATTGCAGCCCAGGA GAAGCAGGGAACCGCCAGGGCCACAGGGGC AGAGACCAGGGCCAGGGTCCCCCTGCAGCC CCTTAGCCCACCCCCTCCCAGTAAGCAGGGC TGCTTGGCTGGCTTCCTTTGCTACAGACCTG CTGCT | - |
| *OCT4* left homology arm | ACTAGTAACGGCCGCCAGTGTGCTGGAATTC GCCCTTTTCCTGCAGGTCCCACCTGCACAGA TATGCAAAGCAGAAACCCTCGTGCAGGCCCG AAAGAGAAAGCGAACCAGTATCGAGAACCGA GTGAGAGGCAACCTGGAGAATTTGTTCCTGC AGTGCCCGAAACCCACACTGCAGCAGATCAG CCACATCGCCCAGCAGCTTGGGCTCGAGAA | GCTCTCCC ATGCATTC AAAC/tgagg |

| | | |
|---|---|---|
| | GGATGTGAGTGCCATGTCTCTCTGCGGGCTC CATCTCTTTCCCCTGTCACCACCTCGCTTTCC CTAGCTCTGGCTCCTCCAACTGCTCTAGGGC TGTTGGCTTTGGACAGAATGTCCAAGCAGTC AGGCCTGTCTCAGCTCATTCTCTAATGTCCTC CTCTAACTGCTCTAGGGCTGTTGGCTTTGGA TAGAATGTCCAAGCAGAGTCAGGCCCGTCTC AGCTCATTGTCTAATGTCATTCTCCTTTCTGT CATTCACTTGCAGGTGGTCCGAGTGTGGTTC TGTAACCGGCGCCAGAAGGGCAAGCGATCA AGCAGCGACTATGCACAACGAGAGGATTTTG AGGCTGCTGGGTCTCCTTTCTCAGGGGGACC AGTGTCCTTTCCTCTGGCCCCAGGGCCCCAT TTTGGTACCCCAGGCTATGGGAGCCCTCACT TCACTGCACTGTACTCCTCGGTCCCTTTCCCT GAGGGGGAAGCCTTTCCCCCTGTCTCCGTCA CCACTCTGGGCTCTCCCATGCATTCAAAC | |
| OCT4 right homology arm | GGTGCCTGCCCTTCTAGGAATGGGGGACAG GGGGAGGGGAGGAGCTAGGGAAAGAAAACC TGGAGTTTGTGCCAGGGTTTTTGGGATTAAG TTCTTCATTCACTAAGGAAGGAATTGGGAACA CAAAGGGTGGGGGCAGGGGAGTTTGGGGCA ACTGGTTGGAGGGAAGGTGAAGTTCAATGAT GCTCTTGATTTTAATCCCACATCATGTATCAC TTTTTTCTTAAATAAAGAAGCCTGGGACACAG TAGATAGACACACTTATCTTGGTTTGTCCTTC AGTTACTGAGGTAGGGATGGGAATATCCAAT GCTCATACCCAAGTGACCCTGAAACTAAGGT GCCATTTACACTCCTTAAGGTCACACAACATC AGAGGGAGAGCTGGGATTGCAGCCAAGTTTA TTTGTACAGGGCCCTGTGATAGGCTAGTTCC CAAAAGCCTGTGATGCAAGAACTTTTGCCCA TAGACTCAGTCACCATGTAGCTGTTACCTGTT CAGAGCTGGCTTTTTGCTTTCCCACCCTACTC TGGAATTCTTAAATGGCTTTATACTTAGAAAT CATCTTATTTCTGTTGAACCTAGATCACCCCA ACCAGAAACTTCTATTAATACTTTGTGCTTTCT TGATACCAGGGTCTATTTGGTTTCCACTTAAG GTTTTTGCATACTCTGCCCATAAGTGACTCAT TAGTTACGGCCGGCCAAAAGGGCGAATTCTG CAGATATCCATCACACTG | gctctcccatgc attcaaactga GG |

**Supplementary Table 8**: List of primers used for NHEJ and integration assays. (eCFP: enhanced Cerulean fluorescent protein)

| Gene target | Primer | Region of amplification | Size | Sequence |
|---|---|---|---|---|
| WNT5A | F | 5' | | GCAGAAGTATTGGGGCTTGA |
| WNT5A | R | 5' | | GCTTGGCTGGACGTAAACTC |
| WNT5A | F | 3' | | ATTGCATCGCATTGTCTGAG |

| | | | |
|---|---|---|---|
| WNT5A | R | 3' | CAAAAGCAGAGGCAAACAAA |
| NKX6.1 | F | Cut site | CACGCTTTAGCAGCCTGAG |
| NKX6.1 | R | Cut site | CTAACTGGTGTGATTCCGGC |
| INS | R | 3' | TTGCTACAGACCTGCTGCTC |
| OCT4 | F | Cut site | GTACTCCTCGGTCCCTTTCC |
| OCT4 | R | Cut site | CAAAAACCCTGGCACAAACT |
| eCFP | F | Cloning | CTAGGGAATTAATTCACAGCCACC |
| eCFP | R | Cloning | CGCGTAGAGTCGCGGTGATCTAGA |

**Supplementary Table 9:** List of primers used for qRT-PCR shown in Figure 5 and Supplementary Figure 5

| Gene | Forward primer | Reverse primer |
|---|---|---|
| OCT3/4//POU5F1 | tgggctcgagaaggatgtg | gcatagtcgctgcttgatcg |
| NANOG | ttgggactggtggaagaatc | gatttgtgggcctgaagaaa |
| REX1/ZFP42 | aggatctcccacctttccaa | caggtagcacacctcctgc |
| LIN8 | acccttccatgtgcagctta | tgtaagtggttcaacgtgcg |
| SOX2 | agaaaaacgagggaaatggg | gtcatttgctgtgggtgatg |
| TERT | atcagccagtgcaggaactt | agctgacgtggaagatgagc |
| FOXA2/HNF3B | gctactcctccgtgagcaac | gggctcatggagttcatgtt |
| SOX17 | ggcgcagcagaatccaga | ccacgacttgcccagcat |
| GATA4 | caggcgttgcacagatagtg | cccgacacccaatctc |
| HNF4A | catagcttgaccttcgagtgc | cgtggtggacaaagacaaga |
| CXCR4 | caccgcatctggagaacca | gcccatttcctcggtgtagtt |
| BRA/TBX1 | gctgaagtgcatccctgc | agcgaggaggaagggaac |
| GATA2 | gccataaggtggtggttgtg | cacaagatgaatgggcagaa |
| RUNX1 | caatggatcccaggtattgg | ccactccactgcctttaacc |
| HAND1 | aatcctcttctcgactgggc | ccttcaaggctgaactcaaga |
| BRCA1 | gaaggccctttcttctggtt | agagtgtcccatctgtctgga |
| BRCA2 | acaaatagacgaaaggggca | gcccttcacttcagcaaat |
| TUBB3 | agtcgcccacgtagttgc | cgcccagtatgaggagat |
| NESTIN | tctttgctcccagtcctgag | gggctctgatctctgcatct |
| PAX6 | tgtgtgctctgaaggtcagg | cctggagctctgtttggaag |
| RAD51 | ctggtggtctgtgttgaacg | ctgagggtaccttaggcca |
| RAD52 | gctgtcacgtcctccaaga | cttccccttgtccatagcct |
| XRCC4 | tggactgggacagtttctga | tcagttcaccaacatatttccc |
| XRCC5 | tgaggaagcgagtaaccagctcataaat cacatc | atgcagtctatgctcttcataaaata cg |
| XRCC6 | agcactcagcaggttaaagctgaagctc aac | gattataaatgcccacagagatca ctat |
| LIG4 | agcttgcccgaggccagttaaacga | aaaaggaacgtgagatgcaaca |
| TBP | gttctgaataggctgtggggg | acaacagcctgccaccttac |
| ACTB | ccttgcacatgccggag | gcacagagcctcgcctt |
| 18s rRNA | tcggaactgaggccatgatt | ctttcgctctggtccgtctt |

**Supplementary Table 10:** List of antibodies and dilutions used in Figure 5

| Antigen | Supplier, species, catalog number | Dilution |
|---|---|---|
| OCT3/4 | SCBT, mouse, sc-5279 | 1:100 |
| SOX2 | SCBT, goat, sc-17320 | 1:100 |

| | | | |
|---|---|---|---|
| TRA1-81 | Abcam, mouse, 16289 | | 1:200 |
| SSEA-4 | DSHB, mouse, MC-813-70 | | 1:50 |
| SOX17 | R&D Systems, goat, AF1924 | | 1:500 |
| FOXA2/HNF3B | Millipore, rabbit, 07-633 | | 1:250 |
| PDX1 | R&D Systems, goat, AF2419 | | 1:500 |
| CTNT | Abcam, rabbit, ab45932 | | 1:500 |
| NKX2.5 | DSHB, mouse, PCRP-NKX2-5-3B4 | | 1:100 |

**Supplementary Table 11:** Synchronized cells have similar gene expression patterns to unsynchronized cells. qPCR values shown for three replicates

| Gene name | # | Category | H1 UT | H1 DMSO | H1 Noco | H1 ABT | HUES8 UT | HUES8 DMSO | HUES8 Noco | HUES8 ABT | DE | CP | NP |
|---|---|---|---|---|---|---|---|---|---|---|---|---|---|
| OCT4 | 1 | pluripotency | 10.2 | 11.5 | 9.8 | 10.5 | 13.4 | 12 | 10.1 | 11 | | | |
| | 2 | pluripotency | 10.64 | 11.68 | 9.9 | 10.1 | 14.15 | 12.5 | 10.3 | 10.95 | | | |
| | 3 | pluripotency | 10.04 | 11.2 | 9.23 | 9.89 | 13.89 | 11.67 | 10.44 | 12.05 | | | |
| NANOG | 1 | pluripotency | 7.3 | 5.6 | 5.85 | 6.2 | 9.5 | 7.1 | 7.45 | 8.8 | | | |
| | 2 | pluripotency | 6.8 | 5 | 5.57 | 6.2 | 9.15 | 7.93 | 8.22 | 8.73 | | | |
| | 3 | pluripotency | 6.56 | 4.8 | 5.72 | 6.89 | 8.66 | 7.44 | 7.72 | 8.16 | | | |
| REX1 | 1 | pluripotency | 4.1 | 4.56 | 3.99 | 4.53 | 4.6 | 4.4 | 4.73 | 4.21 | | | |
| | 2 | pluripotency | 4.22 | 4.67 | 4.13 | 4.38 | 4.17 | 4.84 | 5.01 | 4.44 | | | |
| | 3 | pluripotency | 4.78 | 4.8 | 4.42 | 5.01 | 4.06 | 4.25 | 4.66 | 4.92 | | | |
| GDF3 | 1 | pluripotency | 7.28 | 7.8 | 7.94 | 7.1 | 6.54 | 6.32 | 7.04 | 6.22 | | | |
| | 2 | pluripotency | 6.9 | 8.12 | 7.28 | 6.92 | 6.93 | 7.93 | 8.11 | 8.73 | | | |
| | 3 | pluripotency | 6.67 | 7.74 | 7.92 | 6.19 | 7.05 | 7.44 | 7.72 | 8.16 | | | |
| LIN28 | 1 | pluripotency | 9.3 | 9.6 | 8.85 | 10.2 | 9.5 | 9.1 | 8.45 | 8.8 | | | |
| | 2 | pluripotency | 9.8 | 9 | 8.97 | 11.1 | 9.15 | 8.93 | 8.11 | 8.73 | | | |
| | 3 | pluripotency | 8.56 | 8.4 | 7.52 | 8.89 | 8.66 | 8.77 | 9.72 | 8.16 | | | |
| FOXA2 | 1 | endoderm | 1.23 | 1.5 | 1.2 | 1.74 | 1.45 | 1.67 | 1.74 | 1.85 | 159.1 | | |
| | 2 | endoderm | 1.11 | 1.22 | 1.05 | 1.64 | 1.43 | 1.84 | 1.6 | 1.65 | 147 | | |
| | 3 | endoderm | 1.37 | 1.56 | 1.15 | 1.8 | 1.69 | 1.62 | 1.63 | 1.48 | 161 | | |
| GATA4 | 1 | endoderm | 1.6 | 1.32 | 1.3 | 1.1 | 1.1 | 0.96 | 0.76 | 1.03 | 53.8 | | |
| | 2 | endoderm | 1.41 | 1.25 | 1.45 | 1.02 | 0.98 | 0.88 | 0.64 | 1.14 | 58.1 | | |
| | 3 | endoderm | 1.66 | 1.28 | 1.27 | 1.16 | 1.24 | 1.19 | 0.75 | 0.99 | 56.7 | | |
| CXCR4 | 1 | endoderm | 2.4 | 3 | 2.68 | 2.11 | 1.43 | 1.77 | 1.94 | 1.88 | 66.1 | | |
| SOX17 | 1 | endoderm | 1.8 | 2.5 | 2.89 | 1.5 | 2.4 | 2.71 | 2.89 | 2.16 | 127.8 | | |
| | 2 | endoderm | 1.96 | 2.08 | 2.89 | 1.75 | 2.2 | 2.65 | 2.51 | 2.37 | 144.3 | | |
| | 3 | endoderm | 2.08 | 2.89 | 2.41 | 1.88 | 2.11 | 2.43 | 2.18 | 2.02 | 125 | | |
| BRA(T) | 1 | mesoderm | 1.96 | 2.28 | 2.14 | 2.35 | 1.67 | 1.99 | 2.12 | 2.11 | | 43 | |
| | 2 | mesoderm | 1.85 | 2.49 | 2.15 | 2.22 | 1.59 | 2.11 | 2.36 | 2.18 | | 48.11 | |
| | 3 | mesoderm | 2.15 | 2.51 | 1.99 | 2.21 | 1.42 | 1.94 | 2.56 | 2.74 | | 51.95 | |
| TPNT | 1 | mesoderm | 1.45 | 1.85 | 1.77 | 2.19 | 2.05 | 1.78 | 1.81 | 1.99 | | 85.64 | |
| | 2 | mesoderm | 1.84 | 2.00 | 2.14 | 1.68 | 1.93 | 2.39 | 2.3 | 1.73 | | 81.5 | |
| | 3 | mesoderm | 1.42 | 1.67 | 1.94 | 1.99 | 1.72 | 2.19 | 2.3 | 2.16 | | 81.44 | |
| NKX2.5 | 1 | mesoderm | 0.18 | 1.23 | 1.11 | 1.21 | 0.83 | 1.25 | 0.15 | 1.33 | | 83.8 | |
| | 2 | mesoderm | 1.15 | 1.37 | 1.25 | 1.34 | 1.11 | 1.03 | 0.08 | 1.18 | | 88.1 | |
| | 3 | mesoderm | 1.02 | 1.55 | 1.13 | 1.13 | 0.08 | 0.85 | 0.92 | 1.13 | | 88.1 | |
| PAX6 | 1 | ectoderm | 1.11 | 1.50 | 1.41 | 1.81 | 1.50 | 1.15 | 1.41 | 1.01 | | | 8.11 |
| | 2 | ectoderm | 1.85 | 1.01 | 1.11 | 1.81 | 1.50 | 1.83 | 1.83 | 1.42 | | | 2.861 |
| | 3 | ectoderm | 1.88 | 1.31 | 1.51 | 1.00 | 1.51 | 1.88 | 1.81 | 1.88 | | | 2.011 |
| SOX1 | 1 | ectoderm | 1.81 | 1.81 | 1.83 | 1.51 | 1.81 | 1.88 | 1.51 | 2.58 | | | 1.51 |
| | 2 | ectoderm | 2.51 | 1.11 | 1.81 | 1.81 | 2.05 | 2.00 | 1.83 | 1.51 | | | 15.1 |
| | 3 | ectoderm | 2.22 | 1.41 | 1.15 | 1.85 | 2.04 | 2.23 | 2.33 | 1.88 | | | 8.18 |

## CONTRIBUTIONS

D.Y., M.S., J.C. and M.B. designed the experiments. D.Y., M.S. and J.C. performed all experiments. A.B. provided neural progenitors. R.S. participated in conducting of experiments. D.Y., M.S., J.C. and M.B. wrote the manuscript.

## ACKNOWLEDGEMENTS

We thank Katrina Wamble for excellent technical support. We also thank Drs. Feng Zhang and Rudolf Jaenish for sharing plasmids though Addgene as well as Shoen Kume from sharing KO plasmid. We are also grateful to Drs. Amar Singh and Stephen Dalton for generously sharing FUCCI-H9 cell line. We thank Dr. Vera Gorbunova for providing HR and NHEJ reporter constructs. Dr. Mirjana Maletic-Savatic generously provided NPCs. We also thank all members of M.B. laboratory and Catherine Gillespie for insightful comments. This study was funded by McNair Medical Institute Scholar Award and NIH (P30-DK079638) to M.B. This project was supported by the Cytometry and Cell Sorting Core at Baylor College of Medicine with funding from the NIH (P30 AI036211, P30 CA125123, and S10 RR024574) and the expert assistance of Joel M. Sederstrom and Amanda White.

## REFERENCES

1. Takahashi, K. *et al*. Induction of pluripotent stem cells from adult human fibroblasts by defined factors. *Cell* **131**, 861–72 (2007).
2. Yu, J. *et al*. Induced pluripotent stem cell lines derived from human somatic cells. *Science* **318**, 1917–20 (2007).
3. Thomas, K. R. & Capecchi, M. R. Targeting of genes to specific sites in the mammalian genome. *Cold Spring Harb Symp Quant Biol* **51 (Pt 2)**, 1101–13 (1986).
4. Thomas, K. R., Folger, K. R. & Capecchi, M. R. High frequency targeting of genes to specific sites in the mammalian genome.*Cell* **44**, 419–28 (1986).
5. Capecchi, M. R. Gene targeting in mice: functional analysis of the mammalian genome for the twenty-first century. *Nat Rev Genet* **6**, 507–12 (2005).
6. Zwaka, T. P. & Thomson, J. A. Homologous recombination in human embryonic stem cells. *Nat Biotechnol* **21**, 319–21 (2003).
7. Hockemeyer, D. & Jaenisch, R. Gene targeting in human pluripotent cells. *Cold Spring Harb Symp Quant Biol* **75**, 201–9 (2010).
8. Merkle, F. T. *et al*. Efficient CRISPR-Cas9-mediated generation of knockin human pluripotent stem cells lacking undesired mutations at the targeted locus. *Cell Rep* **11**, 875–83 (2015).
9. Lombardo, A. *et al*. Gene editing in human stem cells using zinc finger nucleases and integrase-defective lentiviral vector delivery. *Nat Biotechnol* **25**, 1298–306 (2007).

10. Zou, J. et al. Gene targeting of a disease-related gene in human induced pluripotent stem and embryonic stem cells. *Cell Stem Cell* **5**, 97–110 (2009).
11. Hockemeyer, D. et al. Genetic engineering of human pluripotent cells using TALE nucleases. *Nat Biotechnol* **29**, 731–4 (2011).
12. Yang, H. et al. One-step generation of mice carrying reporter and conditional alleles by CRISPR/Cas-mediated genome engineering. *Cell* **154**, 1370–9 (2013).
13. Wang, H. et al. One-step generation of mice carrying mutations in multiple genes by CRISPR/Cas-mediated genome engineering. *Cell* **153**, 910–8 (2013).
14. Jinek, M. et al. A programmable dual-RNA-guided DNA endonuclease in adaptive bacterial immunity. *Science* **337**, 816–21 (2012).
15. Heyer, W. D., Ehmsen, K. T. & Liu, J. Regulation of homologous recombination in eukaryotes. *Annu Rev Genet* **44**, 113–39 (2010).
16. Koledova, Z., Kramer, A., Kafkova, L. R. & Divoky, V. Cell-cycle regulation in embryonic stem cells: centrosomal decisions on self-renewal. *Stem Cells Dev* **19**, 1663–78 (2010).
17. Di Domenico, A. I., Christodoulou, I., Pells, S. C., McWhir, J. & Thomson, A. J. Sequential genetic modification of the hprt locus in human ESCs combining gene targeting and recombinase-mediated cassette exchange. *Cloning Stem Cells* **10**, 217–30 (2008).
18. Yu, C. et al. Small molecules enhance CRISPR genome editing in pluripotent stem cells. *Cell Stem Cell* **16**, 142–7 (2015).
19. Maruyama, T. et al. Increasing the efficiency of precise genome editing with CRISPR-Cas9 by inhibition of nonhomologous end joining. *Nat Biotechnol* **33**, 538–42 (2015).
20. Lin, S., Staahl, B. T., Alla, R. K. & Doudna, J. A. Enhanced homology-directed human genome engineering by controlled timing of CRISPR/Cas9 delivery. *Elife* **3**, e04766 (2014).
21. Chu, V. T. et al. Increasing the efficiency of homology-directed repair for CRISPR-Cas9-induced precise gene editing in mammalian cells. *Nat Biotechnol* **33**, 543–8 (2015).
22. Tsakraklides, V., Brevnova, E., Stephanopoulos, G. & Shaw, A. J. Improved Gene Targeting through Cell Cycle Synchronization. *PLoS One* **10**, e0133434 (2015).
23. Vasquez, R. J., Howell, B., Yvon, A. M., Wadsworth, P. & Cassimeris, L. Nanomolar Concentrations of Nocodazole Alter Microtubule Dynamic

Instability *In Vivo* and *In Vitro*. *Mol Biol Cell* **8**, 973–85 (1997).
24. Hande, K. R. *et al*. The pharmacokinetics and safety of ABT-751, a novel, orally bioavailable sulfonamide antimitotic agent: results of a phase 1 study. *Clin Cancer Res* **12**, 2834–40 (2006).
25. Thomson, J. A. *et al*. Embryonic stem cell lines derived from human blastocysts. *Science* **282**, 1145–7 (1998).
26. Cowan, C. A. *et al*. Derivation of embryonic stem-cell lines from human blastocysts. *N Engl J Med* **350**, 1353–6 (2004).
27. Maehr, R. *et al*. Generation of pluripotent stem cells from patients with type 1 diabetes. *Proc Natl Acad Sci USA* **106**, 15768–73 (2009).
28. Sakaue-Sawano, A. *et al*. Visualizing spatiotemporal dynamics of multicellular cell-cycle progression. *Cell* **132**, 487–98 (2008).
29. Singh, A. M. *et al*. Cell-cycle control of developmentally regulated transcription factors accounts for heterogeneity in human pluripotent cells. *Stem Cell Reports* **1**, 532–44 (2013).
30. Pauklin, S. & Vallier, L. The cell-cycle state of stem cells determines cell fate propensity. *Cell* **155**, 135–47 (2013).
31. Mao, Z., Bozzella, M., Seluanov, A. & Gorbunova, V. Comparison of nonhomologous end joining and homologous recombination in human cells. *DNA Repair (Amst)* **7**, 1765–71 (2008).
32. Adams, B. R., Hawkins, A. J., Povirk, L. F. & Valerie, K. ATM-independent, high-fidelity nonhomologous end joining predominates in human embryonic stem cells. *Aging (Albany NY)* **2**, 582–96 (2010).
33. Hou, Z. *et al*. Efficient genome engineering in human pluripotent stem cells using Cas9 from Neisseria meningitidis. *Proc Natl Acad Sci USA* **110**, 15644–9 (2013).
34. Roccio, M. *et al*. Predicting stem cell fate changes by differential cell cycle progression patterns. *Development* **140**, 459–70 (2013).
35. Chetty, S. *et al*. A simple tool to improve pluripotent stem cell differentiation. *Nat Methods* **10**, 553–6 (2013).
36. Kannagi, R. *et al*. Stage-specific embryonic antigens (SSEA-3 and -4) are epitopes of a unique globo-series ganglioside isolated from human teratocarcinoma cells. *EMBO J* **2**, 2355–61 (1983).
37. Scholer, H. R., Ruppert, S., Suzuki, N., Chowdhury, K. & Gruss, P. New type of POU domain in germ line-specific protein Oct-4. *Nature* **344**, 435–9 (1990).
38. Boyer, L. A. *et al*. Core transcriptional regulatory circuitry in human embryonic stem cells. *Cell* **122**, 947–56 (2005).

39. Kim, J., Chu, J., Shen, X., Wang, J. & Orkin, S. H. An extended transcriptional network for pluripotency of embryonic stem cells. *Cell* **132**, 1049–61 (2008).
40. Levine, A. J. & Brivanlou, A. H. GDF3, a BMP inhibitor, regulates cell fate in stem cells and early embryos. *Development* **133**, 209–16 (2006).
41. Richards, M., Tan, S. P., Tan, J. H., Chan, W. K. & Bongso, A. The transcriptome profile of human embryonic stem cells as defined by SAGE. *Stem Cells* **22**, 51–64 (2004).
42. Lee, C. S., Friedman, J. R., Fulmer, J. T. & Kaestner, K. H. The initiation of liver development is dependent on Foxa transcription factors. *Nature* **435**, 944–7 (2005).
43. Arceci, R. J., King, A. A., Simon, M. C., Orkin, S. H. & Wilson, D. B. Mouse GATA-4: a retinoic acid-inducible GATA-binding transcription factor expressed in endodermally derived tissues and heart. *Mol Cell Biol* **13**, 2235–46 (1993).
44. Nair, S. & Schilling, T. F. Chemokine signaling controls endodermal migration during zebrafish gastrulation. *Science* **322**, 89–92 (2008).
45. Kanai-Azuma, M. *et al.* Depletion of definitive gut endoderm in Sox17-null mutant mice. *Development* **129**, 2367–79 (2002).
46. Faial, T. *et al.* Brachyury and SMAD signalling collaboratively orchestrate distinct mesoderm and endoderm gene regulatory networks in differentiating human embryonic stem cells.*Development* **142**, 2121–35 (2015).
47. Lyons, I. *et al.* Myogenic and morphogenetic defects in the heart tubes of murine embryos lacking the homeo box gene Nkx2-5.*Genes Dev* **9**, 1654–66 (1995).
48. Li, H. S., Yang, J. M., Jacobson, R. D., Pasko, D. & Sundin, O. Pax-6 is first expressed in a region of ectoderm anterior to the early neural plate: implications for stepwise determination of the lens.*Dev Biol* **162**, 181–94 (1994).
49. Pevny, L. H., Sockanathan, S., Placzek, M. & Lovell-Badge, R. A role for SOX1 in neural determination. *Development* **125**, 1967–78 (1998).
50. Serrano, L. *et al.* Homologous recombination conserves DNA sequence integrity throughout the cell cycle in embryonic stem cells. *Stem Cells Dev* **20**, 363–74 (2011).
51. Cong, L. *et al.* Multiplex genome engineering using CRISPR/Cas systems. *Science* **339**, 819–23 (2013).

# CITATION

## CHAPTER 1

Jorge Angel Ascacio-Martínez and Hugo Alberto Barrera-Saldaña (2012). Genetic Engineering and Biotechnology of Growth Hormones, Genetic Engineering - Basics, New Applications and Responsibilities, Prof. Hugo A. Barrera-Saldaña (Ed.), ISBN: 978-953-307-790-1, InTech, DOI: 10.5772/38978.

## CHAPTER 2

Mark W Ronsyn, Jasmijn Daans, Gie Spaepen, Shyama Chatterjee, Katrien Vermeulen, Patrick D'Haese, Viggo FI Van Tendeloo, Eric Van Marck, Dirk Ysebaert, Zwi N Berneman, Philippe G Jorens and Peter Ponsaerts, "Plasmid-based genetic modification of human bone marrow-derived stromal cells: analysis of cell survival and transgene expression after transplantation in rat spinal cord," BMC Biotechnology20077:90, DOI: 10.1186/1472-6750-7-90.

## CHAPTER 3

Laura Riolobos, Roli K Hirata, Cameron J Turtle, Pei-Rong Wang, German G Gornalusse, Maja Zavajlevski, Stanley R Riddell and David W Russell, "HLA Engineering of Human Pluripotent Stem Cells," Molecular Therapy (2013); 21 6, 1232–1241. doi:10.1038/mt.2013.59.

## CHAPTER 4

Kalpana Dulal, Benjamin Silver, and Hua Zhu, "Use of Recombination-Mediated Genetic Engineering for Construction of Rescue Human

Cytomegalovirus Bacterial Artificial Chromosome Clones," Journal of Biomedicine and Biotechnology, vol. 2012, Article ID 357147, 10 pages, 2012. doi:10.1155/2012/357147.

## CHAPTER 5

Kazushi Inoue, Elizabeth Fry, Dejan Maglic and Sinan Zhu (2013). Genetically Engineered Mouse Models for Human Lung Cancer, Oncogenesis, Inflammatory and Parasitic Tropical Diseases of the Lung, Prof. Jean-Marie Kayembe (Ed.), ISBN: 978-953-51-0982-2, InTech, DOI: 10.5772/53721.

## CHAPTER 6

Shahbakhi, M. , Far, D. and Tahami, E. (2014) Speech Analysis for Diagnosis of Parkinson's Disease Using Genetic Algorithm and Support Vector Machine. Journal of Biomedical Science and Engineering, 7, 147-156. doi: 10.4236/jbise.2014.74019.

## CHAPTER 7

Armand de Gramont, Sarah Watson, Lee M. Ellis, Jordi Rodón, Josep Tabernero, Aimery de Gramont and Stanley R. Hamilton, "Pragmatic issues in biomarker evaluation for targeted therapies in cancer," Nature Reviews Clinical Oncology 12, 197–212 (2015), doi:10.1038/nrclinonc.2014.202.

## CHAPTER 8

Yang, D. *et al.* Enrichment of G2/M cell cycle phase in human pluripotent stem cells enhances HDR-mediated gene repair with customizable endonucleases. *Sci. Rep.* 6, 21264; doi: 10.1038/srep21264 (2016).

# INDEX

## A

adeno-associated virus (AAV)  64, 65
amplification refractory mutation system (ARMS)  195
antibiotic resistant  100, 102
antidystrophin  81
autoantigens  81

## B

Bacterial artificial chromosome (BAC)  97
bioanalytical method  173, 174
biomarker  171, 172, 173, 174, 175, 177, 179, 181, 182, 183, 184, 187, 188, 189, 190, 191, 192, 193, 194, 196, 197, 198, 199, 200, 201, 203, 208, 212, 278
biomarkers  171, 172, 173, 174, 175, 181, 182, 183, 184, 188, 190, 191, 192, 193, 194, 195, 197, 199, 201, 207, 211, 215
biopsy  182, 183, 192, 194, 197, 203, 211
bioreactor  17, 19, 21, 24, 29
Biotechnology  1, 13, 28, 29, 30, 277
bovine  5, 6, 8, 9, 24, 29
bovine chorionic somatomammotropin (BCS)  9
brain-derived neurotrophic factor (BDNF)  33
Breast Cancer International Research Group (BCIRG)  193

## C

Canine Growth Hormone (CFGH)  7
caprine  9, 24
Caprine Growth Hormone (CHGH)  6
carcinomas  117, 118, 119, 121, 123, 129, 138, 139, 153
chorionic somatomammotropin (CS)  2
chromosomes  158, 159, 160
Clara Cell Secretory Protein (CCSP)  120
Clinical Laboratory Improvement Amendments (CLIA)  189
clone  97, 99, 100, 101, 102, 104, 106, 108, 112
complete culture medium (CCM)  51
cyclosporin  39
Cytogenetics  83
cytoplasm  10

## E

embryoid bodies (EBs)  75
embryonic stem cell (ESC)  63
embryonic stem cells (ESCs)  64
enhanced green fluorescent protein (EGFP)  31, 33, 34, 48, 51
enzyme-linked immunosorbent assays

(ELISAs) 174
Epidermal growth factor (EGF) 130
Equine Growth Hormone (ECGH) 7
European Medicine Agency (EMA) 173

## F

Feline Growth Hormone (FCGH) 7
flow cytometry 34, 58
fluorescence activated cell sorting (FACS) 52
fluorescence in situ hybridization (FISH) 179

## G

ganciclovir 65, 68, 81, 84
genetic algorithm (GA) 155, 157
Genetic Algorithm (GA) 157
genome 119, 154, 218, 219, 220, 239, 240, 242, 244, 273, 274, 275, 276
genomes 97
glial cell-derived neurotrophic factor (GDNF) 33

## H

hematoxylin-eosin (HE) 38, 42, 45, 47, 56
heterozygosity 68, 81, 92, 93
histocompatible 63, 81
homologous recombination (HR) 218
homology directed repair 217, 218
homozygosity 65, 80, 92
Human Chorionic Somatomammotropin (HCS) 4
human cytomegalovirus (HCMV) 97, 114
Human cytomegalovirus (HCMV) 97
human leukocyte antigens (HLA) 64
hybrid search (HS) 156
hygromycin 65, 68, 69, 83, 84

## I

inhibitors 49

inversion 158

## L

laboratory-developed tests (LDTs) 189

## M

mapping 160
marrow-derived stromal cells (MSC) 31
microtomy 192
Monte Carlo search (MCS) 156
murine cytomegalovirus (MCMV) 99
mutagenesis 99, 100, 102, 111, 112, 113, 114, 116
Mutants 100
mutation 119, 122, 124, 128, 130, 139, 145, 153, 158, 159

## N

nerve growth factor (NGF) 33
neural progenitors 217, 242, 244, 246, 272
neuroendocrine (NE) 121
neurons 156
neurotrophin 31, 33, 34, 48, 59, 60
non-homologous end joining (NHEJ) 217, 218
non-small-cell lung cancer (NSCLC) 118

## O

oncogene 121, 125, 126, 127, 140, 145, 147, 148

## P

Parkinson's disease (PD) 155, 156
Parthenogenesis 80
periplasm 10
phycoerythrin (PE) 52
polymerase chain reaction (PCR) 5
porcine 7, 9, 24, 28
probabilistic neural network (PNN) 156
probes 68, 71, 84
progesterone receptor (PR) 194

## Q

quadratic programming (QP)  161

## R

recombineering  97, 100, 115, 116
region of interest (ROI)  100, 101, 103, 104

## S

silver in situ hybridization (SISH)  192
somatolactin (SL)  2
somatomammotropin  2, 9, 21
somatotrophs  2
somatotropin  5, 28
support vector machine (SVM)  155, 166
Support Vector Machine (SVM)  157, 160

## T

Teratoma assays  83
tumorigenesis  120, 122, 123, 128, 129, 130, 131, 132, 135, 136, 137, 139, 142, 146, 147, 148, 150, 151, 152, 153
tumorigenicity  50

## X

xenogeneic hosts  48